EARTHQUAKE ENGINEERING
SIXTEENTH REGIONAL EUROPEAN SEMINAR

PROCEEDINGS OF THE SIXTEENTH REGIONAL EUROPEAN SEMINAR ON
EARTHQUAKE ENGINEERING / STARÁ LESNÁ / 6-12 OCTOBER 1991

Earthquake Engineering

Sixteenth European Regional Seminar

Edited by
EMÍLIA JUHÁSOVÁ
Institute of Construction and Architecture
Slovak Academy of Sciences, Bratislava

A.A.BALKEMA / ROTTERDAM / BROOKFIELD / 1992

ORGANIZING COMMITTEE
E. Juhásová (Chairman)
M. Tichý (Secretary)
O. Fischer (CS-AEE)
G. Martinček
A. Tesár

Y. Koleková
I. Motlik
M. Vrabec
M. Chudik
E. Andrejková

ORGANIZER'S ADDRESS
Institute of Construction and Architecture, Slovak Academy of Sciences, Dúbravská cesta 9, 842 20 Bratislava, Czecho-Slovakia

SPONSORING ORGANIZATIONS
Slovak Academy of Sciences
European Association for Earthquake Engineering
CS Society for Mechanics
Czecho-Slovak Association for Earthquake Engineering
CS Committee of Unesco
Škoda Works

CIP-DATA KONINKLIJKE BIBLIOTHEEK, DEN HAAG

Earthquake

Earthquake engineering: proceedings of the 16th Regional European Seminar on Earthquake Engineering, Stará Lesná / Czecho-Slovakia / 6-12 October 1991 / ed. by Emília Juhásová. – Rotterdam [etc.]: Balkema. – Ill.
Co-prod. with Ister Science Press.
ISBN 90 5410 006 0 bound
Subject heading: earthquakes.

All rights reserved. No part of this publication may be reproduced, stored in a retrieval system, or transmitted in any form or by any means, electronic, mechanical, photocopying, recording or otherwise, without the prior written permission of the copyright owner.

Published outside Czecho-Slovakia and Poland by
A.A. Balkema, P.O. Box 1675, 3000 BR Rotterdam, Netherlands
A.A. Balkema Publishers, Old Post Road, Brookfield, VT 05036, USA

ISBN 90 5410 006 0

© 1992 Ister Science Press Ltd.
Printed in Czecho-Slovakia

CONTENTS

Preface 1
Executive Committee of EAEE activity 3
The address of the CS representant at the EAEE 5

1. PAST EARTHQUAKES. THE STRUCTURES WHICH HAVE SURVIVED THE PAST EARTHQUAKES 7

D. PROCHÁZKOVÁ: Properties of earthquakes in central Europe   8
I. BROUČEK: Historical earthquakes on the territory of
   Slovakia 16
M.R. MAHERI: Behaviour of large engineering structures
   during Manjil, Iran earthquake of June 1990 26
S.G. SHAGINIAN: Spitak earthquake effects and problems of
   strengthening of buildings 36

2. SEISMIC DESIGN OF STRUCTURES. THE WAYS OF SEISMIC STRENGTHENING OF EXISTING STRUCTURES 45

O. FISCHER: Dynamic background to quasistatic seismic solution to structures 46
R. FOLIČ, B. SIMEONOV: Energetic aspects of behaviour of
   hollow brick panels strengthened by reinforced concrete
   grid 52
K. PILAKOUTAS: Ductility design of reinforced concrete
   members 62
A.T. MINASIAN, M.G. MELKUMIAN, E.E. KHACHIAN: Some aspects
   of the estimation of buildings seismic resistance by the
   results of vibration tests 72
D. UKLEBA: Design of the multistoreyed building on a seismic
   effect with the account of elasto-plastic deformation   78

3. SEISMIC RESPONSE REDUCTION SYSTEMS AND SEISMIC ISOLATION 83

E. JUHÁSOVÁ: Optimization of base isolation systems for
   important structures 84
B. KIRIKOV: Systems of seismic protection of old structures 98
G. MARTINČEK: The vibroisolation effect of barriers in
   subsoil 105

R. MASOPUST: Base seismic isolation for important building structures 116
J.M. EISENBERG, A.M. ZHAROV, E.I. GERASIMOVA, B.K. GAYAROV, A. MELENTIEV: Seismic response reduction by means of inelastic deformations and failure. Mechanism of structures control and seismic isolation 126
Y. KOLEKOVÁ, E. JUHÁSOVÁ: The seismic isolation effects of different soil layers combinations 133

4. SOFTWARES AWAILABLE FOR SEISMIC CALCULATION AND DESIGN 141

D. ANIČIČ, D. RADIČ: Software for seismic design of masonry buildings 142
A. CHOLEWICKI, J. PERZYŇSKI, E. WADECKI: Correction of internal forces for coupled shear wall structures at postelastic stage 148
L. GALANO, A. VIGNOLI: Review and comparison of the method for the evaluation of inelastic design spectra 159
J. GLÜCK, M. STERNIK: An equivalent column in lateral load analysis of pile-structure interaction 169

5. RECOMMENDED STRUCTURAL SOLUTIONS FOR MORE SEISMIC RESISTANT STRUCTURES. SEISMIC CODES 179

A. CASTELLANI: Eurocode 8 in a relation to Italian Seismic Standard 180
R. FLESCH: Earthquake resistant design of bridges 193
J.P. NEWELL, G.P. ROBERTS: Aseismic desig methods for industrial plant 203
L. PEČÍNKA, J. ŽĎÁREK: Seismic fragilities for nuclear power plant risk studies 213
A.G. TYAPIN: Influence of embedment on the first resonace of soil-structure system 220

6. SPECIAL THEMES 227

A.M. ANSAL, H. YILDIRIM, M.A. LAV: Geotechnical aspects in earthquake engineering 228
J. BENČAT: Static and dynamic engineering soil parameters evaluated by means of box tests 240

B. SVOBODA: Liquefaction and change of microstructure of the soils under dynamic loading 247

J. MUBARAKOV, H. SAGDIEV: Dynamic methods for earthquake resistance study of underground structures 255

R. CIESIELSKI, W. KOWALSKI, E. MACIAG, T. TATARA: Seismic forces induced by mining shocks in prefabricated panel building 261

H. SAGDIEV, E. JUHÁSOVÁ: Seismic effect of underground explosions 271

Z. ZEMBATY: The effect of strong motion duration on first passage failure of structures 280

List of authors (A) and participants (P) 291

PREFACE

The host organization - Slovak Academy of Sciences had the honour to organize the 16th European Regional Seminar on Earthquake Engineering in 1991 which was held in the House of Scientist of SAS in Stará Lesná, in High Tatras.

The custom of European Association for Earthquake Engineering to organize these Seminars, as the opportunity to bring together engaged leading experts in earthquake engineering and young scientist, students and engineers, is welcome and popular among those which are tied in some way with the problems of seismic effects and improvement of seismic resistance of structures. In order to concern attention on the chosen important questions of earthquake engineering the scientific topics were stated as follows:

1. Past earthquakes. The structures which have survived the past earthquakes.
2. Seismic design of structures. The ways of seismic strengthening of existing structures.
3. Seismic response reduction systems and seismic isolation.
4. Software awailable for seismic calculation and design.
5. Recommended structural solutions for more seismic resistant structures. Seismic codes.
6. Special themes.

The presented volume contains the contributions to these topics.

The staff of organizing committee appreciate the valuable work of authors of papers as well as the soul of the Seminar. We could really say that during the Seminar the barrier between the elder and younger, between the skilful experts and beginners really fell down, discussions after the presented lectures or during the breaks were wide and thoughtful.

We are indebted to sponsoring organizations for their support.

The common work of authors and other participants is the contribution for the next development of earthquake engineering and structural dynamics, for mutual cooperation and strengthening of participants personal contacts.

The reader of papers can meet chosen chapters from the new researches wich are connected with the topics of this Seminar and create the contents of this volume.

We must pass most credits to each of authors for their work as the very basis of the scientific level of Seminar and the Proceedings.

                                        Emília Juhásová

EUROPEAN ASSOCIATION FOR EARTHQUAKE ENGINEERING

Central Office: Institute of Civil Engineering Rakusina 1, P. O. Box 165
41000 Zagreb, Yugoslavia

President: Nikolai N. Skladnev, USSR
Vice-Presidents: Ricardo Duarte, Portugal, Rainer Flesch, Austria
Secretary-General: Dražen Aničić, Yugoslavia (Central Office)
Secretary: Alberto Castellani, Italy
Executive Committee Members: delegates of France, Greece, Turkey and United Kingdom
Honorary Members: G. Brankov, Bulgaria, S. Bubnov, Yugoslavia, J. Ferry Borges, Portugal, R. Yarar, Turkey

R E P O R T

on the Central Office activities for the period from September 1990 to September 1991 (prepared for the EAEE Executive Committee meeting held in Stara Lesna, Czecho-Slovakia, on October 9th, 1991)

As decided at the EAEE General Assembly, Moscow, 1990, the Central Office continued its activities located in the Civil Engineering Institute of Croatia in Zagreb, Yugoslavia, with Drazen Anicic acting as secretary-general for the 1990-1994 period. Following decisions of the EAEE General Assembly (Moscow, September 13, 1990), the Central office undertook several action:

1. National delegates were asked to prepare their Directories of earthquake engineering research. Following countries fulfilled this task: USSR, UK, Portugal, Czecho-Slovakia, France, Yugoslavia, Greece and Austria. Directories were or will be published in EAEE Bulletins. No response reached Central Office one year after the questionnaire was distributed from Turkey, Spain, Bulgaria, Rumania, Hungary, Germany, Norway, Cyprus, Iran, Denmark, Egypt, Iceland, Israel, Italy and Switzerland.

2. The letter of gratitude was sent to the Soviet Committee for Participation in International Organizations on Earthquake Engineering for the organization of the 9th ECEE in Moscow.

3. Mr.Igor Konstantinov as representative of the EAEE President dr. N.N.Skladnev visited Central Office from November 21st to 25th, 1990. Future activities of Presidents office were discussed. Secretary-general informed in detail his guest what should be done for better contacts between European and Soviet scientists.

4. The offer was made to the Spanish Association (dr.A.Lopez-Arroyo) to organize the 17th Regional Seminar on EE for young scientists in Madrid in July 1992 before or after the 10th World Conference. No response. I suggest that dr. R.Duarte as first Vice-president may contact dr. Lopez-Arroyo about this matter.

5. The offer was made to the Israeli Association for EE (dr.Avigdor Rutenberg) for organization of the 18th Regional Seminar on EE in Israel in September 1993. No response yet. I suggest to renew this invitation to the Israeli Association.

6. Close contacts were established with organizers of the 16th Regional Seminar on EE (dr.Emilia Juhasova and Mr.Milan Tichy) and procedure applied in organization of previous seminars was clarified.

7. The pamphlet "EAEE Basic Information" was updated and published in the Bulletin (Vol. 11, No.1).

8. Close contacts with French Association of EE, which is one of the most active in Europe last years, resulted in valuable collection of

French Bulletins, Technical Notes (Cahier Technique) and special conference reports.

9. Following suggestion of dr.Philippe Bisch, French delegate, Secretary-general invited in written the Algerian Association for Earthquake Engineering (dr.Belazoughi) to join EAEE. In personal contacts in Trieste in December 1990 this invitation was repeated one more time. No response from Algeria.

10. French Association organized the International Conference on Buildings with Load Bearing Concrete Walls in Seismic Zone in Paris, June 13-14,1991. EAEE was declared as supporting organization.

11. Central Office contacts with:
- UNESCO-NGO (Non-Governement Organization) Standing Committee supplying some data about EAEE,
- Professor Kenzo Toki, Disaster Prevention Research Institute, Kyoto University, Japan, about possible data exchange in the frame of International Decade for Natural Disaster Reduction,
- many convention organizers contacted Central Office exploring possibilities to organize next European Conference or other meetings in collaboration with EAEE (UK,Holland,Spain,Austria,France,Italy),

12. New Rules for EAEE Working groups were distributed for public opinion. Final version was given for balloting by EC members in February 1991. Final text of these Rules are proposed now to the Executive Committee for final approval.

13. Central Office tried to activate EC members to propose new Working Groups activity plan. No initiative reached Central Office. WG convenors were asked to nominate WG members and propose WG plans. Only dr.N.N.Ambraseys and dr.M.Dolce sent their proposals until this report was completed.

14. In 1990-1991 period three issues of the EAEE Bulletin were published (Vol.10, No3, 1990, Vol.11, No1, April 1991 and Vol.11, No2, September 1990) under editorship of R.T.Duarte and D.Anicic.

15. Following written agreement - N.Skladnev, R.Duarte, R.Flesch and D.Anicic were nominated in the editorial board of the new Soviet scientific-technical journal "Earthquake Engineering" which was planned to be published by Serviceprogress in Moscow. There are no information if this journal actually started.

16. Turkish National Committee for Earthquake Engineering published recently Proceedings of Thirteenth Regional Seminar on Earthquake Engineering, Sep. 14-24, 1987, Istanbul, Vol.1-2,1160 pages, Istanbul, Dec.1990. Congratulation to the Turkish Committee for the nice job. There is now continous editing of EAEE Seminars Proceedings.

Secretary - General:
Drazen Anicic

The Address of
the Czecho-Slovak Representative at the EAEE:

Ladies and gentlemen, dear colleagues,

I have the honour of opening the Regional European seminar on Earthquake engineering that takes place in the beautiful mountain part of Slovakia. Stimulation of these Seminars forms one of the important activities of the European Association on Earthquake Engineering. This institution has had its rise at Skoplje in 1964 with the aim of bringing together the effort of European and Mediteranean countries in protection their territories against the scourge of earthquakes. The number of countries participating in this association grew since that time from 9 to 23, and the activity in this branch as well as the number of interested specialists has always been increasing.

The earthquake is one of the classical natural elements that wexes human being since the beginning of its existence and is actual till present time. And also since ancient time the man tries to avoid the destructive consequences of earthquakes, of course only in recent decades the reasonable and scientific base could be given to this effort. This activity can be understood as the fulfilment of the principal commandment, that the first human beings, Adam and Eva, have received in the Paradise from God: Be fruitful and increase in number; fill the Earth and subdue it (Gen. 1,28) - that implicates: learn to know the Earth, learn to use its riches and learn to protect yourselves against its dangers, of course also against the earthquakes.

Newertheless, in spite of all efforts, of all conferences, of all computers and thousands of publications, there are always human victims and material damages caused by earthquake - for example there were several hundreds of people killed by the earthquakes in Northern Peru, Costa Rica and Western Caucassus during the month of April this year. It must be of course said that most of the victims are not due to the lack of knowledge or of technical development. Those victims rather have the economical, or even political reasons, due to which the technical knowledges have not been able to reach all the parts of the world. Thanks God, the principal ideological barier, obstructing international contacts, has recently fallen, but other obstacles - the ignorance of real seismic risc and insufficiency of financial means - have stil remained in some parts of the world. It is an evident fact, that no government has money enough, but such a government, that spends money to keep its power instead of spending them for the real welfare of the population including its seismic safety, such a government commit a mortal sin against its own country and will be held responsible for all the victims by the Last Judgment.

We have come together just to promote the knowledges of the earthquakes and to help each other to work more effectively on the field of Earthquake engineering.

We are opening the 16th in the row of seminars. This row has started in 1973 in Varna - Bulgaria, the scope of these seminars was to afford fundamental informations about Earthquake engineering to young civil engineers and scientists; later on, as Earthquake engineering has already broken into normal University education, these seminars became to be also a forum for the exchange of knowledges and of new informations between young specialists. This trend can be of course understood, I nevertheless recommend most urgently that the original aim of the seminars, namely the fundamental forming of earthquake engineering specialists, would remain as principle contens of these seminars.

We have met here people from several countries together. Our program includes about 33 registered papers, but I hope that there will moreover remain some time for aditional discussions and personal contacts during our free time. I think, this is also the right place to express our gratitude to Slovak Academy of Sciences, which has offered this nice place for our meeting, and to all the organisers. On the first place it is the iniciator and the head of Organising Committee Mrs Emilia Juhásová, the secretary of the committee Mr Milan Tichý and the whole staff that, besides the normal administration of the meeting like this, managed to face all the administrative reversals and difficulties arising from the economical changes in our country. With full gratitude shoud be also mentioned the sponsors, namely UNESCO and Škoda-Works, whose financial contribution is being highly appreciated.

Let me welcome all of you once again, open the seminar and wish it much success.

                                                           Ondřej Fischer

T1

PAST EARTHQUAKES. THE STRUCTURES WHICH HAVE SURVIVED THE PAST EARTHQUAKES.

PROPERTIES OF EARTHQUAKES IN CENTRAL EUROPE

D. Procházková [1]

INTRODUCTION

In earthquakes, the complex problems of the possibility of predicting them and determining the seismic risk of a specific locality from both the practical and scientific points of view are of paramount interest to us. The prerequisite is the study of earthquakes and the regularities of their space-and-time distribution, and last but not least also the study of genetic connections between the foci of strong and weak earthquakes with horizontal and vertical inhomogeneities of the geological medium in the Earth's crust and the upper mantle.

Each earthquake originates as a result of certain physical causes, however, its origin is also connected with a number of other conditions which because of their pecularities and great variety are the sources of random features of the process.

The application of uniform metodology (Procházková 1984, 1988a, 1990) allows establishing common and different properties of single earthquakes as well as of individual regions in respect of quantitative and qualitative characteristics of seismic regimes. Attention must be directed towards both the study of events that form the majority and fit statistical laws, and events that distinctly differ from these used laws. In case that anomalous events (such as e.g. different duration time of aftershock sequences, prominently weak or prominently strong aftershocks, etc.) are well documented, we investigate the causes of their differences and consequently we usually obtain new findings on earthquake occurrence. Although the causes of all differences have not been identified as yet, establishing their existence is a valuable contribution to the knowledge of objective laws and regularities.

The present paper summarizes the properties of the macroseismic fields, the seismic sources, the magnitude-frequency distribution and the seismic regime (space distribution, time distribution, space-time distribution of earthquake foci).

DESCRIPTION OF THE REGION UNDER INVESTIGATION

From the geological point of view, the area of Central Europe is mainly formed by the Hercynides and Alpides. Central European Hercynides lie mostly on the outer margin of the Alpine-Carpathian foredeep. Their principal outcrops are the Bohemian Massif, Black Forest, Vosges, the Rhine slaty mountain range. They are hidden below the platform sediments between Munich and Berlin, and partly between the Oder and Vistula lineaments. Central European Alpides comprise the region south of the Alpine-Carpathian foredeep; they are divided into the Alps (Western, Eastern, Southern), the Carpathians (Western, Eastern, Southern) and the Dinarides. Part of the Alpide region are also central "median" massifs, e.g. the Pannonian Massif. The development in the course of the Upper Tertiary shaped the tectonic skeleton of the present Central European relief (Procházková,

---

[1] GEOSCI, Praha, ČSFR

Roth 1991).

The specific feature of the region under study is low seismic activity on its greater part. Therefore, we had to study historical and present earthquakes and to use all available sources of information and data (lists of more than one hundred references are contained in Procházková 1984, 1991). In the catalogues used, macroseismic data predominate over instrumental ones.

The map of epicenters of earthquakes with epicentral intensity $I_o \geq 5°$MSK-64, Fig. 1, documents that the earthquake epicenters in the study area are unevenly distributed. The map contains data of the last five to eight centuries; the set of shocks in the map begins being regionally homogeneous since as late as the beginning of the 19th century when cultural differences in Central Europe became balanced. The map points to clustering of shocks into regions.

A comparison of regional findings on the geologically recent orientation changes and the relative size of principal stresses in the crust, geologically and geomorphologically established young movements of the Earth's surface, geodetically established recent movements with historical and recent seismicity allowed the determination of the features of the focal regions in which stronger earthquakes occur. It shows that the seismicity of the region under consideration is connected with the current tectonic movements (Procházková, Roth 1991).

PROPERTIES OF EARTHQUAKES

The analysis of the quantitative earthquake characteristics (Procházková 1984, 1988a, 1990) shows that even in a single focal region, earthquakes differ in size, mechanism, focal depth and source dimension. The earthquake foci are usually associated with faults (many earthquake foci lie on the crossing of faults) that separate units characterized by different geologico-tectonic processes in the recent time. In the period under study only some parts of the faults are active. If the focal regions are connected with fault crossings, we often observe the earthquake foci to be related to one or another fault, one of them being preferred in the recent time as far as earthquake occurrence is concerned. In some places, a shock connected with one fault system is followed by a shock connected with the other system (e.g. in 1976 in the Friuli region, the 1985-86 earthquake swarm in Western Bohemia, in 1984 in the Semmering region, etc.).

In greater part of the study area, earthquake foci occur in the upper parts of the Earth's crust, i.e. $h \leq 10$ km; characteristic focal depths in the majority of the focal regions range between 5 and 8 km. In the south of the study area, the seismoactive layer is thicker, reaching a depth of as much as 25 km; two highly seismoactive parts were identified here round depths of 5 and 17 km. A pronounced difference can be found in the Vrancea region only, in which the foci are in deeper parts, in particular round depths of 42 and 128 km.

The shapes of macroseismic fields depend on the focal region; there are great differences between individual focal regions. While the elongation of isoseismals in the near zone points to a system of faults in the focal region and to the focal mechanisms that triggers the motion of the system of blocks

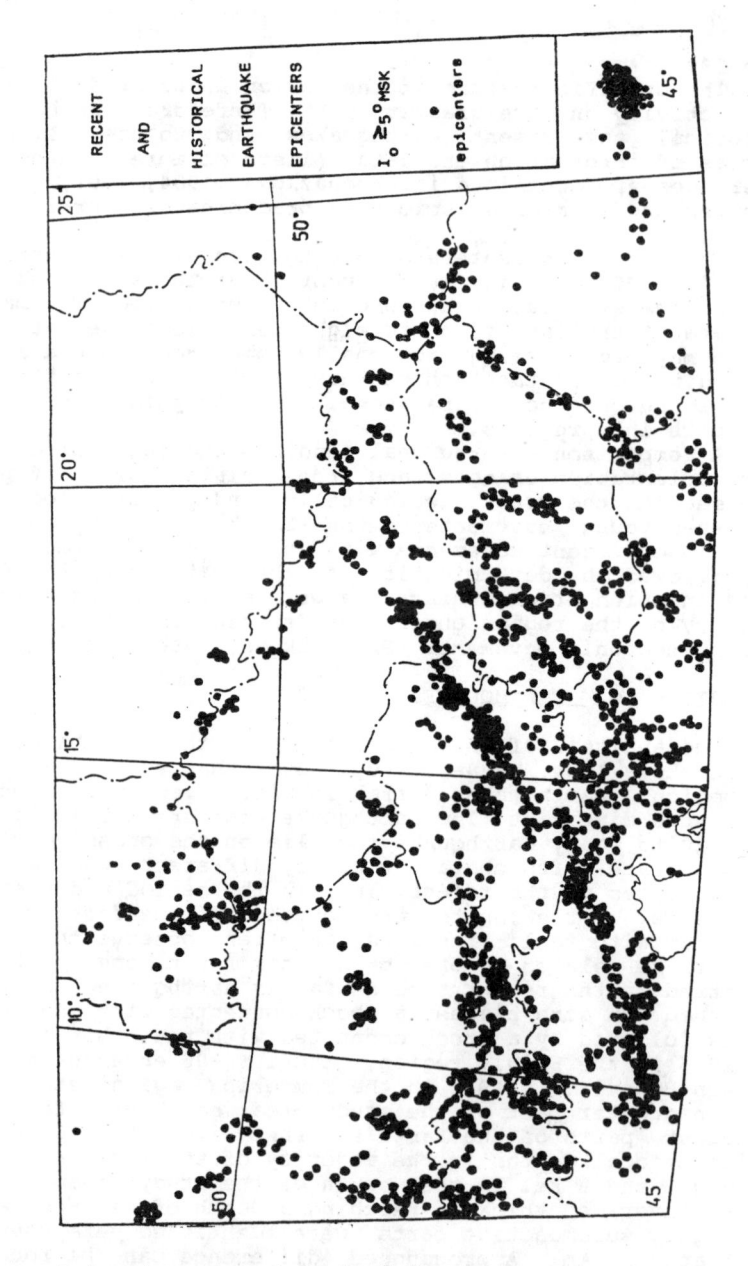

Fig. 1. Epicenter map of Central Europe.

under an earthquake, the elongation of more distant isoseismals (with mean radius r>2.5 h, h being the focal depth in km) depends on the structure of the Earth's crust and the upper mantle, through which seismic waves propagate. Small intensity attenuation is observed in the Bohemian Massif, in the Moesian and East-European Platforms, i.e. in earlier geological units. Great intensity attenuation is observed in young geological units (Western Carpathians, Pannonian Basin), mainly in the vicinity of interfaces with earlier units. Details are given in works (Procházková 1984, Procházková et al. 1986).

In the study area, the size of macroseismic fields is directly proportional to the earthquake size and the focal depth, and indirectly proportional to the attenuation coefficient; the mean statistical dependence is given by the relation (1).

$$\log S_n = (2.6 \pm 0.11) + (0.53 \pm 0.03) I_o + (0.10 \pm 0.05) h - (0.60 \pm 0.02) K_n - (13.00 \pm 0.19) \beta_2 ,$$
$$K_n = 3 - 9^o \text{ MSK-64}, \qquad (1)$$

where S_n is the plane (in km^2) of the isoseismal with intensity K_n; I_o is the intensity in the epicenter; h the focal depth in km; and β_2 the attenuation coefficient (Koevestligethy formula) in the distant zone.

Appart from the general character of intensity distribution (intensity values decrease with growing distance from the epicenter), we can observe a local increase or decrease of intensities with respect to the general character of the field. In some cases we observe these phenomena systematically; for instance, in earthquakes in the Eastern Alps, a systematic intensity increase is observed, e.g. in the České Budějovice Basin and the surroundings of Pardubice (see example in Fig. 2); in earthquakes in the Hronov-Poříčí region a systematic intensity increase is observed around Jablonec and Tanvald. This phenomenon corresponds with the local geological structure of the concrete region (Procházková, Drimmel 1983).

The numerical values of the parameters a,b of the empirical function $\log N_c = a - b I_o$ which describes the distribution of the number of earthquakes (N_c - is cummulative frequency) according to the intensity in the epicenter (I_o), are given in papers (Procházková 1984, 1990). They vary within rather broad ranges, depend on the focal region and on the time interval. In the period of last 80-130 years it holds

$$a = (0.15 \pm 0.39) + (7.95 \pm 0.79) b,$$
$$\sigma = 0.65, \quad r_k = 0.864 \qquad (2)$$

where σ is the mean square error and r_k is selected correlation coefficient.

The time pattern of earthquake regime is not frequently depicted by Benioff's graphs (see examples in Fig. 3). On their basis, we single out active and quiescent periods in a focal region. 57 Benioff s graphs for the focal regions in the study area together with a detailed discussion are given by Procházková (1984, 1988a). Synthesis of these results shows that:
- individual active periods of one focal region can differ in their character;
- individual focal regions differ in the duration of active and

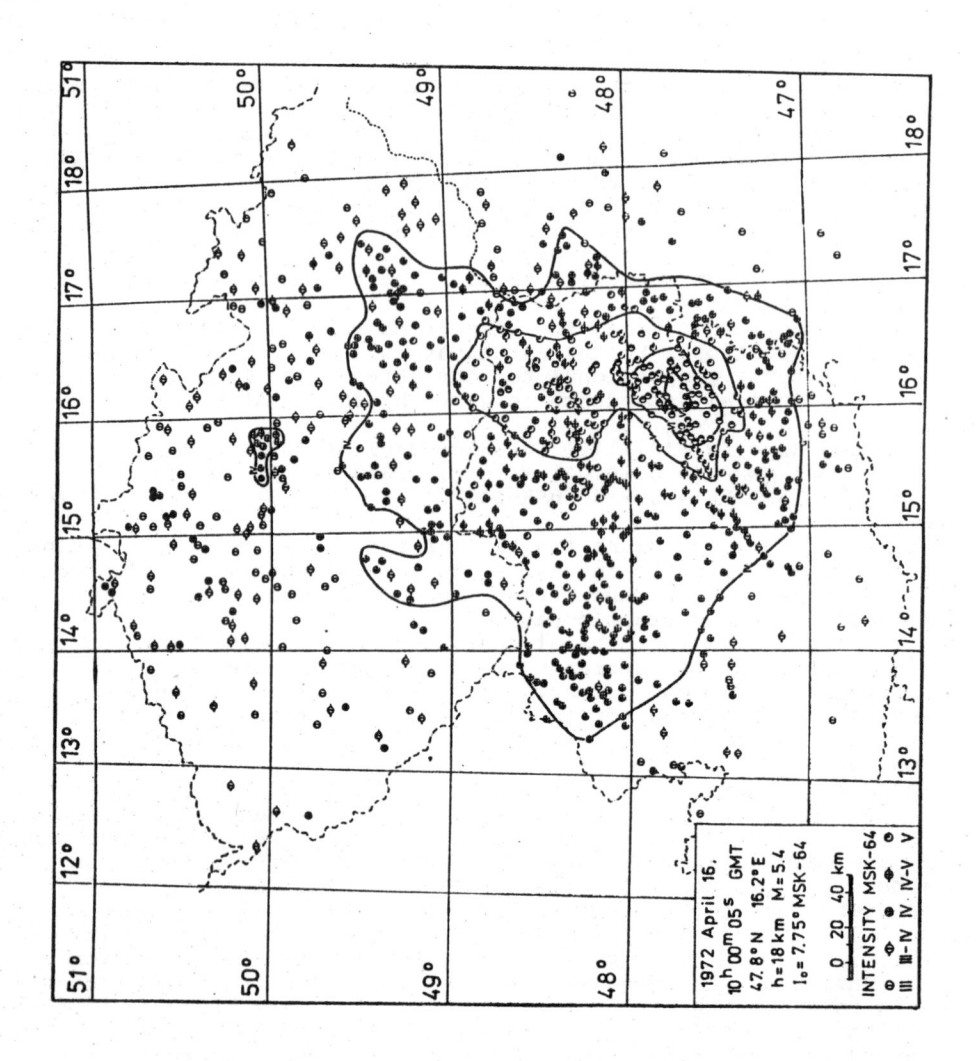

Fig. 2. Isoseismal map of Alpine earthquake of April 16, 1972.

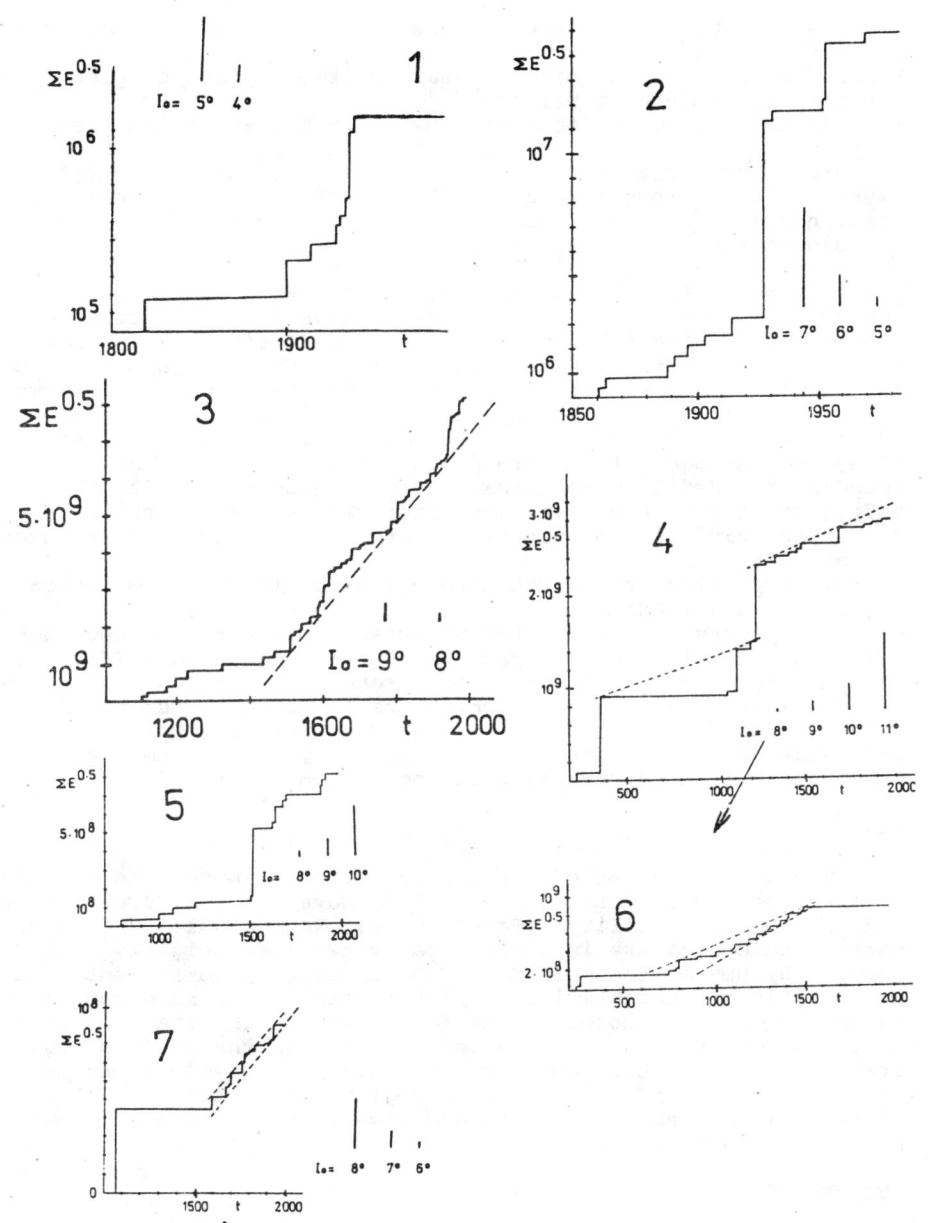

Fig. 3. Benioff's graphs : 1 - Southern Bohemia, 2 - Bakony Forest, 3 - Vrancea (h=i), 4 - Brescia, Lago di Garda, 5 - Slovenia, 6 - Verona, Padua, 7 - Bavaria. In graphs individual earthquakes are not denoted by means of independent $E_i^{0.5}$ [$J^{0.5}$]. In each time unit [year], we give the sum of energy of all earthquakes which originated since the beginning of the selected time period to date.

13

quiescent periods (since several years to several hundreds years);
- also in the domain of microearthquakes we can singled out active and quiescent periods.

Earthquakes also tend to cluster in the short-time range, there are groups of earthquakes differing from one another by a different structure and different properties (Procházková 1984, 1988a, 1990). Strong earthquakes ($I_o \geq 8°$ MSK-64) are usually the main shocks of multiple shock groups (two or more shocks of comparable intensity with aftershocks or with foreshocks and aftershocks). Shocks with epicentral intensity $7° \leq I_o < 8°$ MSK-64 have generally foreshocks and aftershocks. Earthquakes with epicentral intensity $6° \leq I_o < 7°$ MSK-64 have always aftershocks. Shocks with epicentral intensity $I_o < 6°$ either have aftershocks or they, themselves, are aftershocks or foreshocks of stronger earthquakes; they may also belong to earthquake swarms. A great majority of the dependences derived for foreshocks, aftershocks and multiple shock groups (Procházková 1984, 1988 a, 1990) are analogous to the relations derived for the area surrounding the Mediterranean Sea, Japan, Greece, etc.; they only differ quantitatively, i.e. by numerical values of the parameters. The original findings of the study of the area under investigation are :
- the properties of earthquake swarms in Western Bohemia (Procházková 1988b);
- the discovery of two different types of aftershock sequences in the same place of one focal region (Procházková 1984, 1987).

Analysis of earthquake occurrence in space and time (Procházková 1984, 1988a) shows that within a certain focal region, and also in a wider area comprising more regions, earthquake foci migrate, and the activity in one region often means quiescence in the neighbouring region.

CONCLUSION

Though character of endogenic tectonic movements in the area under study in the last 5-10 Ma have a considerable time and regional stability (Procházková, Roth 1991) we do not observe this feature in short-term view; i.e. Chigarev (1980) found rhythms of tectonic movements lasting several centuries. An analysis of characteristics of the seismic regime given above has shown that in none of the focal regions in the area under study nor in the entire area under study is the seismic regime stationary. It is the consequence of various physical processes taking place in individual focal regions connected with changes of tectonic processes in space and time in the time scale comparable with human life.

REFERENCES

Chigarev N.V. (1980): Seismogenez i blokovoe stroenie zemnoj kory (na primere Srednej Azii). Doklady AN SSSR, 255, 313 (in Russian).

Procházková D. (1984): Analýza zemětřesení ve Střední Evropě. Doctor s Thesis. Geoph. Inst. Czechosl. Acad. Sci., Praha (in Czech).

Procházková D. (1987): Properties of Earthquakes in the Mur-

Muerz-Leitha-Little Carpathians Region. Publ. Ser. Swiss. Seismolog. Service, no 101, Zuerich, 147.
Procházková D. (1988a): Earthquakes in Central Europe. Contr. Geoph. Inst. Slov. Acad. Sci. 1985, Veda, Bratislava, 18, 62.
Procházková D.(1988b): The 1985/86 Earthquake Swarm in Western Bohemia. Seismolog Research Letters, 59, 71.
Procházková D. (1990): Seismicity of Central Europe. Publ. Inst. Geoph. Pol. Acad. Sci., B-14 (231). PAN Warszawa-Lódz, 96p.
Procházková D. (1991): Catalogue of Earthquakes for the Territory of Bohemia and Moravia for the Period 1981-89. Manuscript.
Procházková D., Roth Z. (1991): A Complex Study of the Process of Earthquake Origin in Central Europe. Manuscript.
Procházková D., Dudek A., Mísař Z., Zeman J. (1986): Earthquakes in Europe and Their Relation to Basement Structures and Fault Tectonics. Rozpravy ČSAV, 96. Academia, Praha, 86p.

TWO HISTORICAL EARTHQUAKES ON THE TERRITORY OF SLOVAKIA.

Ivan Brouček[1]

The last published catalogue of earthquakes on the territory of Czechoslovakia appeared in the year 1957 [1]. The problems connected with the estimating of seismic hazard enforce a revision of this catalogue based on a new study of the original documents. On example of two historical earthquakes the results of such revision are demonstrated.

THE 1443 EARTHQUAKE IN CENTRAL SLOVAKIA.

According to historical sources, a strong earthquake occured in the Western Carpathians in the year 1443. The large extent of the macroseismic field and the discrepances in data led the authors of seismological catalogues to the opinion that it might be the question of several earthquakes in different regions dated about the year 1443 [1].

INPUT DOCUMENTS AND DATA.

In respect to reliability and trustworthiness, reports on the earthquake can be divided into three categories:

A) Hand-written reports by the contemporaries of the earthquake. There exist six original documents on the earthquake, written at the time the witnesses to the event were still alive.

B) Reports by professional historians and chroniclers of the later time.

Reports in the documents of this category are based on earlier, mostly unknown documents. In some cases, the origin of a report can be established on the basis of comparison of its contents, sometimes the authors themselves refer to their predecessors. It can be generally presumed that the data become less reliable with the time distance from the event unless, of course, their origin is given.

C) Data from the catalogues of natural disasters and from the documents of local importance.

Catalogues of extraordinary natural events began appearing roughly at the beginning of the 18th century. Their authors were usually scholars concerned with natural sciences. Many of them devoted a lot of time and effort to the work, they lacked the experience of professional historians needed for a critical evaluation of historical data and of the reliability of the authors of early chronicles.

---

[1] Geophysical Institute of Slovak Academy of Sciences, Dúbravská cesta 9, 842 28 Bratislava, ČSFR

THE DATE OF THE EARTHQUAKE

All the original documents (i.e.category A) agree in the date 5th June 1443, even though the form of presentation are different. The documents of category B present the date in an analogous way.

A detailed analysis of the sources and a comparison of texts revealed that other dates in connection with this earthquake had turned up by mistake.

The hypothesis of only one earthquake is also backed by virtually identical data on the hour of the event. With a wiew to the unreliable methods of determining time, the data dispersion is adequate (between 7 and 10 o`clock), the most probable time being 8-9 o'clock in the morning (i.e. 7-8 h UT).

REPORTS ON THE EARTHQUAKE EFFECTS.

A detailed report on the earthquake is contained in the Codex of Kremnica. The data do not make it clear, however, which locality the report concerns. We can presume it was in Kremnica. The damage caused by the earthquake of June 5th 1443 is not described in detail, the report only mentions that the earthquake afflicted towers and buildings. Important information is contained in the next part of the document in which mention is made a long aftershock series ("the earthquake lasted throughout the next year"), and some more shock are described. The descriptions of the aftershock series and the other shocks show that the epicentre was close to the place where the notes were written.

A relatively great volume of data on effects in Banská Štiavnica has been preserved. The entry in the municipal protocol (dated 1501) mentions serious damage in the town; buildings collapsed, mine working were heavily damaged, shafts were filled. The concusion of the entry "the most solid hills suddenly turned into deep valleys" suggests that in the area of the town landslides occured. The chronicler M.Bél treats the earthquake in great detail. His account obviously rests upon the entry of the municipal protocol. He writes about the damage to fortifications and his report continues with description of the reconstruction of the destroyed town. Two passages of his text mentions about losses of lives: "it seems that (the earthquake) buried together with it (i.e. the town) the inhabitants, as well", "the number of inhabitants, reduced by repeated disasters (i.e. by the fire in 1442 and the earthquake)". He continues writting that the ruins of the old town, which was situated on a step slope, collapsed into the valley lying beneath, and therefore the inhabitants built up a new town in another place fearing the disaster might repeat.

More recent archeological finds are in agreement with these documents.Unearthed objects, mainly ceramics, provide indirect evidence of the extensive disaster which afflicted in the middle of the 15th century. According to the finds dated to the 13th-14th centuries, the initial character of structures in the study locality was distinctly defensive as opposed to the finds of the 15th-16th centuries which point to the existence of buildings of primarily economic character (roasting kiln for ore, assay furnace). It can thus be assumed that in the course of the 15th century, architecture forcibily ceased to serve its original purpose and was replaced by new architecture [2].

After historical sources, another locality intensively afflicted was "castle Libec in the Zvolen District, which fell into ruin but for one cellar, and ower 30 persons were buried". Judging by the name, one would think it refers to the town of Ľubietová, in its surroundings, however, there was no castle and therefore some historians think it concerns town Slovenská Lupča. This question has not yet been satisfactorily solved.

The chronicle quotes other localities where more serious damage occured: "a castle in the Prievidza district" (most probably Castle Bojnica), and the town of Prievidza itself, in which a church collapsed.

Damage is also mentioned in the two Polish sources (Dlugosz, Calender of Krakow), the former writes explicitly about Krakow-St.Catherine's Convent. The latter only gives a general description of cracks in walls and vaults. The earthquake also afflicted towns in Zips area. Factual data are only available on the town of Levoča. Of interest were the inscriptions in the cathedrale giving important events in the life of the town in the period between 1241 and 1506. The inscriptions have not been preserved, but its text is known. The first 0f them was very brief: "In 1443 an earthquake". The other, a more detailed one, had the following text related to the earthquake: "On June 5th 1443 A.D., a general earthquake occured which destroyed a lot of buildings". We can assume that unless they originated successively with the events described, the inscriptions were made at the beginning of the 16th century at the latest, and by their nature they actually belonged to category A.

The next reports pertain to places farther from the epicentre. Two Viennese chronicles do not explicitly mention observations in Vienna, but they are quoted by the Czech chronicler Lupacius. He writes about observations made in Bohemia naming the town of Hradec Králové and its environs, and about the effects in Moravia (towns Olomouc and Brno). Anomalous effects are described by Frühauf in his chronicle,after which many a corn-loft and barn collapsed in the surrounding of the town of Jihlava due to the earthquake.

According to an other chronicle, in the environs of Moravská Třebová in the year 1443 "a violent earthquake was felt, which struck people with unusual terror". Despite the fact that the chronicle was written as late as at the close of 17th century, from the following text it can be inferred that the reports are reliable.

The reports contained in Silesian chronicles speak about a strong earthquake, even about ruined buildings, factual data, however, only exist from Wroclaw (effect are not given), Opava (houses shaking, tower bells chiming) and from Briegk (church vault damaged).

MACROSEISMIC FIELD.

The largest number of reports on the most serious damage concentrate into areas of Banská Štiavnica and Ľubietová. The intensity in Prievidza seems to be somewhat lower. The centre of these three observations lies nearby Kremnica towards the south, which backs the hypothesis on the epicentre being in the vicinity of the town. Under this assumption and with the view of other reports, we find the macroseismic field to be strikingly asymetric. From the regions south and sutheast of the afflicted area, reports the totally missing. As opposed to it, in the northwest direction data are available from a relatively great distance (Wroclaw, Hradec Králové, Jihlava). Similar asymetry, though perhaps not so pronounced, is also observed in other cases.

For that reason, the asymetric shape of the macroseismic field is not in contradiction with the facts established so far. Apart from that, the paucity of reports may have been due to other causes too, first of all to wars and the longlasting Turkish dominance in these regions, under which the documents were destroyed or transferred far from the place of their origin.

The epicentre coordinates were computed from the observations in the epicentral region (i.e. localities with intensity 8° MSK). Since the data on the effects in Nová Baňa are not documented well enough, the observation was not included into the computation, and the centre of the remaining four localities (Prievidza, Kremnica, Ľubietová and Banská Štiavnica) was determined as the probable epicentre. Coordinates 48.71°N and 18.94° E based on this assumption. There are not enough data available for the error in the coordinates to be determined. It can be presumed, however, that in any case the epicentre lay inside the isoseismal of 8°MSK, i.e. roughly, in the radius of 25 km.

The determination of focal depth was far more difficult. As a matter of fact, the uneven distribution of observation points does not allow complete isoseismals to be constructed, isoseismal 7°MSK is totally missing. For this reason, focal depth was computed from only a part of macroseismic field ($h=47$ km according to Koevesligethy's, $h=67$ km according to Blake's formula) and the resultant values have to be accepted with reservations; it can be assumed, however, that the earthquake focus was in comparatively greater depth than the foci of common earthquakes in the Western Carpathians.

It can thus be inferred that June 5th 1443 there ocured

a single strong earthquake in the region of Western Carpathians, the other dates mentioned in this connection do not correspond to real facts.

THE EARTHQUAKE ON 15TH JANUARY 1858 IN ŽILINA.

In the evening hours on 15th January 1858, northwestern Slovakia and the ambient regions were shaken by a strong earthquake, whose effects were manifest in a vast area. Therefore the authors of older catalogues set a evidently overestimated intensity $I_o = 9°MCS$ [1,2].

OBSERVATION MATERIAL.

Of the copious official documents on investigation of the earthquake effects, relatively few were found. The reason are first of all the changes made in the state and administration systems of the afflicted area.

The principal sources of information are the published professional works by four nature sciencists, F.J.Schmidt, H.L.Jeitteles, M.B.Sadebeck and G.Kornhuber. The first three even undertook study trips to the afflicted area in order to see the earthquake effects with their own eyes.[3,4,5,6]

The works mentioned and other documents contain 380 data on the single localities in which the earthquake was observed another 50 data are negative. As there were no intensity scales at that time, in most reports the effects of shocks are classified by general terms, such as "violent, strong, mild, observable", etc., a more detailed description allowing an intensity evaluation is given only if damage incurred, or if from the viewpoint of that time, interesting or unusual phenomena were observed.

BUILDINGS IN AREA OF ŽILINA.

In the region afflicted most intensively, the majority of buildings were rural, one-storeyed wooden houses. The chimneys of only some of them were bricked. In larger villages there were even masoned houses, usually built of worked stones connected with mortar. Roofs were covered with wooden or straw thatches, particularly in villages. Masoned houses were frequently built on earlier foudations, in the course of their existence they were enlarged or annexed to other structures and thus did not constitute a compact unit due to the building material used, for one thing, and for the other, to the quality of work. Churches were an exception, they were more stable thanks to better-quality construction and minimum of subsequent alterations.

After concordant reports, the most intensively afflicted town was Žilina, in the middle of 19th century an important trade and artisan's centre of northwestern Slovakia. At the time of the earthquake, the town had about 2500 inhabitants and 381 houses. The area of the town corresponded to the present historical core, situated on a small elevation formed of limestone, occupying some 25-30 ha. Most of the houses in the town were masoned, built of worked stones connected with mortar, parts of some of them even of bricks as it can be inferred from description of damage. About a quarter of houses were built of wood. These mostly onestoreyed little houses were in the outskirts of the town outside the fortified area. In the centre of the town there was a square lined with renaissance two-floor bourgeois houses preserved until now. Three sides of the square were rimmed with vaulted arcades. The town had three churches, St.Paul with two towers, a Gothic church of the Holy Trinity with one tower (it burned down during the fire in 1848 and had not been rennovated at the time of earthquake), and the Franciscan Church outside the bulwark one tower.

EARTHQUAKE EFFECTS.

The earthquake caused panic throughout the town. People fled from their houses, mothers carrying sleeping children wrapped in blankets. They spent greater part of the night in the open air, some sought shelter in the suburb wooden houses which were only slightly damaged an obviously appeared safer than the damaged masoned buildings. Nobody was killed or seriously injured.

All the four works mentioned give a detailed description of the earthquake effects, three of them on the basis of the author's own experiences, Kornhuber according to official documents issued by the Vice-Regent's Office in Bratislava. For an exact assesment of damage, however, of decisive importance is the list and assesment of damage elaborated by A.Trajczik, a construction expert, in February 1858.[7]

All the houses in the town were damaged. In the masoned buildings cracks were generated, in many cases along the entire height of the building, as many as several cm wide. Plaster dropped off. On the first floors, joists slipped out so that the ceilings either completely collapsed or were hanging down wraped. Gables ceacked and separated from masonry, in one case a gable fell down. Chimneys either collapsed or were heavily damaged. The vaulted arcades in the square were broken, cracks appeared even in pavements beneath. In the churches, statues of saints fell of their pedestals, the organ pipes fell out. In the church at the square (St.Paul) crack formed and plaster dropped off, the massive jack arch broke. The towers did not suffer serious damage. The adjacent building of the episcopal orphanage was heavily damaged, cracks appeared mainly on the first floor. In addition to similar damage, there were also cracks in the tower of the Franciscan Church.

* In five houses the extent of damage was such that they had to be declared as uninhabitable. The damage to each was estimated at more than 1000 ducates. The damage to the orphanage amounted to the highest sum - 2000 ducates. Approximately another 30 houses suffered damage amounting to more than 500 ducates, and their first floors were uninhabitable.

Wooden houses were only slightly damaged, in the official assesment no sums are given; in the conversation with a visitor to the town the author of the report gave average damage to wooden houses a making 3-5 ducates (probably damaged chimneys, stoves or loosened beams). Total damage amounted to 37000 ducates (the sum does not include the damage to churches).

From environs of Žilina reports on more extensive damage are avilable from Kysucké Nové Mesto, but no details are given. Other data only pertain to individual buildings, churches in particular. The reason for the relative paucity of data on damage is first of all the great portion of resistant wooden buildings which constituted a majority of structures in smaller settlements.

Serious damage was caused to the pilgrimage church in Višňové, which had to be temporarily closed. The vaults cracked in many places, the main jack arch was crushed in several places above the choir, bricks loosened throughout the masonry thickness. In manor house Bytčica, vaults of the rooms on the first floor as well as the ground floor were torn, especially in the corners. The manor house in Teplička was damaged likewise. Minor cracks were reported in churches of Terchová, Varín, Belá, Divinka, Stránske, Strečno, Konské and in Gbelany Castle.

Harmful effects afflicted more distant places, too, but the data on damage are from sources, whose trustworthiness could not be reliably verified, e.g. from newspaper reports, therefore, in the evaluation they were considered only in connection with other effects observed in the respective place or its proximity.

MACROSEISMIC FIELD.

In the proximity of the epicentre the macroseismic field is influenced by the local geological structure of the region, therefore, the highest isoseismal is elongated in the direction of the Carpathian arc. In the data processing, the shape of isoseismals was indoubtely also affected by the uneveness of inhabitation and by the type of buildings.

Isoseismals of lower degrees are distinctly elongated in the direction to the northwest reaching far into the Bohemian Massif. While in the Carpathians the observed intensity is due to seismic wawes which propagate in surface layers with a relatively high attenuation, into the Bohemia Massif seismic energy comes from the deeper parts of the Earth's crust where the attenuation is lower. A similar phenomenon is observed in the propagation of the Eastern Alps earthquakes.

The macroseismic epicentre computed from the isoseismal of 7°MSK lay some 7 km east of Žilina (coordinates 49.22°N, 18.85°E). For the computation of the depth, 6 isoseismals were available. With the use of Koevesligethy's formula the focal depth was determined at 12 km, the application of Blake-Shebalin's formula gives h=18 km.

The analysis of damages to the buildings in Žilina gives the intensity I_0=7.5°MSK.

REFERENCES

1. Kárník V., Michal E., Molnár A.: Erdbebenkatalog der Tschechoslowakei. Travaux géophysiques 1957, Praha 1958.

2. Réthly A.: A Kárpátmedencsék Földrengései (455-1918). Budapest 1952.

3. Jeitteles H.: Bericht über das Erdbeben am 15. Jänner 1858 in den Karpathen und Sudeten. Sitzungsberichte der Mathem. Naturwiss. Classe der Wissenschaftlichen Akademie in Wien, XXXV., Wien 1859.

4. Kornhuber A.G.: Das Erdbeben vom 15. Jänner 1858, besonders rücksichtlich seiner Ausbreitung in Ungarn. Verhandlungen für Naturkunde III. 1. 1858., Bratislava 1858.

5. Schmidt J.F.: Untersuchungen über das Erdbeben am 15. Jänner 1858. Mittheilungen der k. k. Geographischen Gesellschaft II. 2. 1858. Wien 1858.

6. Sadebeck B. A.: Das Erdbeben vom 15. I. 1858 und seine Ausbreitung in Preussen. 35. Jahresbericht der schlesischen Gesellschaft für Vaterländische Kultur 1858. Breslau 1858.

7. Trajczik A.: Bericht über die Schäden in Sillein am 15. Jänner 1858. Manuscript, Region archives Žilina.

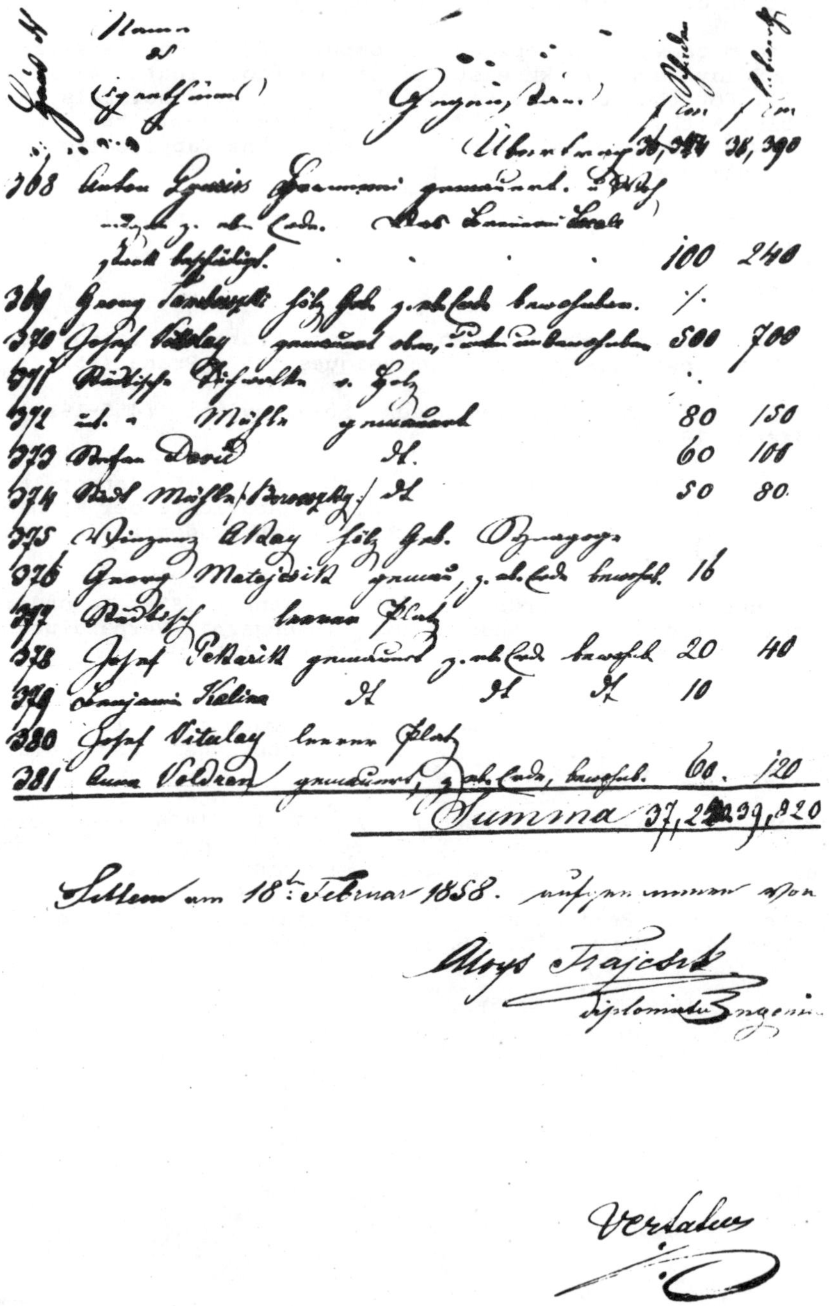

Fig. 1. Facsimile from [7] concerning the damages during Žilina earthquake 1858.

Fig. 2. The old picture which is documenting the earthquake in the town Komárno. Notice the towers of churchs which are falling down.

SEISMIC PERFORMANCE OF LIFELINE STRUCTURES, CASES FROM MANJIL, IRAN EARTHQUAKE OF JUNE 1990

M. R. Maheri[1]

INTRODUCTION

The earthquake of 20 June 1990 struck a densely populated rural area of north-west Iran, destroying a number of towns and hundreds of villages. The area worst affected by the earthquake was centred around the town of Manjil in the mountainous Rudbar region of Gilan province. The official reports of casualties put the number of dead at over 40,000, with half a million homeless. Numerous rock falls and land slides followed the main shock and the stronger after shocks, blocking roads and further damaging the buildings. Considering it's magnitude (reported as between 7.3 and 7.7), the Manjil earthquake is the strongest quake in recent years to strike centres of population. As in many previous earthquakes in Iran, the collapse of un-reinforced brick masonry roof and floor slabs of 1 to 3 storey houses was responsible for the majority of the casualties. The roofs of most of the collapsed buildings were either the traditional brick masonry dome type or the flat slab, steel I-beam and jack arches. Neither type, when un-reinforced, have the ability to withstand the horizontal forces of an earthquake. There were, however, many reinforced concrete or steel-framed semi-engineered and engineered buildings in the area, particularly in the larger villages and towns. These buildings, in general, behaved better and although the majority of buildings in the epicentral area were damaged beyond repair, most maintained their integrity and did not collapse.

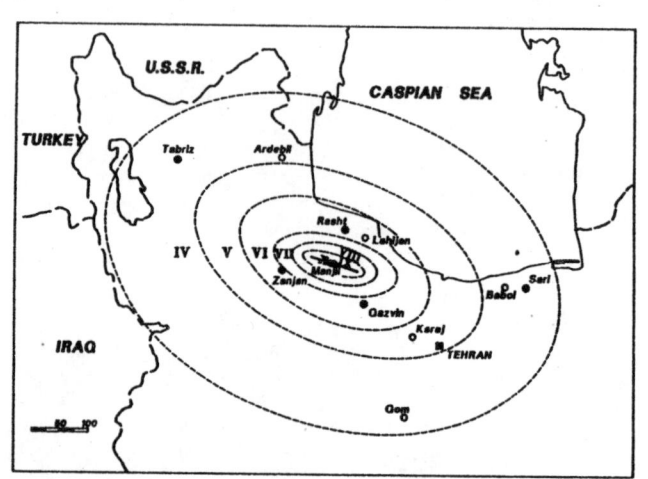

Fig. 1- The isoseismal map of the Manjil earthquake.

1 - WS Atkins Engineering Sciences, Epsom, U.K.

The area devastated by the earthquake of 20 June is a well-watered, historically agricultural, area. Because of it's proximity to the capital, Tehran, in the last few decades it has also developed into a centre for power generation and other industries. Consequently a number of large engineered structures such as dams, power plants, bridges, silos and storage tanks were constructed in the area in recent years. Considering the importance of these structures, not only for the short term need of post-earthquake rehabilitation or the long term economy of the region, but also the consequences of failure in terms of human casualties, their seismic safety should be a prime consideration in design. The majority of lifeline structures in the epicentral area of Manjil earthquake were, however, either not designed or poorly designed for seismic loading. In the following, the earthquake response of a few structures, observed during a post-earthquake visit to the area are discussed.

RASHT WATER-TOWER

The only large engineered structure which completely failed and collapsed in the Manjil quake was a 47m high reinforced concrete elevated water tank. This 20 year old tower (No. 1), situated in the centre of the city of Rasht (some 60km north of the epicentre) was not designed to resist earthquake forces. Two other similar water towers in the outskirts of the city however fared better and, although suffering some damage, did not collapse. The construction of these two towers, identical in design and very similar to tower No. 1, had just been completed and were empty at the time of the quake. This probably accounts for their better behaviour as compared to the ill-fated tower No. 1.

Tower No. 1 was situated in the grounds of the offices of the local water authority, some 6 to 7m west of the main building and only 20 to 30m away from the buildings of a nearby hospital. Fortunately the failed tower was thrown away from the above buildings and collapsed in the central court of the compound (Fig. 2).

Fig. 2- Destruction of the water tank on impact with the ground.

The tower consisted of two thin-walled, prestressed-concrete, cylindrical water tanks with a combined capacity of 1500m^3. The tanks were supported by a 25.50m high, 6.0m diameter and 0.30m thick reinforced concrete shaft, itself placed on a conical, double walled hollow foundation arrangement. The wall of the shaft was reinforced by two sets of 14mm dia.

bars, a reinforcement arrangement hardly adequate to resist horizontal ground loading. The tank was reported to be one third full at the time of the quake. Considering the weight of the tank and the centre of gravity of the tower, the earthquake induced bending stresses in the thin-walled, slender shaft of the tower would have been extremely high. Another factor contributing to the dynamic weakness of the tower under the circumstances lies in its relatively low fundamental frequency of vibration. The strong frequency range of the earthquake in Rasht was also relatively low, leading to high amplifications of the response. Considering the above two factors, the level of ground accelerations required for failure of the shaft and collapse of the tower may not have been very high . The mode of failure of the water tower may be reconstructed from the debris as follows;

i) Possibly at the onset of the earthquake the bending stresses in the shaft exceeded the tensile capacity of the reinforcement bars, leading to bending failure in the shaft. The position of the bending crack, to judge from the remains of the shaft, appears to have occurred at about its mid-height.

ii) The inertia force of the quake then forced the tank and upper section of the shaft westwards, opening the crack in the process and pushing the lower part of the shaft in the opposite direction.

iii) Under the heavy compression and bending forces, the lower half of the shaft crushed and disintegrated into a heap of broken concrete. The tank and upper part of the shaft, meanwhile, followed a free-fall mode , during which time the tank was separated from the shaft, turning over in process.

iv) The final damage to the tank and upper part of the shaft was as a result of impact with the ground. On this impact the half of the upper shaft coming into contact with the ground also crashed whereas the other half remained relatively intact. The fall of the heavy tank on the other hand caused multiple bending, shear and buckling failures in various concrete sections of its structure (Fig. 3).

Fig. 3- **Bending failure in the shaft resulted in collapse of the tower.**

LUSHAN OLD CEMENT FACTORY

This 300 tonne capacity cement factory was constructed some 30 years ago. It is the smaller of two cement factories in and around the town of Lushan some 12 km south of epicentre (the other factory is a new 2100 tonne capacity factory which suffered no serious damage during the earthquake).

Fig.4 Lushan cement factory, damage to the roof and installations.

Although the factory was operational at the time of the visit it had suffered extensive damage particularly in it's associated industrial and residential buildings. Save for the main buildings housing the grinders, the furnace and the large cylindrical steel storage silos, all the other buildings in the factory suffered varying degrees of damage (Fig. 4). The damage in the industrial buildings totalling some 1,500m^2 in area was generally in the form of partial collapse of non load-bearing masonry walls or in collapse of the corrugated steel roofs. According to the factory manager some 1100m^2 office buildings and over 25,000m^2 residential buildings belonging to the workers and staff were damaged between 30% and 70%. The most important of the industrial buildings was the 500m^2 area laboratory within which the majority of the instruments and equipment were damaged either due to the collapsing walls and roof or directly as a result of ground shaking. The damage to the main parts of the factory were as follows;

i) The large rotating 300 tonne furnace was thrown off position along it's long axis for over 1.0m. It did not however roll over it's concrete support as the component of the earthquake in that direction was weaker. The furnace was re-positioned on it's supports shortly after the earthquake. However horizontal cracks could be seen in both of it's concrete supports. These cracks developed as a result of the flexural failure of the 2.0m thick, short, reinforced concrete legs of the supports under the

ii) Although no damage was visible in the cylindrical steel cement silos the thick reinforced concrete foundation bases of these silos had also developed similar cracks as described in (i) above.
iii) Damage to one of the two large cylindrical grinders, which was still out of action at the time of the visit (40 days after the earthquake).

Also the main power supply to the factory was cut during the earthquake as the falling rocks and land slides damaged the power lines and pylons. Because of the importance of the factory, particularly in view of the increased local need for building materials such as cement for the post earthquake reconstruction, the authorities considered it imperative to recommission the factory as soon as possible. As a result, despite loss of life amongst the workers and staff, the partial destruction of accommodation and offices and the damage to the main sections, the factory was operational again within a few days of the main event.

For the reason mentioned above the seismic safety of such plants as cement factories should be a prime consideration and in particular, the secondary response of important elements and installations under the earthquake loading should be given due attention. In design and construction of the 30 year old Lushan cement factory, seismic safety was evidently not a consideration. About two or three miles away from this factory the much larger new cement factory was better equipped to withstand the earthquake loading.

SARAVAN (RASHT) SILO

This 20,000 Tonne capacity grain silo is situated some 40 km north of the town of Manjil and was another large structure which underwent relatively high ground accelerations (possibly a maximum of 0.30g). As is evident in Fig. 5, the structure consists of 27 reinforced concrete cylindrical storage units and two adjoining buildings housing the silo's facilities and offices. The behaviour of this reinforced concrete structure during the earthquake was exceptionally good and no apparent structural damage were evident in the storage units or the associated building. There were however some damage to the systems and equipment which were not either properly isolated or anchored to limit their secondary response. The damage to the facilities nevertheless were minor enough not to affect the functioning of the silo.

Fig. 5- Saravan Silo, no structural damage to this large concrete structure.

SEFID-RUD DAM

The largest and the most important structure in the epicentral area of the Manjil earthquake is the Sefid-rud buttressed concrete dam (Fig. 6). It is situated approximately 2 km north west of the town of Manjil where it collects the waters of the Ghezelozan and Sefid-rud rivers. It is an important source for electricity generation for the region. Sefid-rud dam is a 106m high, 425m long, buttressed gravity dam. There are 23 buttresses, each 5m thick; the width of buttresses at the foundation level is about 100m. The slope of the dam on the downstream face is 1 in 0.6 and on the upstream side 1 in 0.4. It has a vertical crown section 14m high and 10.5m wide. The reservoir was almost full at the time of the main event, the water level being 5m below the maximum level. The water outlets consist of two adjacent intake towers at the west end of the dam and 4 sluice gates at two different levels at both the east and west sides. At the time of the visit the reservoir was being emptied through two sluice gates and the water level was 60 to 70m below the crown. As expected under the circumstances, the drainage of the reservoir started almost immediately after the quake. This was necessary not only to investigate the possible damage to the upstream face of the dam but also to reduce the level of hydrodynamic forces exerted on the dam under subsequent after shocks, as an already weakened dam would be very vulnerable to such after shocks.

Fig. 6- Sefid-rud dam, downstream view.

The Sefid-rud dam was designed in the 1950's, construction began in 1958 and was completed by 1967. Because of the importance of the structure the seismic safety was an important consideration in design. In those days however seismic design of structures was carried out using the equivalent static approach. The dynamic behaviour of such structures was not fully understood and without the modern computational facilities the dynamic and hydrodynamic forces on the dam could not be accurately calculated. The seismic factor adopted for the equivalent static design of this dam was 0.25 [1]. This is a rather high factor compared to other similar designs of the day, reflecting a conservative approach in design.

It is very unlikely that a conservative factor was chosen because of the particular seismic conditions of the site, as there are no references in geological surveys of the day to the presence of a seismic fault crossing the downstream river only 300m north of the dam [2]. Furthermore, if the presence of this fault had been known the dam would almost certainly not have been constructed on this site.

The higher seismic factor enabled the main structure of the dam to resist the high levels of ground acceleration during the earthquake. The fact that the stronger component of the quake happened to be almost parallel to the face of the dam also greatly helped it's behaviour during the quake. Unfortunately, there were no seismographs on or in the vicinity of the dam at the time of the main event to record the level of ground accelerations. The nearest accelerograph at Abbar close to the line of the seismic fault recorded maximum acceleration of 0.65 in the east-west direction and 0.2 in the north-south direction. It can therefore be safely assumed that the accelerations suffered by the dam were in excess of these values.

Damage to the Main Structure

Although the main body of the dam behaved well and retained it's integrity, a number of cracks developed mainly in the buttresses but also in the upstream face of the dam. The structural damage visible can be summarized as follows;

i) Horizontal cracks; These were observed mainly in the upper parts of the buttresses at their intersection with the crown. Most of the 23 buttresses developed these cracks. The cracks were probably extended to the upstream (reservoir) face of the dam where at the same level some horizontal cracks were just visible. The cracks, associated in places with spalling of concrete at the surface to a width of 7 to 10cm, were probably caused by the out-of-plane bending (overturning) action of the crown under the North-South component of the quake. As the drainage of the reservoir was revealing more of the upstream face of the dam, more horizontal cracks could be seen at lower levels. There were also a number of horizontal cracks at the base of the buttresses parallel to the dam.

ii) Diagonal cracks; In a few central buttresses there were also a number of diagonal shear cracks. As the height of the dam in its central sections is more than the end sections the overturning and flexural responses of the structure under the North-South component were higher in those sections, hence resulting in shear failures in the supporting buttresses.

iii) Differential displacements of the dam sections; The dam structure consists of 23 sections separated from each other by construction and expansion joints. Under the horizontal ground motion some lateral differential displacements developed between these sections. An investigation carried out by the engineers from the Ministry of Energy revealed a maximum of 50mm difference in the alignments of the bench marks on some of these sections. Considering the overall dimensions of the dam, such relative displacements would be within the expected range.

iv) Pounding of sections; Evidence of pounding of the dam sections against each other could be seen on the crest in the form of spalling of concrete at the joints. The spalling could be seen in most joints, however, the extent of the pounding damage at the interface of the adjoining sections could not be investigated. The type and size of seismic joints, if they were at all a consideration in design of the dam, were inappropriate to mitigate the damaging effects of pounding.

Non Structural Damage

i) The main visible non structural damage to the dam was in the long unsupported parapet of the north side. Flexural failure at the central section of this reinforced concrete wall in the form of vertical cracks together with horizontal crack at the base due to the bending failure were responsible for the collapse of two relatively large sections of the wall (Fig. 7).

ii) Subsidence of the fill at both ends of the dam adjacent to the concrete section of the dam, caused by compaction of the loose fill under ground vibration.

iii) Overturning failure in the majority of concrete and stone masonry guard blocks. Most of the blocks were thrown off their foundations by up to two meters in the west direction.

iv) Destruction of the guard post at the east side and guard house in the west side of the dam due to rock-fall (Fig. 8). The reinforced concrete guard house was completely destroyed under the falling rocks, some a few meters in diameter, causing some casualties.

v) Rock falls were evident at both east and west sides of the dam, some of which had blocked the access roads to the dam before being moved aside.

vi) Damage to dam's installations and systems.

Fig. 7- Sefid-rud dam, failure of long, un-supported parapet.

Fig. 8- Destruction of dam's guard-house under rockfall.

RUDBAR CONCRETE BRIDGE

The Rudbar concrete bridge (Fig. 9) is a 190m long reinforced concrete bridge spanning across the Sefid-rud river some 3 km north of the town. It consists of 5 piers and six deck sections each 30m long. The end sections are supported directly on two large end piers.

The width of the bridge deck is 10.5m and it's height is about 10m. The piers are T shaped and are approximately 8m high and 10.5m wide at the top. The bridge deck consists of four concrete beams running along the length of each section and simply supported on the piers and a 35cm thick reinforced concrete slab. As far as could be gathered, the bridge deck sections were directly placed on the piers separated only by 20mm thick rubber pads. There were no rubber pads at the vertical gaps between the end sections and the piers.

Fig. 9- Rudbar concrete bridge, west view.

Considering the high level of ground acceleration around Rudbar (5 km north of the epicentre) this concrete bridge behaved reasonably well during the earthquake. No transverse displacements of the simply supported bridge deck were apparent. The structural damage visible in the bridge are as follows:

Fig. 10- Buckling of pavement due to pounding of bridge deck sections.

Fig. 11- Failure in the south-west retaining wall.

i) The spalling of deck concrete at both ends of the bridge, caused by the pounding action of the relatively flexible and free standing bridge deck against the rigid end-piers. The size of the vertical gap between the two sections was evidently very small. Insufficient gap together with lack of rubber pads or similar shock absorbent elements in the joints were the main causes of failure in these locations.

ii) The second form of failure could be seen on the bridge deck at almost every joint between the deck sections. At these joints the pedestrian concrete paving had buckled (Fig. 10). This was caused by the pounding action of the bridge deck sections against each other. Any possible pounding damage to the bridge deck itself could not be verified. It is probable that the presence of bitumen asphalt between the adjoining concrete deck sections prevented damage to the deck.

iii) The collapse of a reinforced concrete retaining wall next to the south end-pier (Fig. 11).

CONCLUSIONS

The seismic-designed structures in the epicentral area of the Manjil earthquake area behaved generally well and save for certain design deficiencies such as inadequate seismic joints (leading to pounding failure), the level of damage to the main structure was limited and repairable. A number of lifeline structures were also present which were not designed for seismic loading. Bearing in mind that these are major structures situated in an area known to be prone to severe earthquakes, lack of seismic considerations in design is surprising. The behaviour of several important structures during the earthquake also illustrated the point that in seismic design of lifeline structures the safety and strength of secondary structures, systems or equipment is as important as the integrity of the main structure itself.

Acknowledgements

The author's visit to the affected area was on behalf of the British Earthquake Engineering Field Investigation Team (EEFIT). The kind cooperation of the International Institute of Earthquake Engineering and Seismology in Iran during the visit in acknowledged.

References

[1] Moinfar,A.A. and Naderzadeh,A., An Immediate and Preliminary Report on the Manjil,Iran Earthquake of 20 June 1990, Building and Housing Research Centre,Pub. No.119, 1990.

[2] Maheri, M. R., Engineering Aspects of Manjil, Iran Earthquake of 20 June 1990, A Field Report by EEFIT, WS Atkins Engineering Sciences, U.K., Report G8630/90/001, 1990.

SPITAK EARTHQUAKE EFFECTS AND PROBLEMS OF STRENGTHENING OF BUILDINGS

S.G. Shaginian[1]

INTRODUCTION

In the paper there are discussed the problems dealing with nature of Spital earthquake of December 7, 1988, visual survey of damaged apartment and civil buildings and recommendations for their strengthening and restoration.

Life loss after the earthquake that occured in Spitak on December 7, 1988 accounts approximately 25 thousands people. Spitak was entirely destroyed. Cities Leninakan, Kirovakan, Akhurian and Stepanavan were severely damaged; a number of cities and settlements were partially destroyed. The earthquake magnitude in the epicentre was 7.

By final reports of State Committee this is considered to be the biggest earthquake of Caucasus recorded instrumentally.

Devasting effects of catastrophic and destroying earthquakes that occured in Armenia are known long ago. Earthquakes of 893 and 1319 totally destroyed capitals of ancient Armenia Dvin (70 000 people died, 20 000 of them died in Yerevan) and Ani.

The temple of Garni was destroyed during the earthquake of 1840 that occured in Ararat; big rock masses were separated from mountain Ararat destroying many villages. Single storey buildings were destroyed after the Leninakan earthquake of 1926, temple of Tatev was destroyed after the Zahnezur earthquakes of 1931 and 1968 etc.

SPITAK EARTHQUAKE

Some experts assure that Spitak earthquake of 1988 by its nature was quite an unusual one taking into consideration the character of building damages seismological indications. It was developed in series of shocks. Three sequential shocks were regis-

---

[1] Arm. NIISA, Gosstroy of Armenian SSR, Yerevan, USSR

tered during the first 30 seconds. The second earthquake with magnitude 5.8-6.5 occured in 4 min. 20 sec. after the beginning of it with epicentre 6-7 km away in southern direction from site of the main earthquake.

The specifity of the earthquake we can discuss on the examples of damaged buildings and structures in Leninakan.

In the city of Leninakan 5-9- storeyed dwelling houses, public stone and frame buildings (including industrial ones) were mainly destroyed whereas low-rise buildings were not severely damaged.

Apparently, alongside with high intensity which was higher 2-4 times than the design seismicity of buildings provided by codes for aseismic buildings, poor quality of workmanship, building materials and designs, the following phenomena were observed:

- resonance vibration with exceptional duration (35-40 sec.) of main shock;
- significant effects of vertical component of earthquake (what in USSR codes for dwellings and civil buildings was not taken into account);
- inhomogenity of geological structure and soil conditions (engineering-geological, hydro-geological etc.) of city.

Investigations of experts from Arm. NIISA and experts of other organizations showed that 9-storeyed buildings of series 111 with design seismicity 7-8 erected in Kirovakan (intensity of earthquake was 8) did not suffer, whereas in Leninakan they were destroyed. This is another striking proof that they were subjected to shocks with intensity exceeding the design one.

The analysis of damaged buildings showed that if the periods of natural vibrations of subsoil are near or coincide with the periods of natural vibrations of buildings it brings the behaviour of structures forward resonance capable to damage the strongest structures.

In fact during vibrations with long periods (0.6 sec. and more) a the duration of main shock was exceeding 30 sec. High-rise buildings entered into resonance taking approximately 30-40 cycles of vibrations.

Thus we can suppose that one of the damage causes of 5-9 storeyed buidings was resonance. In the results of measurements of periods of ground subsoil and buildings in Leninakan Japanese experts came to the same conclusion. Low-rise buildings were not severely damaged. This was explained by the fact that dominant long-period vibrations in the frequency response spectra were not dangerous for them.

The character of damage shows the significant effect of vertical component of shocks on buildings and structures. In Leninakan also architectural monuments, which were strong enough in the horizontal direction were damaged by this vertical shock. By the evidences of experts and restorers, which were near the church during the earthquake, the bell was separated, uplifted and at last fell into the church; the small bell of another church with 12 tons weight was thrown away for about 12 m from the building into the neighbour yard [4].

The ground was shaking under the feet and by the evidences of doctor from hospital after the main shock he was uplifted with his chair. The gas-stove was turned over, a hollow rumble and wild roar were heard. This was mentioned in all questionaires.

High intensity of earthquake to some degree is accounted for by complex soil engineering-hydro-geological conditions (geotectonics) of the city. The decipherment of cosmic thermal (infra-red) and other surveys revealed great number of tectonic faults in different directions.

Methods of strengthening the effect of earthquake 8-30 times taking into account lake deposits are described in reports of American scientists who have carried out instrumental measurements. Some other reports are done by National Committee of Academy of Sciences in Washington by prof. Koff and others.

Intensity of the earthquake was increased due to the high level of underground water in Leninakan. This is the main cause of different intensities in different parts of the city.

In this way the village Azatan located 5 km from Leninakan did not suffer any damages, whereas Akhurian located in the same distance from Leninakan was severely damaged. The analogous phenomenon was observed in San Francisco and Mexico during the earthquakes of 1906 and 1985. Apparently this is the characteristic feature of destroying earthquakes timed to regions of deep fractures.

The survey showed that in the result of the earthquake were damaged stone dwelling buildings of series IA-450, I-451 and their modifications, frame-panel buildings of series AI-451 KP etc.

Stone 5-storeyed buildings were damaged due to the following reasons:

- loss of stability of end walls due to unreliable bonds with longitudinal walls; lack of floor anchorage in bearing walls; absence or bad workmanship of reinforced concrete antiseismic belts; insufficient cross stiffness of end sections; lack of steady joints between floor slabs which in its turn brougth forward horizontal displacement of slabs and failure of joint connection of floor slabs with bearing walls; weakening of bearing walls in the result of built-in recesses, openings and removing of some parts of the walls; insufficient shear stiffness of stone masonry and separation of walls in longitudinal direction; poor quality of workmanship.

Frame-panel buildings were damaged because of:

- absence of stiffening diaphragms in one direction and their poor workmanship; great mass of exterior suspended wall panels not provided in designs which are significantly increasing inertia loads on bearing structures of buildings; poor quality of frame joints and precast reinforced concrete elements.

Large panel buildings in the zone of disaster practically were not damaged. Problems of their strengthening due to the increase of design seismicity from 8 to 9 are not taken into consideration in this report.

METHODS OF STRENGTHENING

Main methods for restoration and strengthening of damaged structures according to seismicity 9 for stone buildings are following:

- erection of additional interior and exterior longitudinal and cross walls including "flexible" ground floor; maintenance of spatial stiffness of floors by laying finegrained concrete of M150 grade with 40-50 mm thickness over the welded mesh;

- spraying of fine-grained concrete over the metal screen of all bearing walls and partitions from both sides; injection of big cracks (wider than 1mm) by cement, polymer and other mortars;

strengthening of narrow partitions (narrower than 100cm) and stone columns by metallic rings-posts, maintenance of reliable bonds between exterior and interior wals by setting tensioning metallic belts in the floors;setting of additional bonds of floors and walls around the edges;replacement of partitions from heavy elements by ligth effective ones including gypsum boards over the wood frame, sectional perlite concrete ones etc.

Concerning the frame-panel buildings we are to take into consideration the following:

- setting of metal casing on columns and cross-bars, additional metal bonds or reinforced concrete diaphragms in both directions, maintenance of approximately same building stiffness in longitudinal and transverse directions, strengthening of staircase walls and elements, additional strong bondage of exterior wall panels with bearing structures, maintenance of strength and stability of partitions and some other methods analogous with those provided for stone buildings.

Necessity for strengthening bases and foundation is determined in accordance with materials of supervising engineering-geological researches and detailed investigation of foundation structure state.

Albums of type decisions are worked out with the help of designers. Expenses of main building materials for restoration and strengthening of structures on 1m² of the total area are estimated.

Thus according to data of preliminary analysis we came to the following conclusion:

1. Intensity of the earthquake was 2-4 times larger than the building design seismicity provided by maps of seismic zoning of the republic and seismic codes whereas in seismic codes of 1982 intensity of cities Kirovakan, Spitak and other regions was baselessly decreased to 8 and 7 respectively. There is an urgent need to revise designs and engineering codes of seismic regions and to make higher demands to designs of seismic loading (taking into account the vertical component of all structural systems of apartment and civil buildings), to work out more symmetrical planning decisions and strict structural demands [7].

2. It is neccessary to begin mass construction of large panel, cast-in-place and stone buildings restricting number of storeys, to develop on a broader scale usage of metal frame, ligth-weight structures and goods in earthquake engineering. Designs of above mentioned buildings are developed by the experts of ARMNIISA.

3. It is neccessary to pay a special attention to engineering-geological, hydro-geological conditions of regions while making maps of seismic zoning and microseismic zoning taking into account spectrum of expected earthquakes.

4. Designs are to be realized basing on 2 factors: first - erected building should resist forces provided by designs (design level is of first order);
second -building should not be destroyed by strong earthquake occurring once in a thousand years, i.e. safety of life must be guaranteed by design (design level is of second order). These are demands of Japanese codes on earthquake engineering [6].
It is neccessary to work out standards for designing hospitals, schools and kindergartens similarly as in US codes.

5. It is neccessary to revise design standards on town planning in the seismic regions.

6. It is high time to organize building quality control services. Building materials like cement, steel etc. are to be used only in case of having properties indicating their main characteristics. Control services should be independent like Gosgeonadzor and some other organizations functioning in the USA, Japan, Hungary and other countries.

7. It is neccessary to equip seismic, engineering-seismometrical stations of the republic with up-to-date equipment including imported ones.

8. Designs are to be fulfilled by computers following the experience of leading organizations paying attention to equipment of the institutes.

9. To organize special training of experts on earthquake engineering in Yerevan Politecnic Institute.

CONCLUSION

Intensity of Spitak earthquake exceeded the design one provided by standards of design and construction. Poor quality of workmanship in the zone of disaster and grave shortcomings in designs redoubled the tragedy.

There is an urgent need of strict control over the units under erection and a qualitative expert of designs.

Population in seismic regions should be thoroughly prepared to act accordingly during the earthquake.

REFERENCES

1. P. RUSSO: Earthquake. (tr. from French). Progress. M., 1966.
2. Express-information (experience of foreign experts), 14th series, 7th issue, M., 1986.
3. S.V. MEDVEDEV and others: Seismic effects on buildings and structures. Stroyizdat. M., 1968.
4. Express-information (experience of foreign experts), 14th series, 5th issue, M., 1986.
5. Engineering aspects of the earthquake in Mexico 19-20. IX. 1985. Report. Tokyo 1985.
6. M. HIROSAVA: Earthquake engineering. Report. Tokyo. 1988.
7. Express-information (experience of foreign experts), 14th series, 10th issue. M. 1986.
8. Sh. SUEHIRO and others: Report of Japanese experts on investigation of damaged buildings in Nothern regions of Armenia. Tokyo. 1989.

Figs. 1-2. The view of destroyed buildings after Spitak Earthquake, 1988.

T2

SEISMIC DESIGN OF STRUCTURES. THE WAYS OF SEISMIC STRENGTHENING OF EXISTING STRUCTURES.

DYNAMIC BACKGROUND OF QUASISTATIC SEISMIC SOLUTION TO STRUCTURES

O. Fischer[1]

1. INTRODUCTION

The main advantage of quasistatic solution to the seismic response of structures is its simplicity and its compability with a standard static solution. That is why this approach has been used in most seismic building codes up to the end of fifties. Later on, when the requirements of high-rise building structures even in seismic zones have become even more urgent, the dynamic character of seismic loading had to be respected. This was done in some cases in such a way, that the main advantage, namely the avoiding of dynamic solution of the structure, had been lost, but the inacuracy of static solution had mostly remained. Using such a method one had to calculate at first the natural vibration of the structure in order to determine the equivalent seismic forces and only then to take advantage of the simplicity of statical solution itself. It is truth, this approach can be accurate, too, but mostly it was introduced above all to rid the civil engineers of the feared dynamic calculation. I belive, this last reason is no more justifiable, but it is nevertheless useful to remind the real meaning of this method of calculation, in order to have knowledge of its possibilities, limitations or further simplifications.

2. THE PRINCIPLE OF QUASISTATIC SOLUTION

The problem of quasistatic solution to seismic resistance of structures is to determinate horizontal static forces, which would produce the same response (displacement, moments etc.) as is the maximum dynamic amplitude of the structure subdued to the most effective real earthquake. Let us neglect the fact, that the safety of the structure is different under one static loading and under several times repeated dynamic loading, even if the maximum response is the same; we simply try to find equivalent static forces in order to obtain the same response as is the maximum response corresponding to exact dynamic solution.

Even in the early sixties these equivalent static horizontal forces were only a certain multiple of the weight of the corresponding part of the structure, and the seismic coefficient was constant for the whole structure [1]. Later on this coefficient increased with the height, either linearly, or in some other way; in [2] it is the coefficient for the i-th mass

---

[1] - CS Academy of Sciences - UTAM, Prague

$$\eta_k = z_k \sum_{i=1}^{n} W_i \bigg/ \sum_{i=1}^{n} W_i \, z_i \qquad (1)$$

where z_i - the height of the i-th mass

W_i - the weight concentrated at the point i

This formula (1) can be used for low buildings, where no higher vibration modes are expected. A more general formula has been used since many years in the Russian and some other codes [3]. The coefficient is different for each natural mode and for each mass; for j-th natural mode and k-th point we find

$$\eta_{(j)k} = u_{(j)}(z_k) \sum_{i=1}^{n} W_i \, u_{(j)}(z_i) \bigg/ \sum_{i=1}^{n} W_i \, u_{(j)}^2(z_i) \qquad (2)$$

where W_i - weight of the i-th mass

$u_{(j)}(z_k)$ - ordinate of the j-th natural mode at the point of the height z_k

The coefficient (2) has been derived from the dynamic solution; as it will be seen later, its use does not bring simplifications with respect to the complete dynamic solution.

3. NATURAL MODE ANALYSIS IN SEISMIC EXCITATIONS

Solution to many problems of dynamics using natural modes is often advantageons due to its objectivity; it often reminds the static solution by means of static influence-lines. It is well applicable if the damping is small and not much different in individual parts of the structure. But even in the case of large and different dampings a similar procedure can be used passing to generalised complex modes in Laplace-image domain (see [4]).

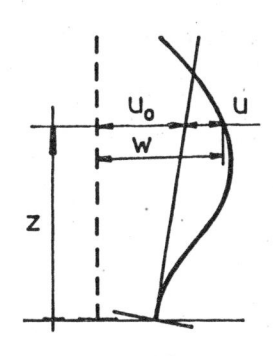

Fig. 1. Seismic excitation

The problem of determination of classical natural frequencies and modes which are used here, can be performed by different methods, well-known from structural dynamics, and there is no reason to argue which of these methods is better. The only requirement is to obtain realistic results not only in displacements, but also in bending moments and other stresses, if necessary. The rest depends on the practice of the engineer, on the computer-type and software he has to his disposal.

Knowing once the natural modes of a structure, the solution procedure is the same with all structure-types, discrete and continuous. Let us demonstrate it on a prismatic beam with uniformly distributed mass: The displacement of any arbitrary point of the structure can be expressed (fig.1)

$$w(z,t) = u_o(z,t) + u(z,t) \qquad (3)$$

where w - total displacement of one point
       u_o - displacement due to the amplitude of seismic excitation
       u - relative displacement due to the deformation of the structure during vibrations
       z - vertical coordinate
       t - time

Supposing viscous damping proportional to the velocity of relative displacement $u(z,t)$ the equation of the motion will be

$$\mu \frac{\partial^2 w(z,t)}{\partial t^2} + 2\mu\omega\beta \frac{\partial u(z,t)}{\partial t} + EJ \frac{\partial^4 u(z,t)}{\partial z^4} = 0 \qquad (4)$$

where μ - beam mass per unit length [kg/m]
       ω - circular frequency of the motion [rad/s]
       β - relative damping (fraction of critical)
       EJ - bending rigidity of the beam [N.m2]

Introducing (3) into (4) and expressing

$$u_o(z,t) = u_o(z) q_o(t)$$
$$u(z,t) = \sum_{j=1}^{n} u_{(j)}(z) \cdot q_{(j)}(t) \qquad (5)$$

where
   $u_o(z)$ - is the static deflection corresponding to the unitary amplitude of seismic excitation (dimensionless)
   $q_o(t)$ - time-history of seismic excitation, dimension [m]
   $u_{(j)}(z)$ - j-th natural mode of the structure (dimensionless)
   $q_{(j)}(t)$ - generalised coordinate of the motion in j-th mode (dimension [m])
   n - number of natural modes taken into account.

Taking also in mind that for the natural vibrations in holds

$$\frac{d^4 u_{(j)}(z)}{dz^4} = \frac{\mu \omega_{(j)}^2}{EJ} u_{(j)}(z) \qquad (6)$$

we can write

$$\mu u_o(z) \ddot{q}_o(t) + \sum \mu u_{(j)}(z) \ddot{q}_{(j)}(t) +$$
$$+ 2 \beta \sum \mu \omega_{(j)} u_{(j)}(z) \dot{q}_{(j)}(t) + \sum \mu \omega_{(j)}^2 u_{(j)}(z) q_{(j)}(t) = 0 \qquad (7)$$

Multiplying (7) by $u_{(k)}(z)$ and integreating each member along the whole structure, with respect to orthogonality of natural modes

$$\int \mu u_{(j)}(z) u_{(k)}(z) dz = 0 \quad \text{for} \quad k \neq j \tag{8}$$

we obtain

$$\ddot{q}_0(t)\int \mu u_o(z) u_{(j)}(z) dz + \ddot{q}_{(j)}(t)\int \mu u_{(j)}^2(z) dz +$$

$$2\beta \omega_{(j)} \dot{q}_{(j)}(t) \int \mu u_{(j)}^2(z) dz + \omega_{(j)}^2 q_{(j)}(t)\int \mu u_{(j)}^2(z) dz = 0$$

or

$$\ddot{q}_{(j)}(t) + 2\beta \omega_{(j)} \dot{q}_{(j)}(t) + \omega_{(j)}^2 q_{(j)}(t) = -\ddot{q}_0(t) P_{(j)} \tag{9}$$

where
$$P_{(j)} = \int \mu u_o(z) u_{(j)}(z) dz : m_{(j)} \quad \text{[dimensionless]}$$
$$m_{(j)} = \int \mu u_{(j)}^2(z) dz \quad \text{[kg]} \tag{10}$$

The equation (9) demonstrates the well-known advantage of natural-mode analysis, giving conditions for generalised coordinates in each mode separately. The seismic excitation is defined by its distribution in space $u_o(z)$, by its magnitude and time-history of acceleration $\ddot{q}_o(t)$.

The value $P_{(j)}$ is the influence-coefficient characterising the contribution of every natural mode to the resulting response.

For seismic excitation in the form of pure horizontal translation

$$u_o(z) = u_o = 1 \quad \text{(const.)} \tag{11}$$

and constant mass the equ. (10) has the form

$$P_{(j)} = 1 \int u_{(j)}(z) dz / \int u_{(j)}^2(z) dz \tag{12}$$

For prismatic cantilever equ. (12) gives

$$P_{(1)} = 1,566 ; \quad P_{(2)} = -0.868 ; \quad P_{(3)} = 0,510 \tag{13}$$

and for shear-type buildings, where:

$$u_{(j)}(z) = \sin(2(j-1)\pi z/(2h))$$
$$P_{(1)} = 1,27 ; \quad P_{(2)} = -0,424 ; \quad P_{(3)} = 0,255 \tag{14}$$

For rocking vibrations, where $u_{(1)}(z) = z/h$

$$P_{(1)} = 1 \cdot \int_0^h (z/h) dz / \int_0^h (z/h)^2 dz = 1,5 \tag{15}$$

Further simplifications of the general expression (12) are possible and can be found in different seismic codes. Mostly the discretised approach has been used, having changed integrals into sums, see (1), (2).

The solutions of (9) depends on the way the excitation

$$a(t) = - \ddot{q}_o(t)$$

has been defined. If it is a deterministic function (a real or artificial accelerogram), the numerical integration will be probably used, introducing the influence-coefficient $p_{(j)}$ according to the natural mode-shape. If the earthquake accelerogram has been considered as a random function, described e.g. by its power spectral density

$$S_{aa}(f) \, , \quad \text{dimension} \quad [m^2/s^3]$$

the generalised coordinate of the response is given by means of its variance

$$\overline{q^2_{(j)}} = \int_0^\infty p^2_{(j)} \, S_{aa}(f) \, |H_{(j)}(if)|^2 \, df \tag{16}$$

where the absolute value of the coplex frequency characteristic (frequency response function, mechanical impedance) is (in the dimension of displacements)

$$|H_{(j)}(if)|^2 = \frac{1}{16\pi^4 f^4_{(j)} \left[\, (1-(f/f_{(j)})^2)^2 + 4\beta^2 \, (f/f_{(j)})^2 \, \right]} \tag{17}$$

In the case the power spectral density can be considered as white noise (i.e. constant, at least in the vicinity of the natural frequency f(j)) having the value of Saa , the integral (16), (17) can be solved in closed form

$$\overline{q^2_{(j)}} = \frac{p^2_{(j)} \, S_{aa}}{64\pi^3 f^3_{(j)} \beta} \tag{18}$$

If the earthquake motion has been described by means of response-spectrum curve, the same aproach can be used for the generalised response-components. The maximum natural mode response will be taken from the given response-curve for either displacement, velocity or acceleration, defined as a function of natural frequency and damping. The acceleration of the generalised j-th component produced by the given earthquake can be for example expressed as

$$\ddot{q}_{(j)} \max = p_{(j)} \, R(acc, \, f_{(j)}, \beta) \tag{19}$$

4. THE RESPONSE RESULTING FROM NATURAL MODES

The resulting response of the given structure (displacements, accelerations, bending moments etc) will be calculated from the natural mode decomposition (5),(3). For the displacement with respect to the ground (relative dispacement) we have

$$u(z,t) = \sum_j u_{(j)}(z) \, q_{(j)}(t) \qquad (20)$$

for absolute acceleration (in a fixed coordinate system)

$$\ddot{w}(z,t) = u_0(z) \, \ddot{q}_0(t) + \sum_j u_{(j)}(z) \, \ddot{q}_{(j)}(t) \qquad (21)$$

for bending moments

$$M(z,t) = \sum_j M_{(j)}(z) \, q_{(j)}(t) \qquad (22)$$

where $M(j)(z)$ is the amlitude of the bending moment corresponding to the vibration in the j-th natural mode (calculated simultaneously with natural modes of the structure).

Such a direct transition from the calculated generalised coordinates to the response of the structure is very simple. Newerthless, mostly another procedure is being used, namely the quasistatic one, consisting of finding the equivalent static loading and subsequent static solution. These equivalent static forces are usually being applied on the discretised structure (lumped-mass system); in fact they are inertia-forces acting at the centre of gravity of each mass m, equal to the product of the mass and corresponding absolute acceleration of this mass (21). Such a system of equivalent static forces is being defined for every natural mode; the static solution of the system gives the j-th component of the response - it should be of course the same as that one obtained from the direct dynamic solution (22).

REFERENCES

[1] Earthquake resistant regulations - a World list 1963. International Association for Earthquake Engineering, Tokyo, 1963

[2] Eurocode No 8: Structures in seismic regions, Part 1, Edit May 1988 ; CEC, EUR 12266

[3] Constructions in seismic regions (in Russian). SNIP II-7-81, GOSSTROJ SSSR, 1981

[4] Naprstek J., Fischer O.: Response of continuons systems to excitation with variable frequency. In: Proc. 8th World Congress on the Theory of Machines, Praha 1991

ENERGY ASPECTS OF THE BEHAVIOUR OF HOLLOW CLAY BRICK
PANELS STRENGHTENED BY REINFORCED CONCRETE GRID

R. Folić[1] B. Simeonov[2]

INTRODUCTION

Panels made of hollow clay bricks with reinforced concrete ribs are vertical bearing elements of the precast system "MONTASTAN" - "1. MAJ", from Bačka Topola. There are three kinds of panels: full panels (FP), panels with door openings (DR), panels with window openings (WD). Panels are industrially produced from hollow clay blocks with reinforced concrete ribs in two orthogonal directions.

Floor structure slabs are also industrially manufactured from clay blocks and reinforced concrete ribs. Vertical panels and floor panels are assembled at the construction site and monolithized, forming reinforced concrete columns of 22×22 cm at the ends of the panel. The floor structures are monolithized by forming reinforced concrete beams-cerclage in the direction of vertical panels.

In order to study the characteristics of bearing and deformability, as well as the energy indicators under loading which simulates seismic influences, these two experimental investigations were realized [6]. They represent a continuation of former analytical and experimental investigations of the precast system "Montastan"-"1.maj", B. Topola.

The results of former investigations are given in papers [1], [2] and [5]. Based on these experimental investigations, the bearing and failure mechanisms of wall panels are analyzed in [7]. In this paper, hysteretic dependances, displacements and ductility, as well as energy indicators, are analyzed on the basis of the experimental investigations of 6 two-storey models under cyclic loading. On the basis of an analysis of the results, the corresponding recommendations and conclusions are given.

TESTED MODELS

Three series of two two-storey models each, in 1:1 ratio i.e. a total of 6 models. The models' dimensions are conditioned by the dimensions of the hollow brick. Before the design of models, an analytic investigation of 5-8 storey buildings was carried out designed by this system. For the non-linear

---

[1] Professor University of Novi Sad, Faculty of Technical Sciences, Institute for Industrial Building, Novi Sad, Yugoslavia
[2] Professor, Institute of Earthquake Engineering and Engineering Seismology, University "Kiril and Metodij", Skopje, Yugoslavia

analysis of a structure's fragments, documentation on the 1966 Parkfield earthquake was used, the N-S component with a maximum acceleration of 0.15 g. It was shown that plastic hinges form at the beams' ends, and later at the first floor of vertical panels as well. A detailed description of results is given in [5].

Two of the six tested models were without openings (FP), and two each with door (DR) and window (WD) openings respectively. Beside the given symbols, the symbols EL 1 and EL2 were added.

The models were tested in horizontal position. First, vertical load was applied at the top of the second floor, which was 10% of the capacity of the cross section for axial forces. The cyclic, horizontal load was applied at the second floor level (Fig. 1).

The integral system of automatic measuring and data gathering was applied during tests.

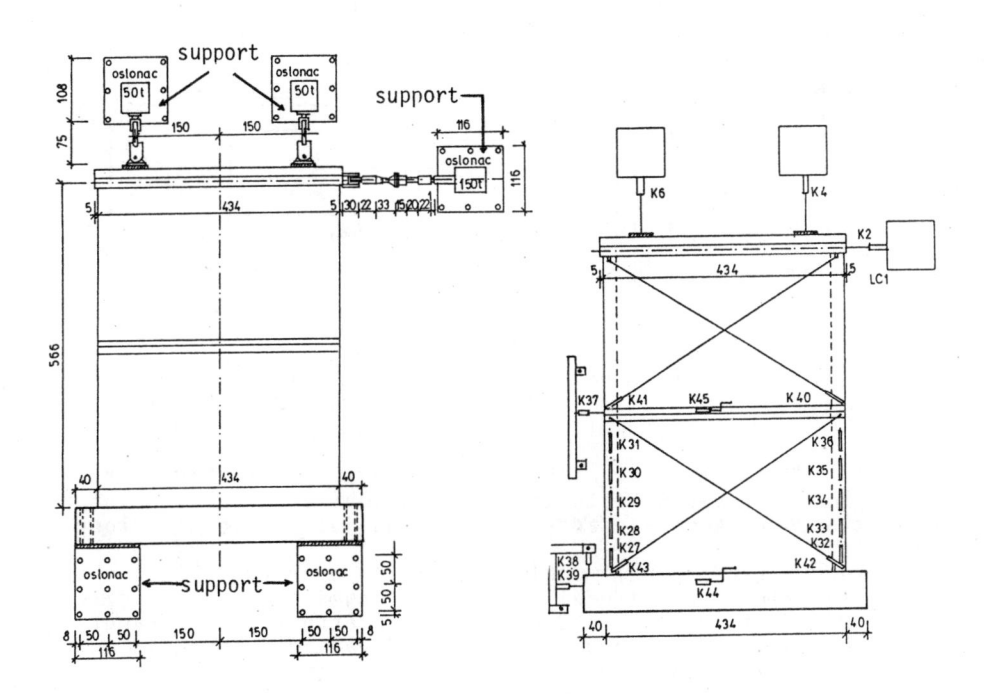

Fig.1 Model and equipment for testing

Fig.2 Outer instrumenting of models - number of channels

Before the testing of models under cyclic loading, the strengths of concrete blocks under pressure were determined (blocks used in making concrete columns of the model), and the mechanical properties of reinforcement in columns were tested the same way. The strength of clay bloks was adopted on the basis of testing performed by ZMRK-Ljubljana, to be 10 MPa.

The instrumenting of models with the marking of instruments for all three series of panels is given in Fig. 2, 3 and 4, for full panels, for panels with door openings and panels with window openings, respectively. The instrumenting was aimed at achieving data on the resistance and deformability of models under cyclic loading. During the testing period all instruments were read, so that various appearances cracks development and the gradual damaging of models were observed.

For both models with full panels the characteristics are the same, only different normal loadings were given, and for EL2 it was 0.84 MPa.

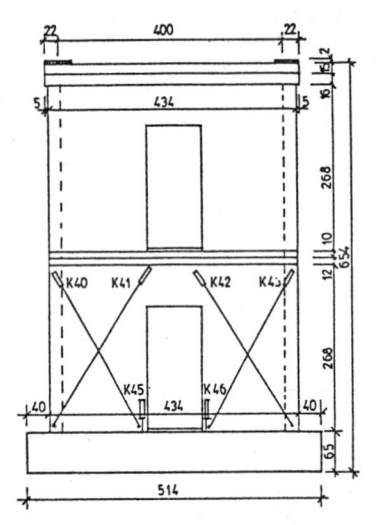

Fig.3 Additional instrumenting of a model with door openings - number of channels

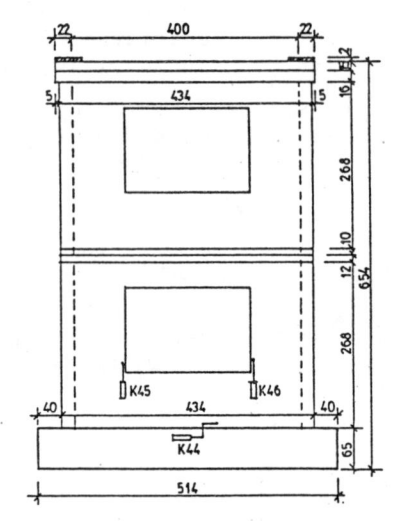

Fig.4 Additional instrumenting of a model with window openings - number of channels

In the first model with door openings, a reinforcement of 2 ⌀ 8 mm is placed next to the opening, while in EL2 a reinforcement of 4 ⌀ 14 mm was added by welding just above the floor structure.

In the first panel model with window openings WD3EL1 the vertical reinforcement next to the opening is 1 ⌀ 8 mm, while in WD3EL2 it is made with 2 ⌀ 14 mm, and anchored in parapet and over above the window. The value of vertical force in both models is 10% of the limit capacity of pressure.

TESTING RESULTS AND THEIR ANALYSIS

Results that will be discussed here are concerned with all three kinds of models, i.e. for full panels, and panels with door and window openings. The results will be assessed on the basis of general hysteretic model behaviour with data on displacements, ductility and energy dissipation.

Displacement and Ductility

The displacement capacities of the tested models are given in Table 1.

Analytical determination of displacement in yielding is based on the capacity for development of the moment in yielding, M_y, and corresponding curve, ϕ_y. The shape of the curve diagram ϕ_y, is taken to be identical to the moment diagram from the elastic analysis of fragments of the structure exposed to horizontal forces. In this way, bending deformations of tested models can be determined.

Table 1 Characteristics of the models´ deformability

	Model	Δ_y^T (mm)	Δ_u^T (mm)	D_Δ	R_y^T (10^{-3})	R_u^T (10^{-3})	ε_{au} (10^{-3})	ε_{bu} (10^{n3})	τ_{max} (MPa)
FP	PN1EL1	14.84	54.09	3.64	2.62	9.56	2.12	2.20	0.38
FP	PN1EL2	12.34	47.75	3.87	2.18	8.44	2.23	2.88	0.53
DR	VR2EL1	16.77	71.26	4.25	2.96	12.59	3.96	2.08	0.80
DR	VR2EL2	16.96	74.67	4.40	3.00	13.19	2.73	1.76	1.01
WD	PR3EL1	15.05	84.25	5.60	2.66	14.89	2.92	1.48	0.88
WD	PR3EL2	15.42	73.77	4.78	2.72	13.03	3.38	1.04	0.91

For determining overall yield displacement, shear deformations are taken into account as well. Since in the yield state cracks in the panel are already formed, it is very difficult to analytically determine the shear deformations of the models.

The yield displacement and the angle of yield deformation, $R_y = \Delta y / h_i$, can be determined from the deformation angle of reinforced concrete diaphragms [3]. From a survey of values of the deformations angle (Table 1), it is evident that it ranges from 2.4×10^{-3} for panels without openings to 3.6×10^{-3} for models with door openings.

The ultimate displacement is an indicator of "ability" of the model to work in a nonlinear area, while the displacement ductility $D\Delta = \Delta_u / \Delta_y$ is defined by the relationship between ultimate and yield displacements.

In the linear area of work, until the appearance of cracks in panels of the first and second floors, the model PN1EL1 behaved satisfactorily. It had symmetric hysteretic loops until the appearance of joint loosening on the bond of beam to column. Due to which a structural improvement of details in reinforcing this bond has been proposed. The middle horizontal yield loading is $Q_{y,sr}^T = 295$ kN, while horizontal displacement of the second floor is 14.84 mm, which, when divided by the height, from the upper edge of the foundation to the middle panel of the second floor, gives the angle $R_y^T = 2.62 \times 10^{-3}$. This angle of yield displacement is similar to the shear angle in reinforced concrete walls of the rectangular cross-section.

Ultimate displacement of the second floor is 54 mm, and the ultimate displacement angle $R_y^T = 9.56 \times 10^{-3}$, which is greater from the maximal angle (6.67×10^{-3}) determined in the GSP Rules (Art.41), i.e. from $h_i/150$. Realized displacement ductility is 3.64, which is satisfactory for models made of full panels.

In the case of the second full model, yield displacement was 12.34 mm, while the shear angle $R_y^T=2.18 \times 10^{-3}$. The greatest displacement of the second floor, at ultimate bearing, was 47.75 mm, while displacement angle $R_u^T=8.44 \times 10^{-3}$. The displacement ductility was $D_\Delta=3.87$. The strain in reinforcement reached 2.23×10^{-3}, which is on the level of yield dilatation (Fig.5). The maximal pressure strain in concrete and reinforcement were 2.88×10^{-3} (Fig.6). Columns are not stressed to the level which concrete and reincorcement can bear. In Fig. 7 the dependance of force-strain - elongation of the model's diagonal is shown.

The hysteretic behaviour of the model with door openings DO2EL1 is shown in Fig.8. The symmetrical enlargement of the horizontal force and displacement is evident. The horizontal yield displacement of the second floor is 16.77 mm, while in its ultimate state it is 71.26 mm, giving the displacement angle $R_u^T=12.59 \times 10^{-3}$. The displacement ductility is 4.25.

In the model DR2EL2, due to a quantity increase in reinforcement next to the door openings, the hysteretic loops are wider than in the previous model (Fig.9). The greatest horizontal displacement is 74.67 mm, which gives the displacement ductility of 4.40. The yield displacement angle is 3×10^{-3}, while in its ultimate state $R_u^T=13.19 \times 10^{-3}$, which is a confirmation of the great displacement capacity of this model.

In the model with window openings WD3EL1 the behaviour in the linear area is symmetric (Fig.10) until cracks appear in the joint of the column to beam joint of the first floor, which is a consequence of the inadequate anchorage of the beam's reinforcement in joining with the column. The greatest displacement of the second floor is 84.2 mm, which gives a ductility of 5.60.

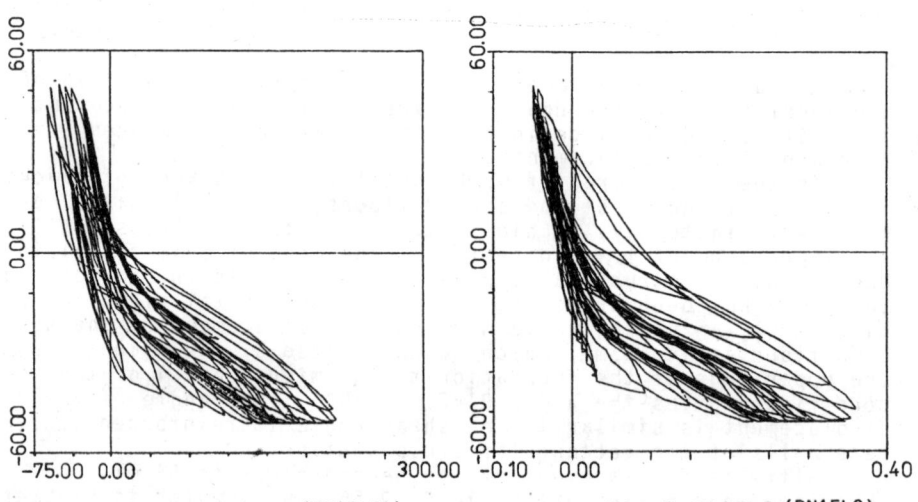

Fig.5 DR2EL2 (PN1EL2)

Fig.6 FP1EL2 (PN1EL2)

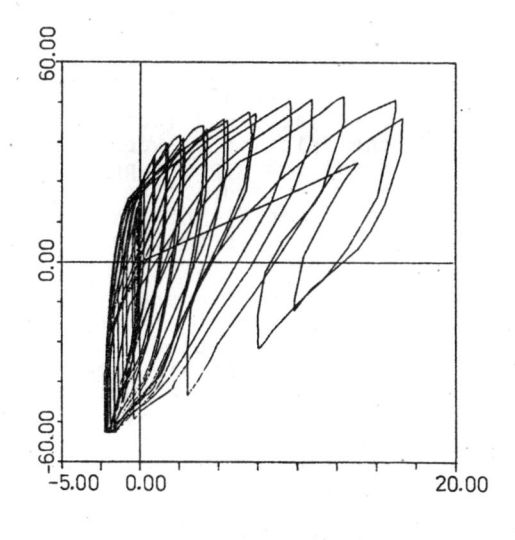

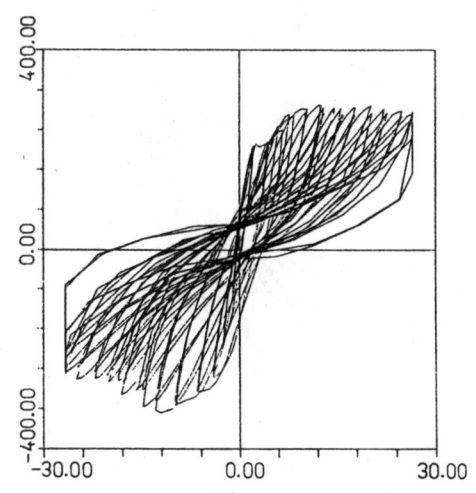

Fig.7 FP1EL2(PN1EL2)

Fig.8 DR2EL1(VR2EL1)

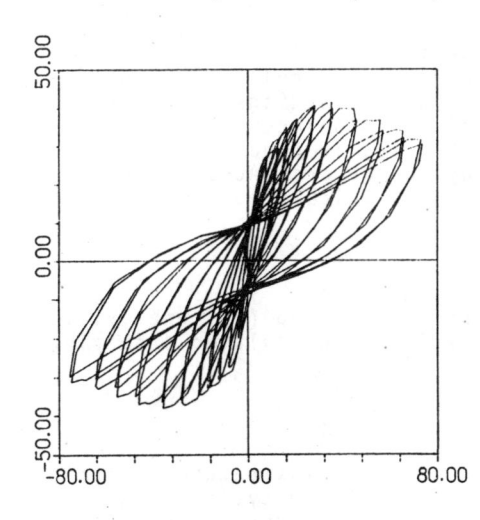

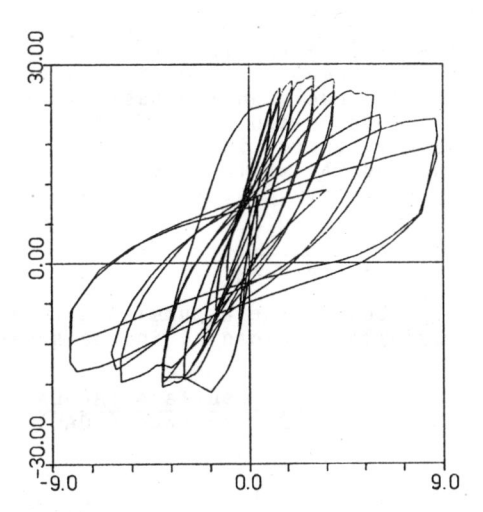

Fig.9 DR2EL2(VR2EL2)

Fig.10 WD3EL1(PR3EL1)

The behaviour of the model FP3EL2 is more favourable than the previous one, which is evident from the Q-Δ relationship (Fig.11), with almost symmetrical behaviour for both directions of loading. The capacity of displacement and ductility does not show greater differences in reference to the model WD3EL1.

Energy Dissipation

The energy transferred to the model during testing equals the total action of external forces. The external forces which

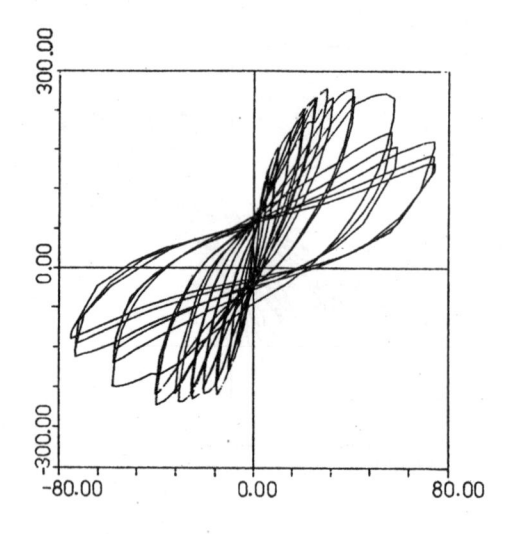

Fig.11 FP3EL2(PR3EL2)

influence the model are the forces of the hydraulic jacks and those of the friction between the model and pipes supporting the model. The work performed by the friction forces is small and negligible.

The internal energy, dissipated during testing, can be calculated by stress and strain integration, for the whole volume of the model and for the overall time of the testing. In paper |9| it is shown that for a two-dimensional model, subjected to bending, shear and axial forces, the internal energy can be calculated by integrating the diagram area of the shear force-displacement from the top of the model.

Accumulated and dissipated energy is calculated for all tested models on the basis of the hysteresis diagram Q-Δ, for the model's top, at different levels of displacement. The accumulated energy in one semi-cycle, Fig.12, is equal to the surface ABDEA, while dissipated energy to the area ACDEA. The energy ductility, D_e, for one semi-cycle is calculated according to

$$D_e = \frac{W_a}{W_a - W_d}$$

Beside the accumulated (W_a) and dissipated (W_d) energy, the equivalent deadening in one semi-cycle was calculated according to

$$h_{eq} = \frac{1}{2\pi} \cdot \frac{\text{surface (ACDEA)}}{\text{surface }(\Delta OAB)}$$

Fig.12

Dissipated energy in one cycle is equal to the sum of the energy in both semi-cycles and normalized by dividing it with the middle value of horizontal displacement in that cycle for comparing, and it is presented in Fig.13. On the basis of these results it can be estimated that the models made of

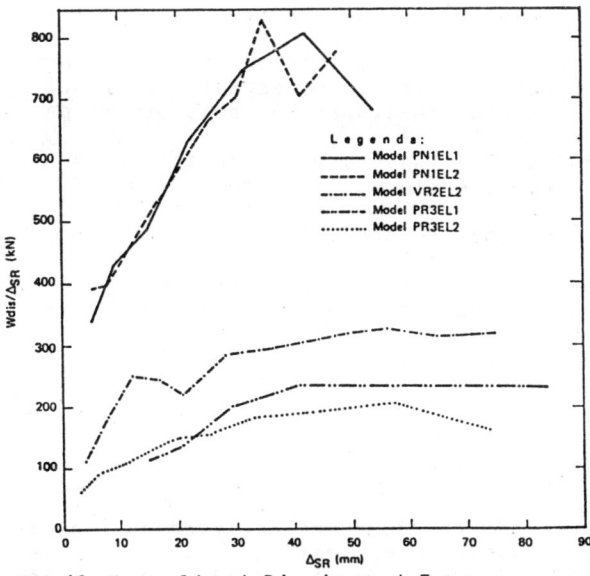

Fig.13 Normalized Dissipated Energy

full panels have an estremely greater capacity for dissipation of energy, in reference to the models with openings for doors and windows, with the same external dimensions. The model DR2EL1 has a smaller capacity for dissipation of energy in reference to other models with openings (it is not shown in Fig.13).

The energy ductility, D_e, is quite high for all models and it is usually greater than displacement ductility, which indicates the favourable energetic properties of tested models and the precast system as a whole.

From the hysteresis dependances $Q-\Delta$ the equivalent damping are calculated for all models, ranging from 0.10-0.25 from critical damping. These values are considerably high, which is to confirm the favourable behaviour of models and bonds under cyclic alternative loading.

CONCLUSIONS

On the basis of the behaviour of models during experimental research and conducted analysis, the following conclusions are reached:
- yield displacement established on the basis of experimental results is similar to displacements in RC diaphragms without openings, so that the rigidity at reinforcement yielding is similar to that of RC diaphragms,
- displacement ductility in tested models ranges from 3.64-5.60, similar to that in RC diaphragms, i.e. it satisfies criteria for aseismic structures,
- the normalized angle of ultimate displacement ranges from 9.0×10^{-3} with full-panel models, to 14.0×10^{-3} with models with window openings. These values are greater than the proposed maximal values of displacement for non-linear dynamical analysis of high buildings, which means that the displacement capacity of models is satisfactory,
- the ultimate displacement of panels, according to tests, is satisfactory. In terms of capacity, it is greater than the

maximal displacements allowed for the designed earthquake level,
- the normalized capacity of dissipated energy is not proportional to the displacement increase, instead it is a function of the opening of cracks and non-linear deformations in panels,
- models made of full panels have a much greater energy dissipation capacity, as compared to models with window and door openings (Fig.13). The model with door openings (VR2EL2), with increased reinforcement by the door, has a greater energy dissipation capacity than model VR2EL1,
- models with windom openings have a smaller energy dissipation capacity, which cannot be neglected in designing buildings in seismic regions,
- equivalent viscous damping of all tested models is very high, while the proposed lowest value should not be less than 10% of critical damping for non-linear dynamic analysis.

S U M M A R Y

The energetic aspects of behaviour are analysed on the basis of experimental and analytic investigation of six wall panel models of the "MONTASTAN" system developed by "1.MAJ", Bačka Topola. Three types with two two-storey models each have been tested in the scale 1:1. Two models are without openings, two with openings for doors and two with openings for windows. Model have been exposed to alternative cyclic loading.

Hysteresis dependence of force-displacement, displacement during reinforcement yield, ultimate displacement, as well as turning angles of such displacements have been analysed. In addition to that, ductility, energy dissipation and damping equivalents for all models have also been analysed.

R E F E R E N C E S

1. Folić, R., Taškov, Lj., Horvat, A.: Eksperimentalno ispitivanje prinudnim vibracijama jednog stambenog objekta izvedenog u sistemu MONTASTAN, četvrti jugoslovenski naučni skup, INDIS '86, Novi Sad, 1986.

2. Folić, R., Simeonov, B., Kaćanski, Dj.: Analitička istraživanja stambenog niza F8/L2 izvedenog u Novom Sadu u sistemu MONTASTAN. četvrti jugoslovenski naučni skup, INDIS '86, Novi Sad, 1986.

3. Simeonov, B.: Experimental Investigation of Strength and Deformation of Reinforced Concrete Structural Walls, Proceedings of the VII ECEE, Vol. 4, Athens, 1982.

4. Simeonov, B., Folić, R., Gjorgievska, E.: Analitička istraživanja montažnog sistema MONTASTAN GIK "1.maj" Bačka Topola, Objekat L-4, Izveštaj IZIIS 87-57, 1987.

5. Simeonov, B., Folić, R., Gjorgievska, E.: Analytic Investigation of the Dynamic Behavior of the MONTASTAN System Buildings, Proceedings of the Eurobuild Conference on De-

sign, Construction and Repair of Buildings in Earthquake Zones, Dubrovnik, Sept. 1987.

6. Simeonov, B., Folić, R., Gjorgievska, E.: Analitička i eksperimentalna istraživanja montažnog sistema MONTASTAN - Eksperimentalno ispitivanje modela, Izveštaj IZIIS 89-96, Skopje, decembar 1989.

7. Simeonov, B., Folić, R., Kaćanski, Dj.: Nosivost i mehanizmi loma zidnih panela od šuplje opeke ojačane armiranobetonskim roštiljem, V kongres Saveza društava za seizmičko gradjevinarstvo Jugoslavije, Bled 24-26. april, 1990.

8. Valenas, J.M., Bertero, V.V., Popov, E.P.: Hysteretic Behavior of R/C Struct. Walls, Report 79/20, EERC - University of California, Berkeley 1979.

9. Wang, T.Y., Bertero, V.V. and Popov, E.P.: Hysteretic Behavior of RC Structural Walls, Report No. 75-23, EERC, University of California, Berkeley, Dec. 1975.

---

* This paper has been made within the project No. 7703 of the Republician Sciences Fund of Serbia.

DUCTILITY DESIGN OF REINFORCED CONCRETE MEMBERS
K.Pilakoutas [1]

INTRODUCTION

In earthquake resistant design, only a fraction of the seismic actions calculated by elastic analysis is required to be resisted by structural elements. In Eurocode 8 [1] the factor utilised to reduce the inertia effects corresponds to the behaviour factor 'q'. This approach is justifiable provided that the structure possesses adequate strength and ductility. For an extreme seismic event, therefore, it is acceptable that a structure is subjected to nonlinear hysteretic deformations provided the maximum deformation capacity of critical members is not exceeded prior to significant strength and stiffness degradation. The concept of ductility design has long been employed in seismic design and is used effectively in widely used codes such as the American Codes ACI 318-83 [2], UBC (1988) [3] and the New Zealand Standard 3101 (1982) [4]. The New Zealand standard additionally incorporates the 'capacity design' philosophy which pays special attention to locations and members likely to undergo inelastic deformations. Nevertheless, the above mentioned codes try to achieve ductile behaviour by prescribing confinement reinforcement and imposing dimensional limitations, without directly quantifying the level of ductility achieved.

A comprehensive procedure arriving at the required confinement for reinforced concrete (RC) members from a chosen behaviour factor, is incorporated in the proposed Eurocode 8. Four stages are involved in the procedure as shown in the following flow chart figure 1 [5].

Choice of the overall behaviour factor 'q' (*from table V, 2.1.4 of Part 1.3*)

↓

Range of member displacement ductility demand μ_Δ

$$q < \mu_\Delta < 0.5(q^2+1)$$

↓

Calculation of the critical section curvature ductility demand

$$\mu_{1/r} = 1 + \frac{[\mu_\Delta - 1]}{3 \lambda_{pl} (1 - 0.5 \lambda_{pl})} = 1 + 3.5(\mu_\Delta - 1)$$

↓

Calculation of the critical section confinement
(For single hoop wall boundaries)

$$\omega_{wd} = 0.9 \sqrt[3]{\mu_{1/r}} \left[0.01 \, \mu_{1/r} + 0.15 \frac{A_c}{A_o} + v_d - 0.4 \right] \geq 0.2$$

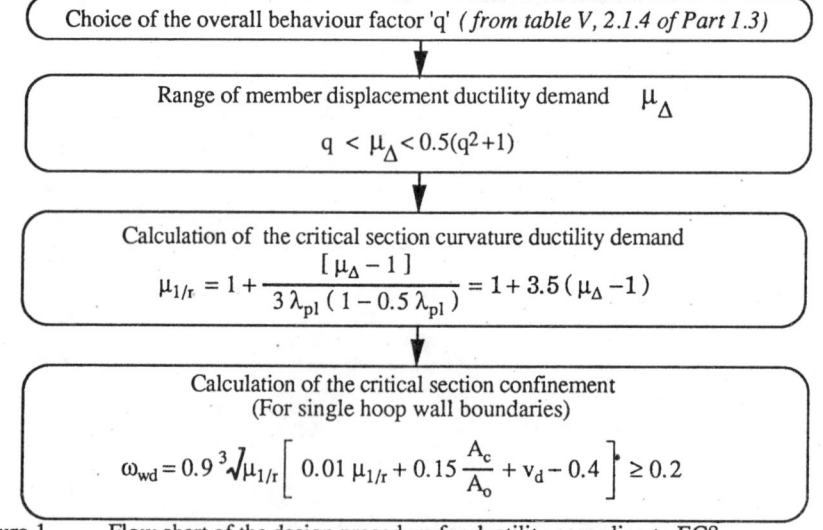

Figure 1   Flow chart of the design procedure for ductility according to EC8

From the four stages involved in this procedure, the ones relating to the overall structural behaviour are currently attracting the bulk of the research efforts. Nonetheless, the derivations on the member level are not being scrutinised to the same extend. In this paper the theory and basic assumptions used by Eurocode 8 are investigated and improvements as well as amendments are proposed. This is achieved with the help of analytical tools which have been developed by taking into account realistic cyclic material models and allowing for the

---

[1] - Department of Civil and Structural Engineering, University of Sheffield, Sheffield, S1 4DU, UK

effects of confinement. The results of parametric studies indicate that the EC8 provisions may be un-conservative in some of the assumptions used, and too simplistic in others.

NONLINEAR RC SECTION ANALYSIS PROGRAM

In order to investigate the variety of parameters that influence the ductility of reinforced concrete members realistic nonlinear models have to be used. A computer implementation of a section analysis program "CRELIC" [5,6] includes cyclic material models for both steel and concrete which are briefly introduced in the following so as to enhance the appreciation of the results.

In order to represent simply the behaviour of the variety of steels used, a tri-linear cyclic model has been formulated as shown in figure 2. Monotonic experiments can be used to obtain the model envelope. A yield level is determined by using the monotonic yield load and the strain hardening stiffness 'E_{s1}'. A maximum stress 'f_{su}' is not to be exceeded at any value of strain. Exceeding an ultimate strain 'ε_{su}' will result in the bar fracture and total loss of strength. Loading and unloading up to the yield level and down to zero follows the initial stiffness 'E_{so}'. On reloading, a stiffness 'E_{sa}' is used. Once the yield level is achieved, stress increases according to stiffness 'E_{s1}'.

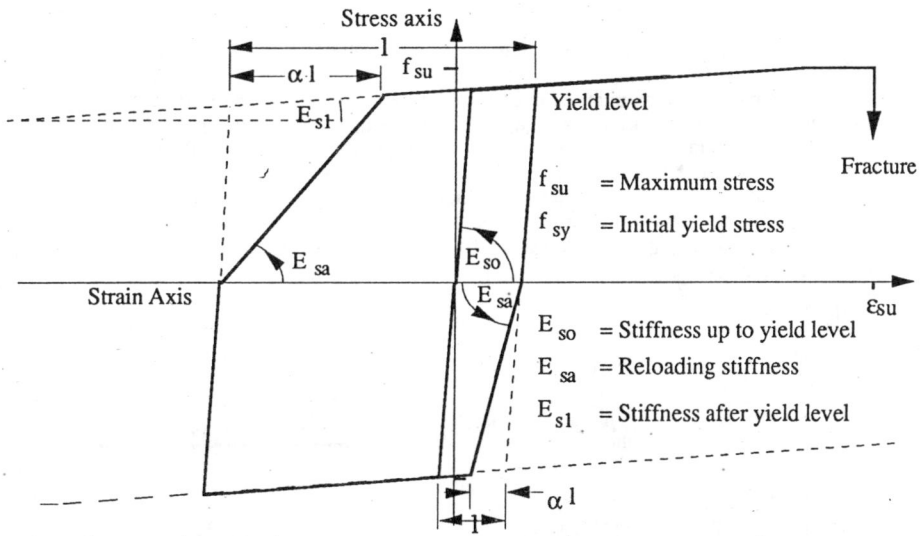

Figure 2   Stress-strain diagram for steel reinforcement [6]

The above simple model is based on a slightly modified form of the Massing model and utilises the stiffness degradation factor 'α' from the work of Santhanam [7]. The implementation of the steel model in a computer program is simple and only the current and previous maximum and minimum permanent strains are required to establish the stress from the current strain.

The concrete model used is based on the work of Mander J.B, Priestley J.N. & Park R [8,9] as shown in figure 3. It was chosen for its direct applicability to the method of sections and was implemented with some modifications. For monotonic tensile loading, concrete is capable of carrying tensile stresses up to a limit of 'f_t'. However, this strength may be affected by initial micro-cracking, and will be lost at the initial stages of cyclic loading. Therefore, in the implementation of the above model the tensile strength has been ignored.

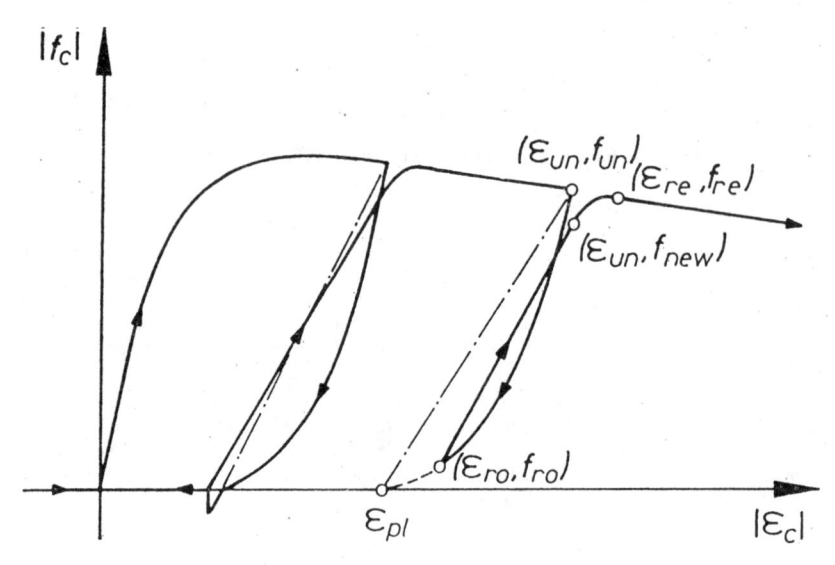

Figure 3   Stress strain cyclic model for concrete [8,9]

The model though uniaxial in its formulation, takes into account the confining stresses. The development of the lateral force in the steel depends on the elastic properties of concrete as well as the axial strain. In order to avoid the elaborate calculations for evaluating the confinement factors, the effectiveness of the confinement is provided to the section analysis program as an input. Simple calculations to account for the effect of hoop pattern and spacing, the maximum enhanced confined stress f_{cc}, the strain at which f_{cc} is achieved and ultimate crushing strain ε_{cc}, are given by Tassios [10] in a background document to Eurocode 8 as shown in figure 4.

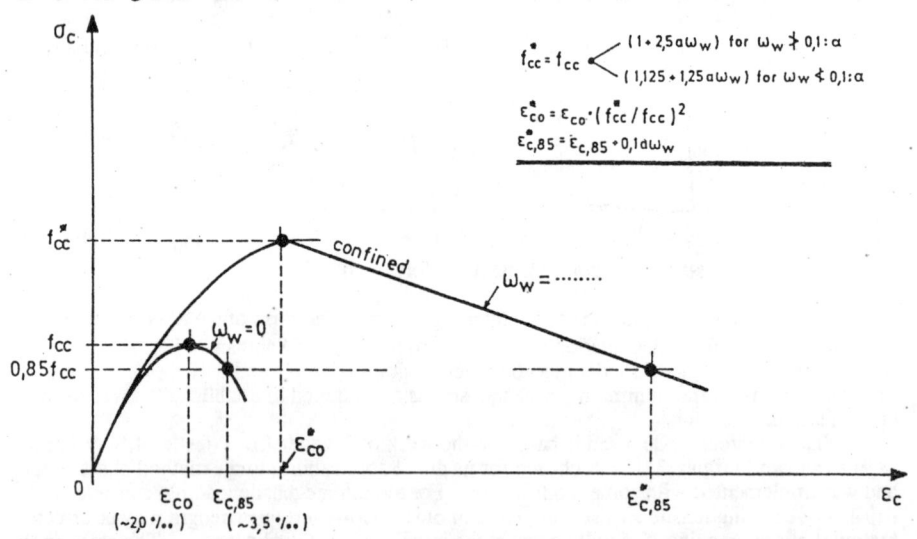

Figure 3   Stress strain model for monotonic loading unconfined and confined concrete [1,10]

64

CURVATURE DISTRIBUTION

The relationship between displacement μ_Δ and curvature $\mu_{1/r}$ ductilities in Appendix "D" of Eurocode 8 is given by (where λ_{pl} is the normalised height of the plastic hinge):

$$\mu_{1/r} = 1 + \frac{[\mu_\Delta - 1]}{3\lambda_{pl}(1 - 0.5\lambda_{pl})} \quad (1)$$

The above expression is based on the curvature distribution shown in figure 5 which considers the curvature to be almost equal to the ultimate curvature within the plastic hinge zone, which seems to be un-conservative when compared with the results from analysis (shown shaded in figure 5). A better upper limit approximation to the expected curvatures[5] can be given by a linear variation of curvatures within the plastic hinge zone as shown in equation 2 below with the value of 'p' equal to one:

$$\mu_{1/r} = 1 + \frac{[\mu_\Delta - (1 + \lambda_{pl})]}{\lambda_{pl}(1.5 - 0.5\lambda_{pl})p} \quad (2)$$

A lower limit approximation to the expected curvatures within the plastic hinge zone can be obtained by using a parabolic curve as shown qualitatively in figure 5. The choise of a parabolic curve can be demonstrated to affect mainly the value of 'p'. For high values of λ_{pl} the value of p can be as low as 0.333.

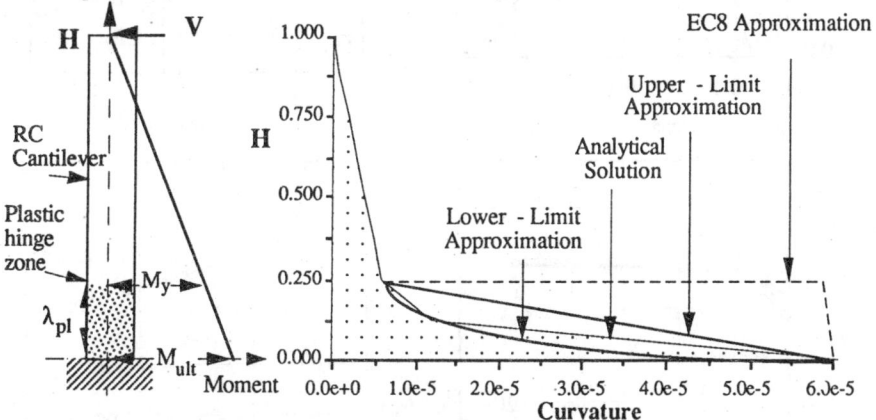

Figure 5   Curvature distribution for a cantilever wall at ultimate load

ESTIMATION OF PLASTIC HINGE LENGTH

Before investigating the relationship between curvature and displacement ductilities, a reliable estimate of the extent of plasticity is required. The plastic hinge height is given by several researchers to be a function of the aspect ratio based on equations proposed by Mattock [11]. However, since the effect of shear is not taken into account in curvature calculations, there are no firm theoretical grounds for the assumption that the plastic hinge height should be related to the section width. By using the similar triangle principle in the moment diagram in figure 5 it is obvious that the normalised length of the plastic hinge zone for a cantilever is given by:

$$\lambda_{pl} = 1 - \frac{M_y}{M_{pl}} \quad (3)$$

Consequently, the plastic hinge length can easily be obtained analytically. The various parameters that affect λ_{pl} have been investigated by making use of the program "CRELIC" [5,6]. Plots of λ_{pl} versus a number of different parameters are shown in figure 6.

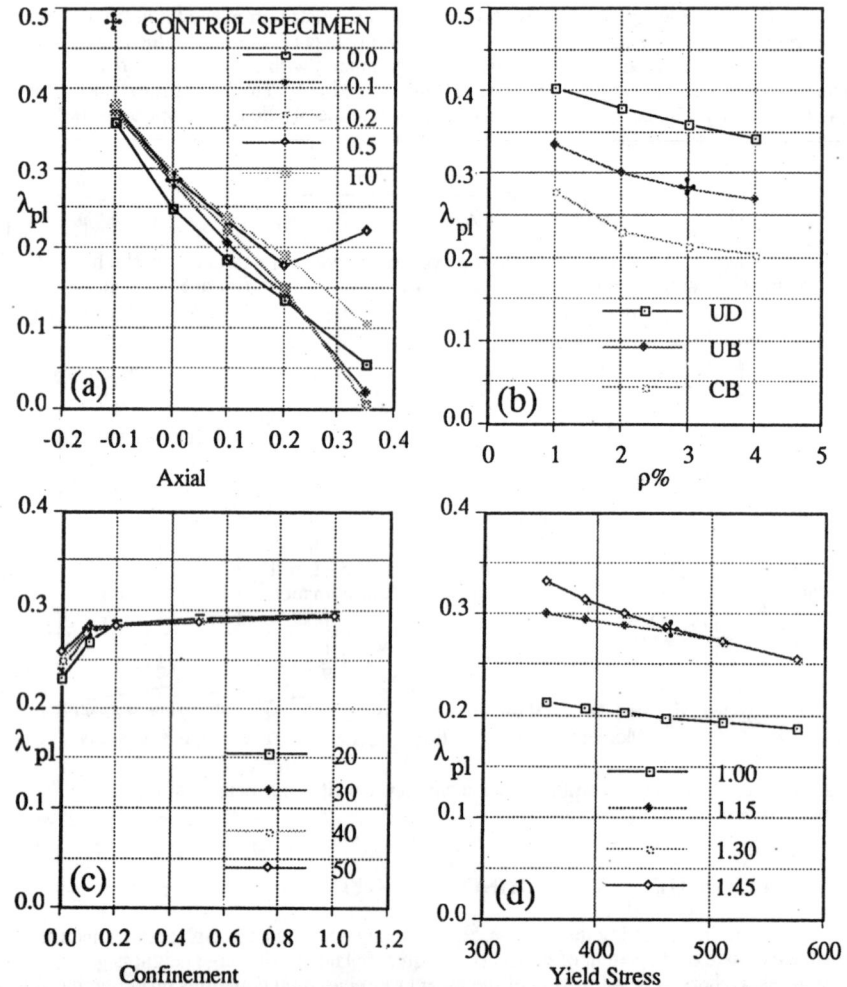

Figure 6 Variation of plastic hinge length with a selection of member parameters

In figure 6 (a) the axial load variation is shown to be of much greater importance in determining the plastic hinge depth than the confinement. Normalised axial loads greater than 0.2 may cause yield in the compressive zone first and hence, since energy dissipation from this mechanism is undesirable, such high loads should be avoided. Confinement reinforcement though it helps in spreading plasticity it does not contribute significantly towards this effect. As expected in figure 6(b) the increasing percentage of the flexural reinforcement reduces the extend of plasticity. The other parameter investigated in figure 6(b) is the effect of different types of steel distribution. Surprisingly, the uniformly distributed reinforcement (UD) indicates a larger plastic hinge zone than when the reinforcement is concentrated in the extremities (CB) even though the latter distribution can be shown to result in higher displacement ductilities. In figure 6(b) the middle curve is for specimens for which the reinforcement is distributed within boundary elements of approximately 10% of the depth of the section (UB). In figure 6(c) the concrete strength is shown not to affect the extend of plasticity. Finally in figure 6(d), the higher yield strength of reinforcement is shown to reduce the extend of plasticity. On the other hand the ratio of ultimate to yield strength indicates that the higher the reserve in strength after yield the higher is the extend of plasticity even though a too high reserve will not be utilised.

ESTIMATION OF CURVATURE DUCTILITY

By using the results of the parametric study employed in the section above, it is possible to compare the accuracy of the three different assumed distributions of curvature shown in figure 4. The plastic hinge length obtained from analysis is substituted in equations 1 and in 2 for values of p =1 & 0.33. The three sets of results are plotted in figure 7 against the curvature ductilities calculated directly by the analytical program.

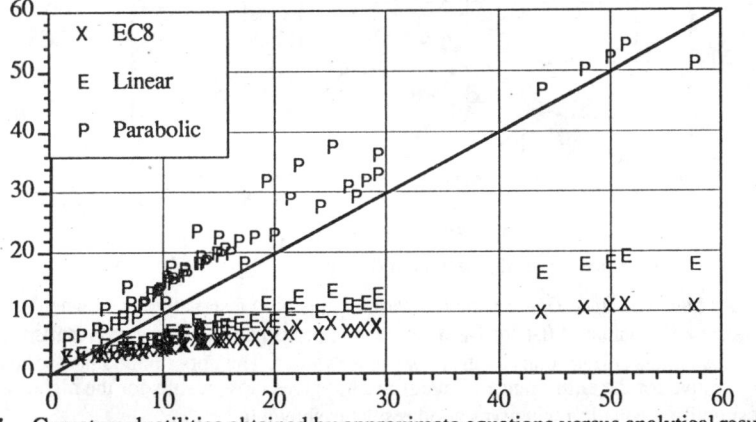

Figure 7 Curvature ductilities obtained by approximate equations versus analytical results

It is clearly demonstrated in figure 7 above that the EC8 approximation underestimates the curvature ductility demand by a factor of about 4. The linear lower bound approximation is slightly better than the Code approximation but still unsafe. The parabolic approximation, however, in general gives a very good upper bound solution.

The above mentioned equations, however, require the plastic hinge normalised length as an input in addition to the displacement ductility and, hence, it is desirable to eliminate that from the equations. From figure 6 it can be seen that the average value of λ_{pl} is about 0.3, and does not fall below the value of 0.15 except in the presence of very high axial loads. Notwithstanding, Eurocode 8 employs the very conservative value of 0.1. Table 1 shows the three pairs of equations arising for two limiting values of λ_{pl} namely 0.1 and 0.3. The graphical representation of these equations is shown in figure 8.

	$\lambda_{pl} = 0.1$	$\lambda_{pl} = 0.3$
EC8 $\mu_{1/r} = 1 + \dfrac{[\mu_\Delta - 1]}{3\lambda_{pl}(1 - 0.5\lambda_{pl})}$	$1 + 3.5(\mu_\Delta - 1)$	$1 + 1.3(\mu_\Delta - 1)$
Linear Approximation $\mu_{1/r} = 1 + \dfrac{[\mu_\Delta - (1 + \lambda_{pl})]}{\lambda_{pl}(1.5 - 0.5\lambda_{pl})}$	$1 + 6.9(\mu_\Delta - 1.1)$	$1 + 2.5(\mu_\Delta - 1.3)$
Parabolic Approximation $\mu_{1/r} = 1 + \dfrac{[\mu_\Delta - (1 + \lambda_{pl})]}{\lambda_{pl}(1.5 - 0.5\lambda_{pl}) \; 0.33}$	$1 + 20.9(\mu_\Delta - 1.1)$	$1 + 7.5(\mu_\Delta - 1.3)$

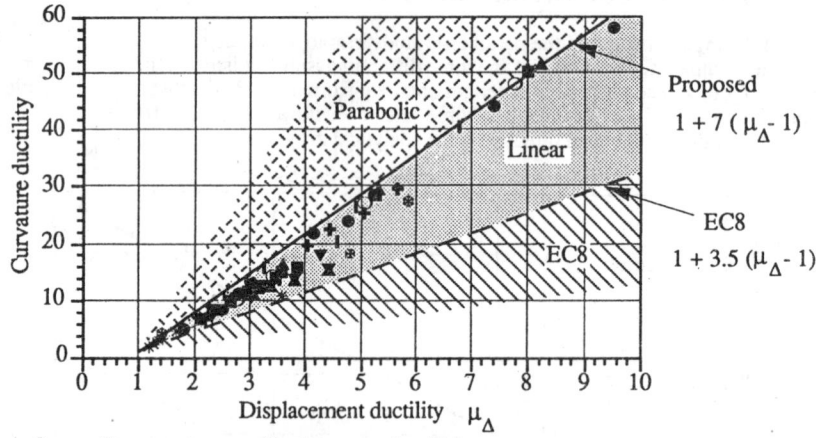

Figure 8   Curvature versus displacement ductilities

As observed from figure 8 the EC8 equation is un-conservative even when using the very conservative value of 0.1 for λ_{pl}. The linear approximation gives good results for low values of λ_{pl} but is still un-conservative for higher values. The upper bound parabolic equation is conservative for the entire range of ductilities but gives close results for the higher values of λ_{pl}. A simplified equation that gives good results is shown to be:

$$\mu_{1/r} = 1 + 7(\mu_\Delta - 1) \qquad (4)$$

The equation proposed above is shown in figure 8 to be conservative for most of the examined members except for the cases with high confinement levels for which estimates are very close to the analytical predictions.

ESTIMATION OF REQUIRED CONFINEMENT

The final stage of converting curvature ductility to confinement is a more difficult task, since the ductility of an unconfined section is influenced by many parameters. It is proposed that a reference curvature ductility $\mu_{1/ro}$ is calculated first for the unconfined section, by establishing the curvatures at yield and ultimate conditions, based on accurate estimates of the neutral axis depth. Thereafter, the controlling curvature parameters at yield and at ultimate are the steel strain ε_{sy} and the concrete ultimate strain ε_{cu}. In which case, the unconfined curvature ductility is given by:

$$\mu_{1/ro} = \frac{\varepsilon_{cu}(1-\xi_y)}{\varepsilon_y \xi} \geq \frac{\varepsilon_{cu}(1-1.5\xi)}{\varepsilon_y \xi} \tag{5}$$

The value of ξ_y is always larger than ξ (up to 40% higher) and hence a conservative assumption would be to use the value of 1.5ξ at ultimate conditions in order to by-pass the extra calculation required for the evaluation of ξ_y. The above equation is also valid for the confined section provided the value of ε_{cu} is substituted by ε_{cc}. In fact it has been shown analytically [6] that the value of the neutral axis depth at ultimate conditions is not affected significantly by variations in confinement and hence the values of ξ at yield and ξ' at ultimate are similar. Therefore, the ratio of the confined to unconfined curvature ductilities can be given by:

$$\frac{\mu_{1/r}}{\mu_{1/ro}} = \frac{\varepsilon_{cc}(1-1.5\xi)}{\varepsilon_y \xi} / \frac{\varepsilon_{cu}(1-1.5\xi')}{\varepsilon_y \xi'} = \frac{\varepsilon_{cc}}{\varepsilon_{cu}} \tag{6}$$

By substituting the equation for ε_{cc} as given by the Eurocode 8:

$$\frac{\mu_{1/r}}{\mu_{1/ro}} = \frac{\varepsilon_{cc}}{\varepsilon_{cu}} = \frac{\varepsilon_{cu} + 0.1\,\alpha\,\omega_w}{\varepsilon_{cu}} \approx 1 + \frac{\alpha\,\omega_w}{10\,\varepsilon_{cu}} \tag{7}$$

Hence, by rearranging:

$$\omega_w = \frac{10\,\varepsilon_{cu}}{\alpha}\left(\frac{\mu_{1/r}}{\mu_{1/ro}} - 1\right) = \frac{1}{\kappa\,\alpha}\left(\frac{\mu_{1/r}}{\mu_{1/ro}} - 1\right) \tag{8}$$

This final equation for calculating the confinement includes the ratio of ductility demand to $\mu_{1/ro}$, as well as the confinement pattern. The value of parameter 'k' equal to 28.5 should correspond to a value of ε_{cu} of 0.0035. The graph below shows the vaules of 'k' obtained for the members analysed by using program 'CRELIC'. The value of k of 28.5 seems to be close to the average for the confinement values examined which validates the simplified assumptions made previously. Nonetheless, a value of 25 would cover specimens with higher concrete strength.

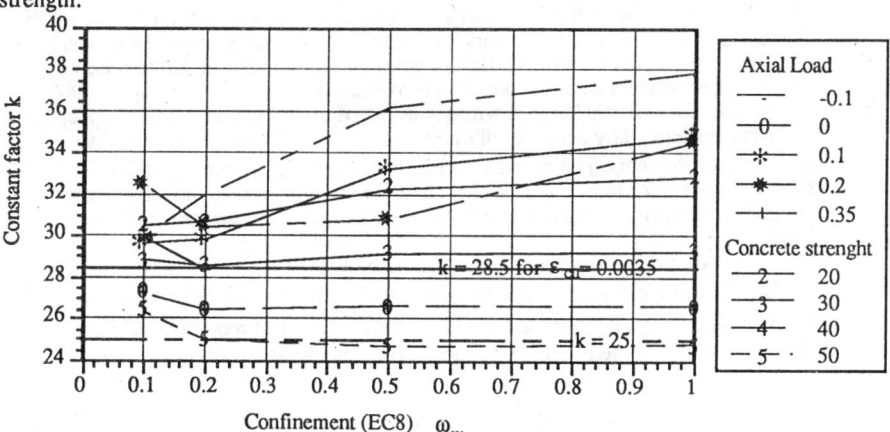

Figure 9   Variation of factor 'k' versus confinement

Combining the above developed equations leads to an improved design procedure for designing for ductility as shown in figure 10 below. The novelty of this approach is in accounting for the pertinent section properties in the determination of the required level of confinement. Results obtained from this procedure are in general in good agreement with analysis, however, they are relatively conservative when the estimates of the initial curvatures are low. This is in contrast to the results from Eurocode 8 which are too un-conservative and too often governed by the minimum limits on confinement.

Choice of the overall behaviour factor 'q' (*from table V, 2.1.4 of Part 1.3*)

Range of member displacement ductility demand μ_Δ

$$q < \mu_\Delta < 0.5(q^2 + 1)$$

Calculation of the critical section curvature ductility demand

$$\mu_{1/t} = 1 + 7(\mu_\Delta - 1)$$

Calculation of available curvature ductility

$$\mu_{1/ro} = \frac{\varepsilon_{cu}(1-\xi_y)}{\varepsilon_y \xi} \geq \frac{\varepsilon_{cu}(1-1.5\xi)}{\varepsilon_y \xi}$$

Calculation of the critical section confinement

$$\omega_w = \frac{10\,\varepsilon_{cu}}{\alpha}\left(\frac{\mu_{1/t}}{\mu_{1/ro}} - 1\right) = \frac{1}{\kappa\,\alpha}\left(\frac{\mu_{1/t}}{\mu_{1/ro}} - 1\right)$$

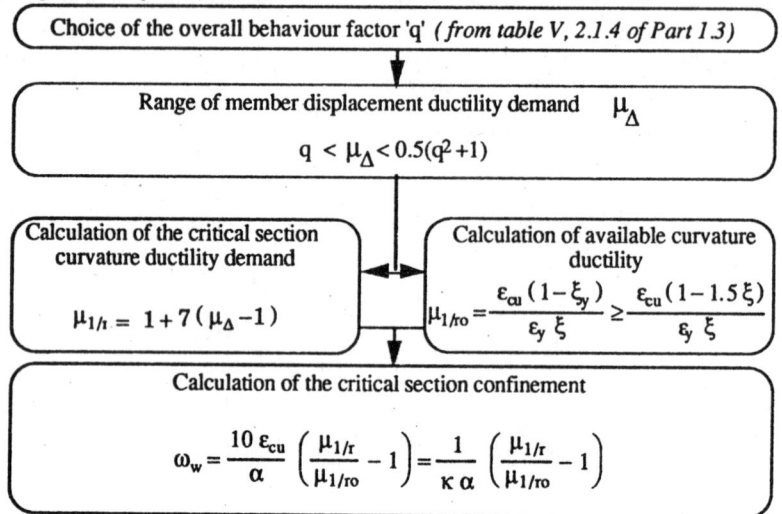

Figure 10   Proposed flow chart for the design of ductility of RC members

CONCLUSIONS

(a) The EC8 provides a complete framework for designing and detailing for ductility the critical members in accordance with a chosen behaviour factor.

(b) The EC8 equation for calculating curvature from displacement ductilities is based on the un-conservative assumption that curvature ductility is constant within the plastic hinge zone. A parabolic distribution for the curvatures within the plastic hinge zone was demonstrated to yield an upper bound approximation.

(c) The height of the plastic hinge is wrongly assumed to be a function of the aspect ratio and a can be derived by using equilibrium calculations for statically determinate structures or elements. The normalised height of the plastic hinge λ_{pl} varies significantly but for normally loaded walls a lower value of 0.15 can be assumed.

(d) An improved equation for calculating curvature from displacement ductility is proposed.

(e) The Eurocode 8 equations for providing confinement reinforcement do not take into account the available ductility of unconfined RC members $\mu_{1/ro}$. A new approach which requires estimating $\mu_{1/ro}$ prior to calculating the confinement required has been developed by using the Eurocode equations for concrete strength enhancement.

ACKNOWLEDGEMENT

The work described herein was partly funded by the UK Science and Engineering Research Council, and the ESEE section of Imperial College, as part of a research programme on the seismic performance of reinforced concrete structural walls, under the supervision of Dr A.S.Elnashai and Professor N.N.Ambraseys.

REFERENCES

1. EC8, "Eurocode No.8 : Structures in seismic regions - Design ", Part 1, General and building, Report EUR 12266 EN, Industrial Processes, Building and Civil Engineering, Commission of the European Communities, May 1988
2. ACI 318-83, " Building Code Requirement for Reinforced Concrete", American Concrete Institute, Detroit,1983, 111 pp.
3. Uniform Building Code, International Conference of Building Officials, Whittier, California, 1988
4. NZS 3101:1982, Parts 1 &2, "Code of Practice for the Design of Concrete Structures", Standards Association of New Zealand, Wellington, 1982, 283 pp.
5. Pilakoutas K. and Elnashai A.S., "Seismic Design of Ductile R/C members", International Conference on "Earthquake, Blast and Impact', SECED, Manchester, 18-20 Sept. 1991
6. Pilakoutas K., "Earthquake resistant design of reinforced concrete walls", PhD Thesis, University of London, 1990
7. Santhanam, T.K., "Model for Mild Steel in Inelastic Frame Analysis", Journal of Structural Engineering, ASCE, Vol. 105, No. 1, January 1979, pp 199-220
8. Mander, J.B, Priestley, M.J.N. & Park, R., "Theoretical Stress-Strain Model for Confined Concrete", Journal of Structural Engineering, ASCE, Vol. 114, No. 8, August 1988a, pp 1804-1826
9. Mander, J.B, Priestley, M.J.N. & Park R., "Observed Stress-Strain Behaviour of Confined Concrete", Journal of Structural Engineering, ASCE, Vol. 114, No. 8, August 1988b, pp 1827-1849
10. Tassios, T.P., "Specific rules for concrete structures", In Backround Document for Eurocode 8 - Part 1, Volume 2 - Design Rules, Commission of European Communities, 1989, pp 1-123
11. Mattock A. H., Discussion of "Rotational capacity of reinforced concrete beams", by Corley W. G., in J. Struct. Div. Am. Soc. Civ. Eng., 93 (ST2), April 1967, pp. 519-522

SOME ASPECTS OF THE ESTIMATION OF BUILDING SEISMIC RESISTANCE BY THE RESULTS OF VIBRATION TESTS

A.Ts.Minasian[1]   M.G.Melkumian[2]   E.E.Khachian[3]

INTRODUCTION

Considerable, negative consequences of intensive earthquakes have essentially affected the role of correct estimation of seismic resistance and guaranteed reliability of functioning and new buildings under construction in seismically active regions. It has become a problem of utmost importance successful solution of which meets the population demands. But it may be solved only after testing a building and/or its elements for an action imitating, to a certain degree, a real earthquake. For the tests of this kind it is necessary to have preliminarily calculated data on the intensity and character of future earthquakes, as well as on the behaviour of buildings and structures themselves. After many years of theoretical and practical experiments we may say that building vibrations during earthquakes may imitated with sufficient accuracy with the help of powerful vibration machines.

Modern buildings represent, complex, many-fold statical indeterminate, structural systems consisting of a multitude of different bearing and nonbearing elements. In its turn, each element may function under different fixation and undergo different prevailing deformations. All these features combined make the problem of seismic stability evaluation even more complicated. The question is that if we study the behaviour of separate elements and have no integrated information on their operation as a whole, we arrive at incorrect conclusions related to evaluation of the building reliability. An opposite statement is also true. Therefore, in estimating the seismic resistance experimentally, the data on separate simple constructions and that on complex systems tested for seismic stability in their natural size or in the form of a model must supplement one another, as a rule. In their report the authors present their criteria for estimating the building capacity to resist the seismic effect in the full-scale or model vibration tests.

CRITERIA FOR ESTIMATING THE SEISMIC RESISTANCE OF BUILDINGS IN THE VIBRATION TESTS

Testing both separate constructions and full-scale buildings, the authors have performed numerous experiments and found that there are mechanisms governing transition of constructions from one qualitative state to another when reinforced concrete constructions affected by horizontal (seismic) forces undergo this or that type of deformation; parti-

---

1, 2, 3 - ArmNIISS, State Committee on Architecture, Yerevan, Republic of Armenia

cularly, it is especially interesting to consider transition of constructions from the state of intensive plastic deformation development to the stage when the destruction begins. Hence, it is possible to define quantitative parameters corresponding to qualitative changes in the construction, and to consider them as the criteria of the state of the systems under test. Such transition parameters for dynamically loaded buildings are presented by the vibration period change, the storey relative displacement (drift) and the storey inertial shear forces, as well as their derivative values of the storey dynamic stiffness and the dependence "restoring force - drift". To define them, it is convinient to perform the vibration test in the resonance mode and to register horizontal displacements and accelerations of every storey and foundation base and, if necessary, vertical displacements of foundations and some floors.

Experimental values of the storey inertial and shear forces are determined by the formulas

$$S_k^e = m_k \ddot{y}_{k max}, \quad Q_k^e = \sum_{i=k}^{n} m_i \ddot{y}_{i max} \qquad (1)$$

where m_k and $\ddot{y}_{k max}$ are the mass and maximum horizontal acceleration, respectively, measured at the level of the k-th storey. Loading value, obtained during the tests, is not estimated by the seismic force value, but by that of the shear forces Q_k^e because their height distribution for the most buildings is similar to the force distribution during earthquakes. These shear force values are then compared with standard Q_k^s shear force values calculated according to the seismic norms.

As it has already been stated above, the state of transition for buildings may be estimated by the changes of the vibration period value of the first type T_1. Generalized results of numerous model and full-scale tests show that if at the loading level $Q_k^e = Q_k^s$ the vibration period value T_1 increases by no more than 1.3 fold in comparision with its initial value, then the building can resist the calculated standard loads. If at the same loading level, the vibration period increases 1.5 and more fold, the building seismic stability can not be guarateed. If at the loading level $Q_k^e > Q_k^s$ by 1.5-2.0 times, no damage is visible in the bearing elements and if the vibration period increases by no more than 1.5 fold, the building is considered to be reliable and may resist the future seismic effects.

The building deformation characterized by the storey relative displacement (drift) is also very important in estimating the seismic resistance of buildings and is determined from

$$\Delta_k^e = y_{k max}^e - y_{(k-1) max}^e \qquad (2)$$

where $y_{k max}^e$ is the maximum horizontal displacement of the k-th storey. The authors have found extreme displacement values for different structural systems in which plastic deformations start to intensively develop as soon as these values are achieved. Particularly, for reinforced concretr frames it is $\Delta = \frac{1}{160} h$ or $\Delta = \frac{1}{200} h$ depending on the girder stffness, and for bracing frames $\Delta = \frac{1}{300} h$ or $\Delta = \frac{1}{550} h$ depending on the diaphragm (solid or with door openings). But in the storey drift calculations it is necessary to take into account that during vibration tests of buildings, depending on their structural concept, the floors may turn if the whole system is bending. In this case it is necessary to find the share of the total horizontal displacement, determined by the building bending. Assume that the floors remain flat during vibrations, i.e. they are not deformed, but only turn by very small angles Θ_K (fig.1). Now it is easy to prove that the top displacement $y_{n\Theta}$ of an n-storey building, caused by bending, may be determined from the formula

$$y_{n\Theta} = \Theta_1 h_1 + \sum_{K=2}^{n-1}(\Theta_K - \Theta_{K-1}) h_K \qquad (3)$$

where Θ_1 and Θ_K the floor turning angles of the 1st and K-th storey, respectively; h_1 and h_K are the distance between the 1st and K-th storey floors, respectively, and the floor of the n-th storey; and n is the number of storeys.

Another factor to be used in estimating the building seismic stabilty is the storey dynamic stiffness. It is determined from the ratio

$$c_K^e = \frac{Q_K^e}{\Delta_K^e} \qquad (4)$$

and represents an inertial shear force causing the K-th storey drift, equal to a unit. During the vibration tests with a stage-by-stage increase of the vibrator shaft loads, damages gradually grow in their amount and, hence, the storey and the whole building stiffness gradually decreases. The vibration period also grows. Thus, by the analogy with considerations concerning the vibration period described above, it is possible to state that if at the loading level $Q_K^e = Q_K^s$ the averaged storey stiffness decreases more than two-fold, then a construction of this type cannot be considered as quake-resistant.

EXAMPLE OF SEISMIC RESISTANCE EVALUATION FOR A 16-STOREY FRAME BUILDING ACCORDING TO ITS VIBRATION TEST RESULTS

After the Spitak Earthquake on December 7, 1988 there appeared a necessity of testing the 16-storey frame buildings under construction in Armenia for their seismic resistance, in order to provide their large-scale construction (Fig.2). These buildings are erected of columns designed for three storeys at a time and girders designed for one span. Near the frame joints in the columns there are areas 55 cm high with bare reinforcement. In these areas, on the column faces there are flat surfaces, perpendicular to the column longitudinal axes (Fig.3). The stiffness diaphragms consist of precast panels fixed to the columns and girders by reinforcement U-stirrups (Fig.4). The diaphragm location scheme is presented in Fig.5. The building has a basement 2.4 m high and, therefore, in the calculations and tests it is assumed to consist of 17 storeys.

The tests were carried out with the help of the vibration machine BIM-80 by ArmNIISA, mounted on the top floor of the building (Fig.6). Before the tests, periods and decrement of building vibrations, equal to T_1 =1.161s, T_2 =0.278s and δ = =0.19, have been experimentally found by hitting the building top with a heavy load. For these periods, preliminary analysis of the vibrator possibilities showed that vibrations of the first type could not induce the required effect. Therefore the main tests were performed with vibrations of the second type. The given building has been designed for a 7-point earthquake effect. The total shear force (Q^s) diagram is presented in Fig.7a. There were five test stages with gradual increase of loading. The loading level, close to calculated one, was achieved at stage V, and a corresponding diagram of experimental shear forces (Q^e) is shown in Fig.7b. To make the comparison more visible, the both diagrams are combined in Fig.7c. It shows that during the tests the middle storeys proved to be underloaded in comparison with the calculated level, while the upper storeys were overloaded from 3 to 11%. On the whole, judging from the diagram completeness, it may be asserted that the building has been underloaded by 17% in comparison with the calculated 7-point effect. At this loading level, the vibration period also appears considerably changed (Fig.8), increasing by almost 47% in comparison with the initial value. This increase is a result of the storey stiffness fall (Fig.9). Comparison of average values of the 1st and 5th storeys shows and almost three-fold fall of stiffness. Besides, the displacement diagram (Fig.10) shows that at stage V the maximum displacement of the 17th storey is 6.0 mm and that of the 8th storey 5.25mm. Corresponding, calculated displacement values, according to the vibration second type, are equal to -2.95 and +2.91mm, i.e. they are less than the experimental values 1.92 fold on the average. But comparing shear forces, generated only by the second type of vibrations, we see that experimental values exceed the calculated ones 1.61 fold. The building under study has been designed for an elastic stage, and during the tests it also behaved elastically, the displacements should also increase proportionally to the shear force increase. But from the given values it follows that the displacements have grown more than the forces.

Thus, using the criteria given above, and comparing them with the vibration test results for a 16-storey building, we arrive at the conclusion that the building cannot reliably resist intensive seismic effects.

CONCLUSION

The article is devoted to criteria for estimating the building seismic resistance. An example of this estimation, according to the vibration test results, is also given. Experimental results are in good agreement, thus confirming correctness of the adopted criteria and efficiency of vibration tests as a seismic resistance estimating method. It is also found that at experimental values close to calculated ones, the building storey stiffness considerably falls, the vibration period increase totally characterizing this fall. The building frame is characterized by increased deformability causing considerable displacements, due to wrong concepts.

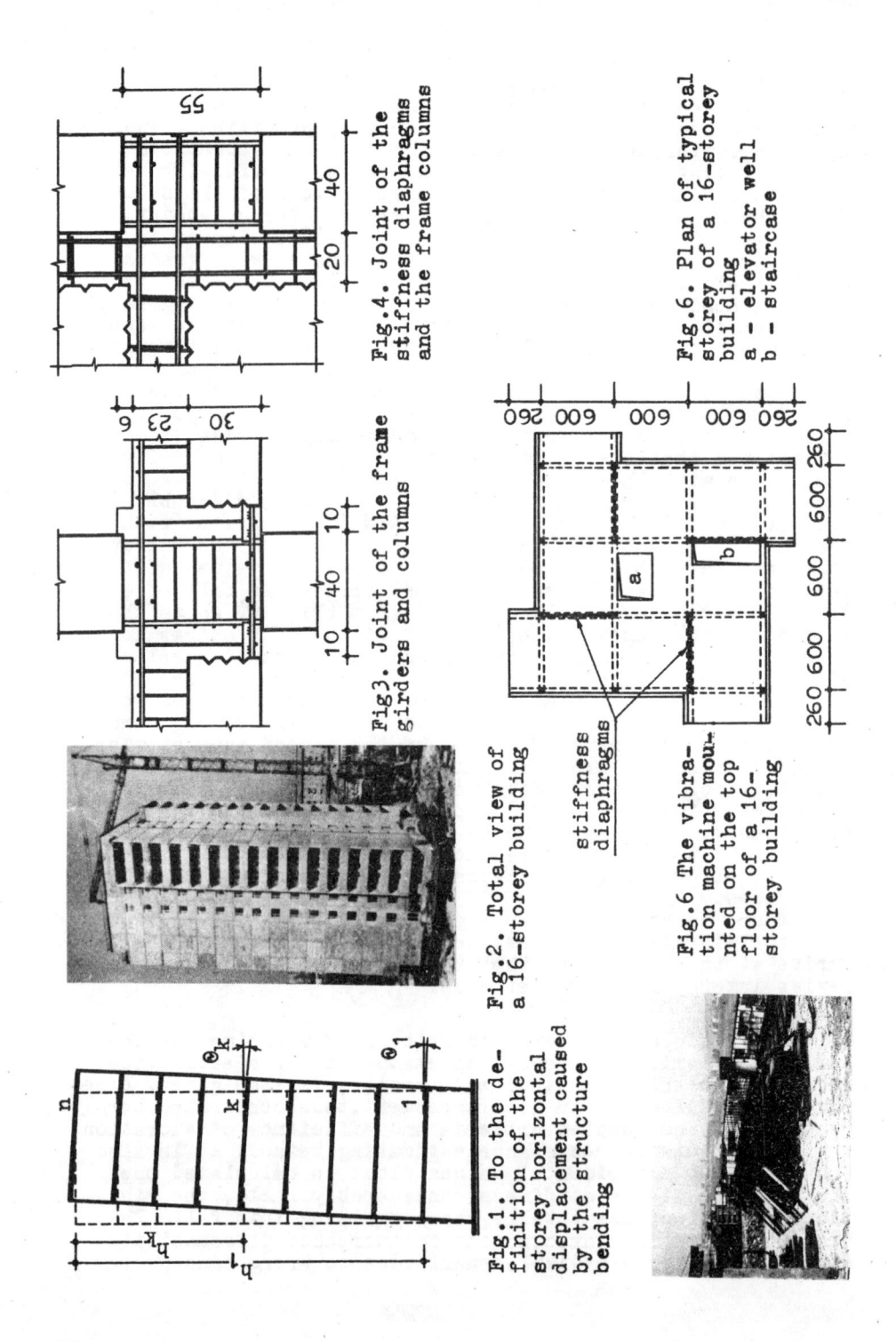

Fig.4. Joint of the stiffness diaphragms and the frame columns

Fig.3. Joint of the frame girders and columns

Fig.6. Plan of typical storey of a 16-storey building
a - elevator well
b - staircase

Fig.2. Total view of a 16-storey building

stiffness diaphragms

Fig.6. The vibration machine mounted on the top floor of a 16-storey building

Fig.1. To the definition of the storey horizontal displacement caused by the structure bending

76

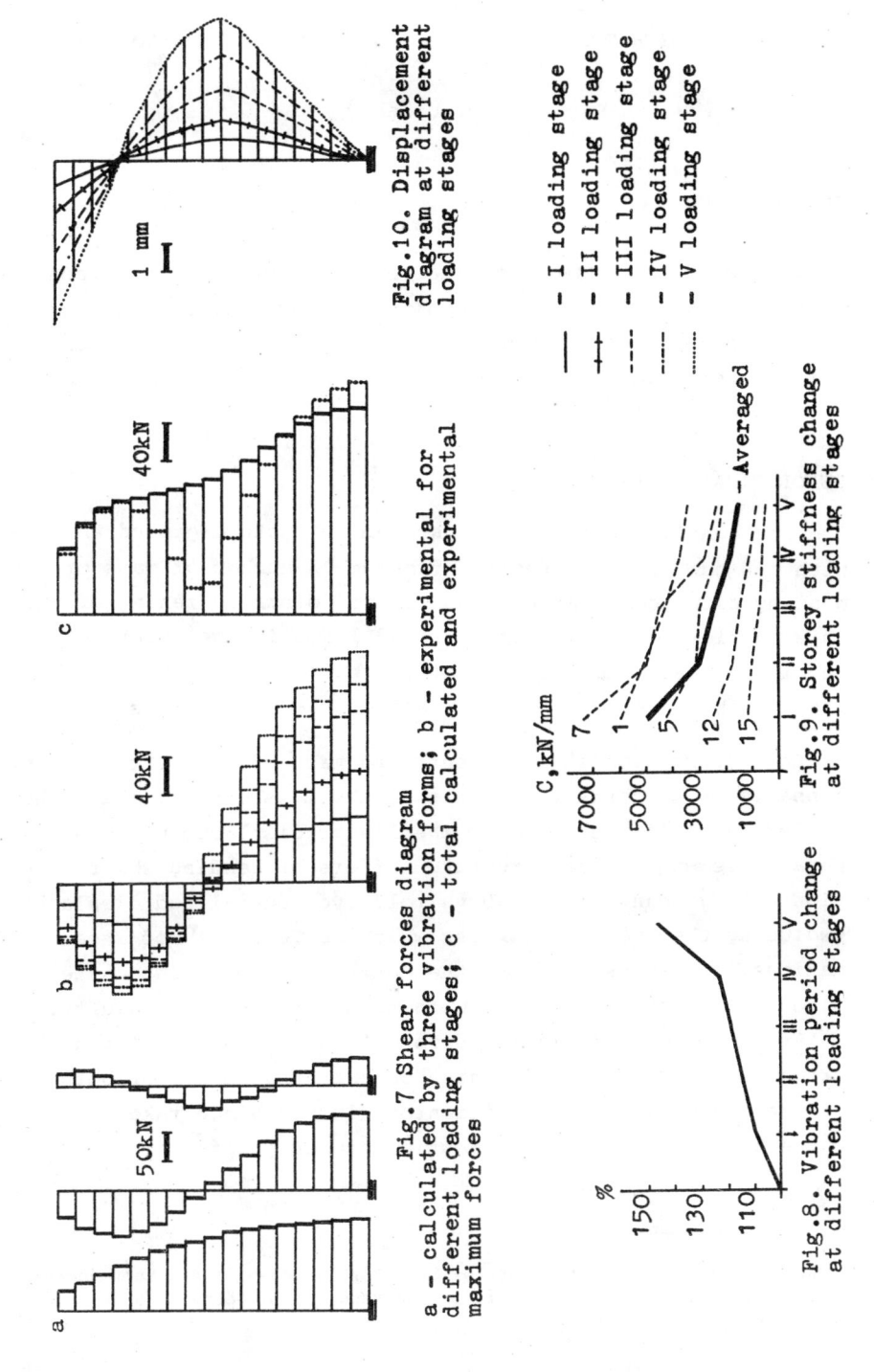

Fig.7 Shear forces diagram
a – calculated by three vibration forms; b – experimental for different loading stages; c – total calculated and experimental maximum forces

Fig.10. Displacement diagram at different loading stages

— I loading stage
— II loading stage
— III loading stage
— IV loading stage
— V loading stage

Fig.9. Storey stiffness change at different loading stages

Fig.8. Vibration period change at different loading stages

DESIGN OF THE MULTISTOREYED BUILDING ON SEISMIC EFFECT WITH THE ACCOUNT OF ELASTO-PLASTIC DEFORMATION

D. Ukleba[1]

INTRODUCTION

The design method of combined elastic deformation of construction and ground-foundation massif on seismic effect is suggested in this paper. Discrete design scheme of multi-mass system "structure-ground thickness" is used. The numerical examples are presented.

THEORETICAL DEPENDENCES

The considered multi-mass system "structure-ground thickness" with n+k number of degrees of freedom (n-number of mass or responding displacements and k-number of joints suffering of turns) is weightless carrying concentrated loads in the joints at the level of intermediate floors of the buildings or structures and also at the level of the interface of the soil layers.

Each conceptional concentrated mass of the system can move linearly as concerns the joints of the mass concentration they suffer turns in addition to the linear displacements. Each separate element of the structure suffers of bending shear deformation with account of the above relative generalized displacements, excluding the ground thickness which suffers only of shear deformations. It is assumed that the centres of non-point masses coincide with joint centres and with those of element rigidity.

At the moment of elastic deformation beginning all plastic properties of the material are concentrated on one or both ends of the element, where formation of plastic hinge takes place, while the element remains elastic between joints.

---

[1] Enginer, Doc. of the Institute of Structural Mechanics and Earthquake Resistance of the Acad. of Sci. of the Republic of Georgia, USSR

It is admitted that the bending moments and normal forces acting at its ends remain constant until the element will go out of this stage of deformation.

The closing of plastic hinge is caused either by changing of velocity mark or by changing velocity mark of relative horizontal displacement of the joint element.

At opening or closing of plastic hinge the element rigidity changes spontaneously and after closing of plastic hinge it requires the initial stiffness.

Vibration of the construction is described by the set of differential equations and their solution is performed with use of Runge-Kutta method with automatic choice of step integration.

Damping of the energy of vibration is considered according to Voigth hypothesis.

Differential equation of the motion of the whole system "structure-ground thickness" for any joint j without accounting for damping can be written as

$$m_j \ddot{u}_j - K_{j-1}(u_{j-1} - u_j) + K'_{j-1}(\Theta_{j-1} + \Theta_j) + K_j(u_j - u_{j+1}) - K'_j(\Theta_j + \Theta_{j+1}) = - m_j \ddot{x}_o,$$

$$I_j \ddot{\Theta}_j + C_{j-1}\Theta_j - K'_{j-1}(u_{j-1} - u_j) + C_j\Theta_j - K'_j(u_j - u_{j+1}) + C'_{j-1}\Theta_{j-1} + C'_j \Theta_{j+1} = 0,$$

$(j = 1, 2, \ldots n; \; \Theta_i = 0, \text{ where } i = k+1, k+2, \ldots n).$

where:

m_j - concentrated loads mass of the system;
K_j and K'_j - rigidity of separated structural elements; at relative displacements of the ends;
C_j and C'_j - rigidity at the end turn of the elements;
u_j, $\dot{u}_j$ and $\ddot{u}_j$ - relative linear displacement, velocity and acceleration of mass m_j;
Θ_j - angle of the turn of the joint of the constructed structure;
I_j - inertia moment of the isolated masses of adjacent structural elements relative of rotation axes perpendicular the drawing plane;
$\ddot{x}_o$ - seismic acceleration of the base soil.

In a particular case we consider the whole system "structure-ground thickness" with n+1 number of freedom (n-linear displacement and 1-foundation turn), it is also assumed that the underground part, the yielding of the base soil, can perform the shear as well as the turn in the vertical plane (Fig. 1).

It is also assumed that each isolated horizontal section of the structure above foundation suffers of the turn for the same angle as the foundation itself and the joints of the concentrated masses of the structure can perform only mutual shear displacements. The ground thickness is a multi-mass system, which works only for a horizontal shear.

As the example the results of calculations of 9 storeyed frame building, which was damaged during Spitak earthquake in Armenia in 1988 is given in this paper. The calculations were made with the account of elastic-plastic deformation possibility for both: building and ground foundation. The building is spread by thick three layers ground of foundation, its depth is 9m and it has real geological data.

N-S - horizontal component of Spitak earhquake accelerogram (Fig. 2) was used as the seismic distribution, which was recalculated for the rock base according to the method developed at the Institute of Structural Mechanics and Earthquake Engineering, Academy of Sciences of the Republic of Georgia.

Fig. 3 shows the plots of displacements in time of joint with the mass m_{10} of the system "structure-ground thickness" for elastic-plastic analysis with account (Fig. 3a) and without the account (Fig. 3b) of the foundation turn accordingly.

Gradual accumulation of plastic deformations with the same mark is observed, on account of it, free vibration takes place regarding to new axes (stroked dotted line) which is displaced because of the accumulation of remaining displacements. In this case the remaining displacements achieve 0.81 - 0.95m accordingly.

On the basis of this analysis a conclusion is made that the possibilities of combined elastic-plastic deformation of the construction and the ground foundation influences much the seismic stability of buildings.

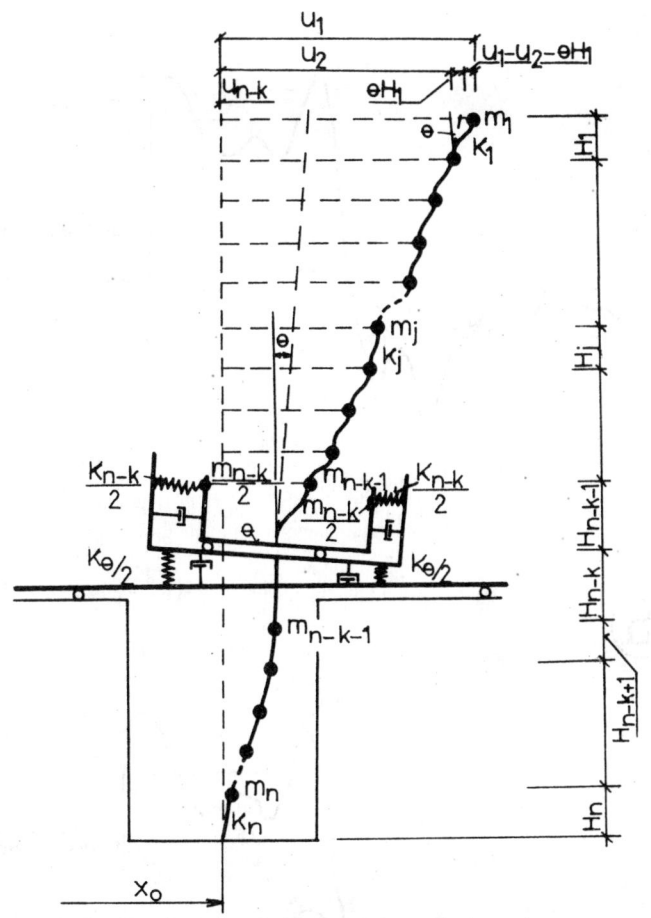

Fig. 1. Design scheme of the system "structure-ground thickness"

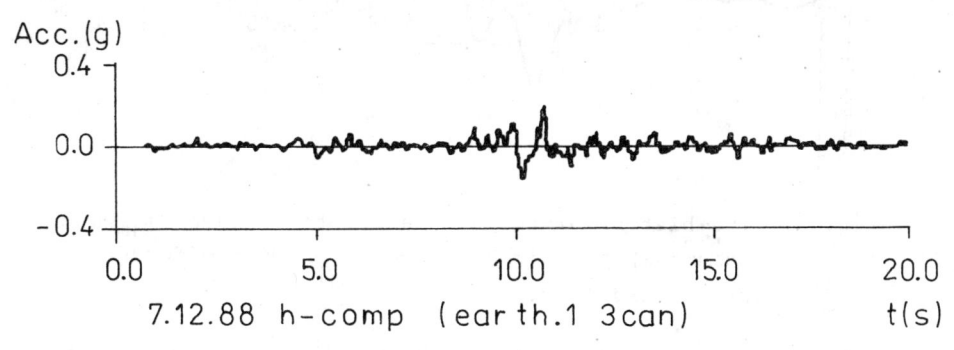

Fig. 2. N-S horizontal component of Spitak /Armenia/ earthquake accelerogram.

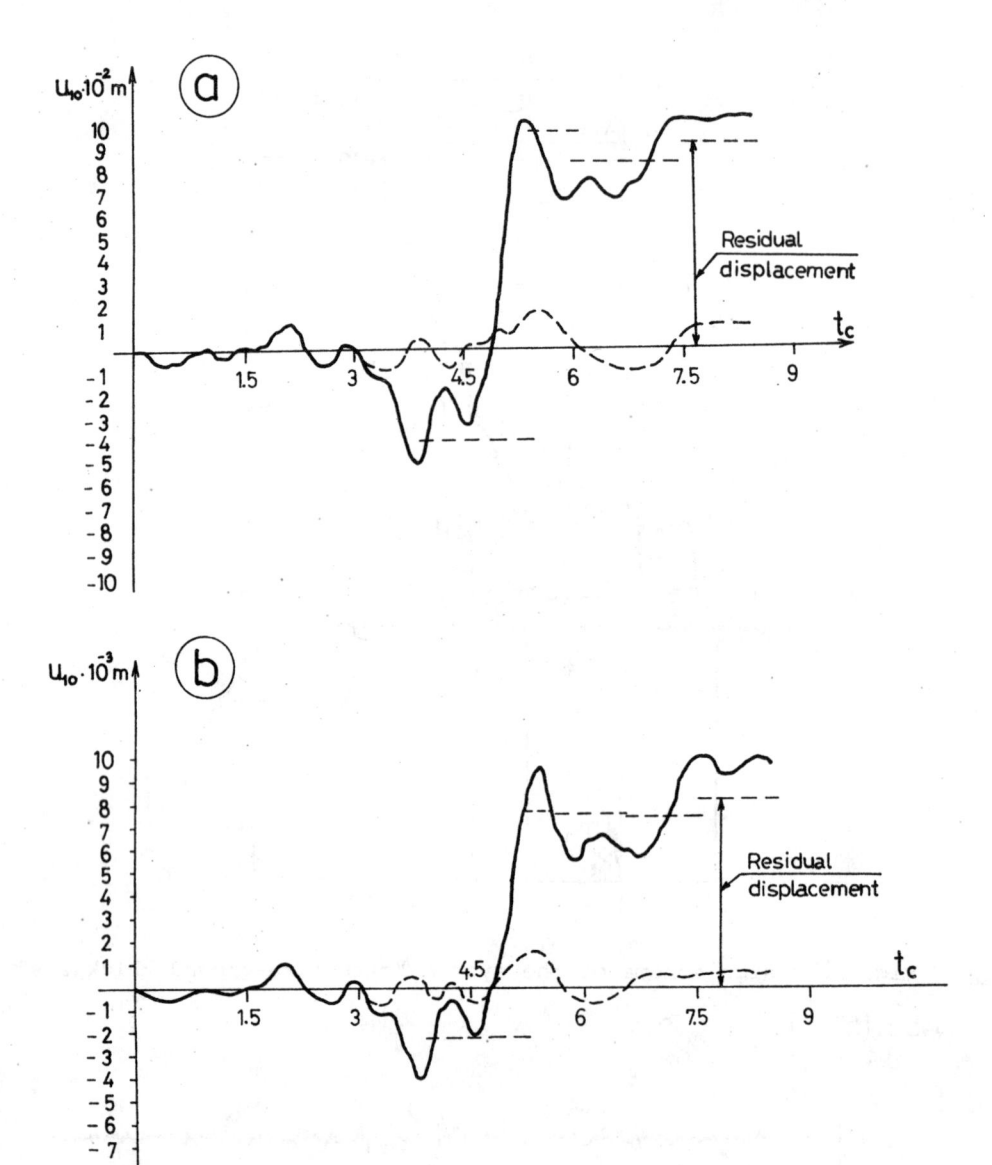

Fig. 3. Time-history of the displacements of mass m_{10} with account (a) and without account (b) of the foundation turn accordingly.

T3

SEISMIC RESPONSE REDUCTION SYSTEMS
AND SEISMIC ISOLATION.

OPTIMIZATION OF BASE ISOLATION SYSTEMS FOR IMPORTANT STRUCTURES

E. Juhásová[1]

INTRODUCTION

The present architectural approaches in the design of structures include a wide variety of used materials, structural systems and the solution of load-carrying and fulfiling elements and details. Some of them are more or less inconvenient for using in seismic areas. Namely the irregularities of the system can cause that the structure, even if being designed according to seismic standards, is not seismic-resistant. The problems arise also when the results of seismic response analysis show that the seismic resistance of the respective structure is not satisfactory either in the case when taking into account the non-linear reserve in the load-carrying capacity of structural elements. Then we should solve the question how to help the structure withstand the effects of strong seismic motions. One of the ways how to improve the behaviour of structure is the application of base isolation systems. The paper is analysing the behaviour of base isolators with different solution of details comparing the individual factors influence on the final isolation effect.

METHODS OF LABORATORY INVESTIGATION

In the case of the investigation of the efficiency of seismic isolators both static and dynamic tests are necessary. The best way is to complete the dynamic tests with the tests on seismic shaking tables. We must say that multi-component controlled shaking tables for laboratory seismic testing are very useful tools in the hands of experienced skilful investigators. Either the possibilities of shaking tables are rather limited, there still exists the sufficiently wide field to analyse the chosen unanswered problems of earthquake engineering by the help of shaking tables.

The laboratory system which we have used for seismic testing was two component shaking table with horizontal and vertical independent x-z motion. The shaking table is driven by electrohydraulic controlled system with the possibility of the simulation of general time history. The transmission of horizontal motion from the actuator to the table is secured by elastic plates system, the vertical motion from the second actuator is secured by another system of elastic plates. The tested model can be loaded either by one component seismic motion chosen either in x or z direction or by two component seismic motion acting both in x and z directions. The seismic response of such system can be measured and registered through accelerometers, strain gauges and pickups

1 - ÚSTARCH SAV, Bratislava, ČSFR

for displacement measurements - usually inductive pickups. While there are nearly no problems concerning the strain and acceleration measurements, the problems rather arise when we want to use the inductive pickups for measurement of relative displacements.

As far as the tested model responds either to one component seismic loading by space vibration, the relative displacements must be precisely measured to obtain the satisfying results. This can be reached through the stiff supporting system which is fixed to the table. The inductive pickups edges should have the connection where the rotation motion or lateral sliding motion are allowed.

Naturally also twice integration of acceleration can be used to receive the displacement quantities. But such results are less precised comparing to displacements measured by inductive pickups. The experimental results in the form of time histories for strains, accelerations, displacements are stored directly using A/D convertors on hard disk or on floppy disks of computers. Optionally multichannel tape recorder can be used too.

SEISMIC RESPONSE OF ISOLATOR PROTOTYPES

Taking into account the knowledges from our previous tests and analysis of seismic isolation properties of different absorbers [1],[2],[3] we have done seismic tests of a few seismic isolator prototypes where was followed the effect of the rubber layers and motions in slide planes. The system was investigated as a stiff structure which rested on two isolators. The chosen systems were denoted as

- shear isolators,
- ring isolators,
- hat isolators.

The measured quantities were accelerations and displacements of the shaking table $\ddot{x}(t)$, $\ddot{z}(t)$, $x(t)$, $z(t)$, absolute accelerations of tested structure $\ddot{u}_1(t)$, $\ddot{w}_{1R}(t)$, $\ddot{w}_{1L}(t)$, and relative displacement of tested structure $u_1(t)$, $w_{1R}(t)$, $w_{1L}(t)$. (Fig. 1.).

The seismic loading was applied succesively like:

- pure x direction excitation,
- pure z direction excitation,
- x-z directions excitation,

in every case like:

- harmonic seismic motion 0.5-20 Hz,
- different seismic accelerograms.

During the tests of shear isolators the different combinations of rubber layers were used, they are denoted as the alternatives A1 to A7 in Table 1.

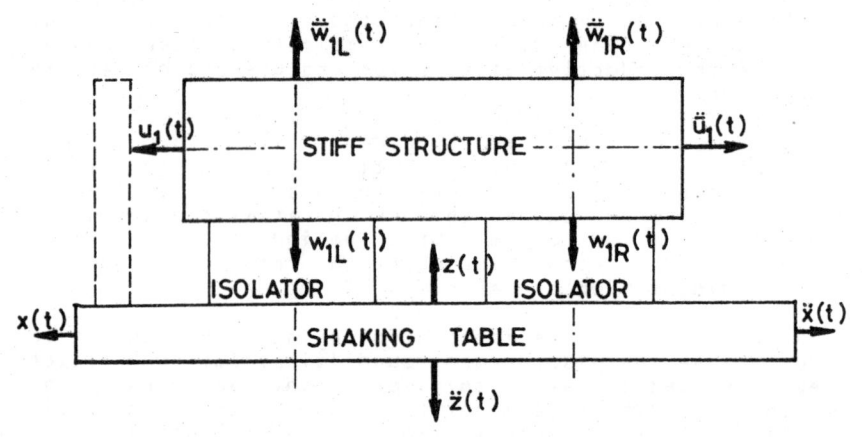

FIG. 1. THE MEASURED QUANTITIES DURING THE SEISMIC TESTS.

The degree of the quality of seismic response was denoted:

N - unsuitable, with large residual displacement,

1 - suitable, with residual displacement not larger than 2-4 mm,

E - excellent, with the proper behaviour during every version of seismic loading, no residual displacement.

TABLE 1. The results of seismic tests of the shear isolator prototypes

Alternative of layers in isolator	Harmonic seismic loading	Different seismic accelerograms
A1	N - unsuitable	N - unsuitable
A2	N - unsuitable	N - unsuitable
A3	1 - suits well	1 - suits well
A4	N - unsuitable	1 - suits well
A5	N - unsuitable	1 - suits well
A6	E - suits excellent	E - suits excellent
A7	E - suits excellent	E - suits excellent

The best alternatives of shear isolators were used as a base for the next group of isolators - ring isolators. The idea of them was to spread up the range of allowable horizontal motion in the isolators and to preserve the limit motion in the isolator against the additional shock effects. The behaviour of such adopted isolators is described in Figs. 2-6 for harmonic seismic excitation and in Figs. 7-8 for the chosen seismic accelerograms of different intensity.

The extremes of the loading seismic acceleration and the transmitted acceleration are in Table 2. Similarly we can follow the changes in relative displacements in both horizontal and vertical directions.

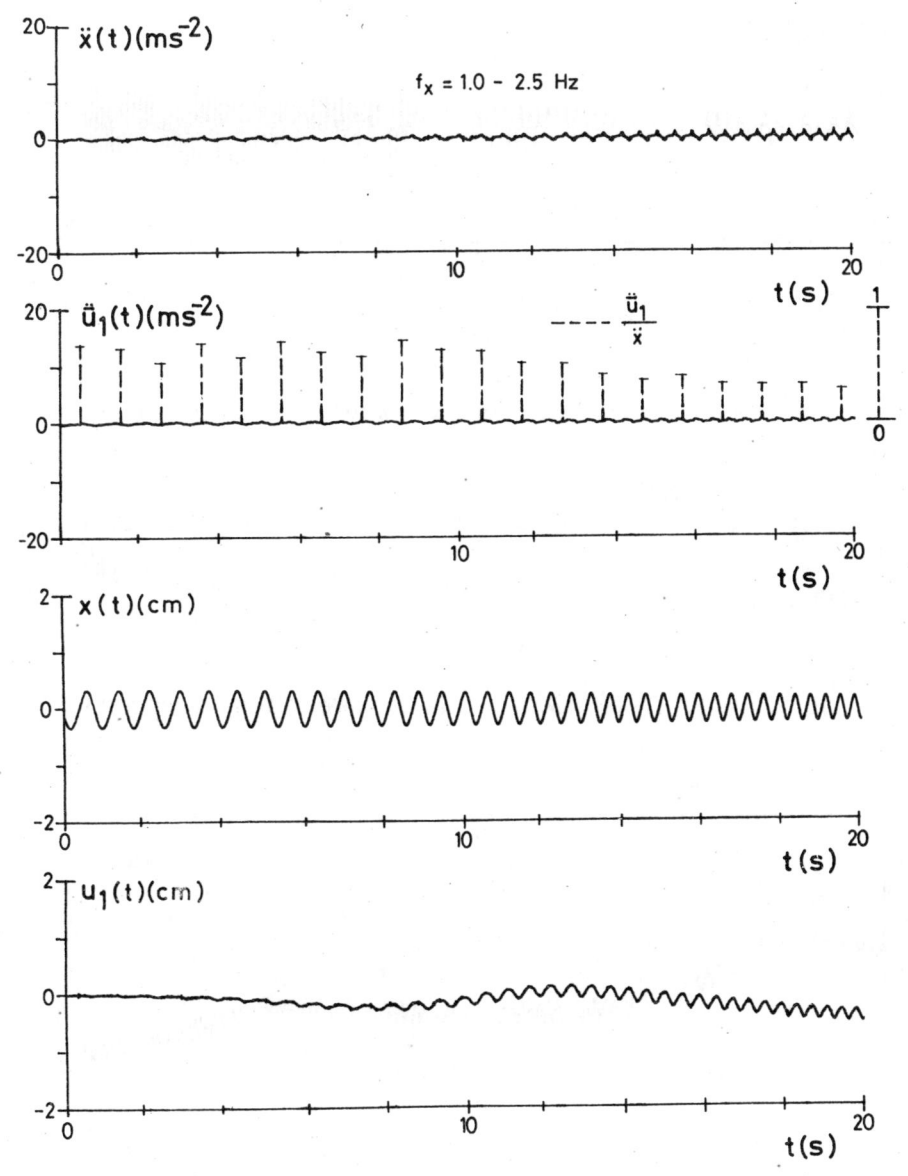

FIG 2. HORIZONTAL HARMONIC SEISMIC LOADING AND THE RESPONSE OF RING ISOLATOR PROTOTYPE $f_x \in (1.0, 2.5)$ Hz

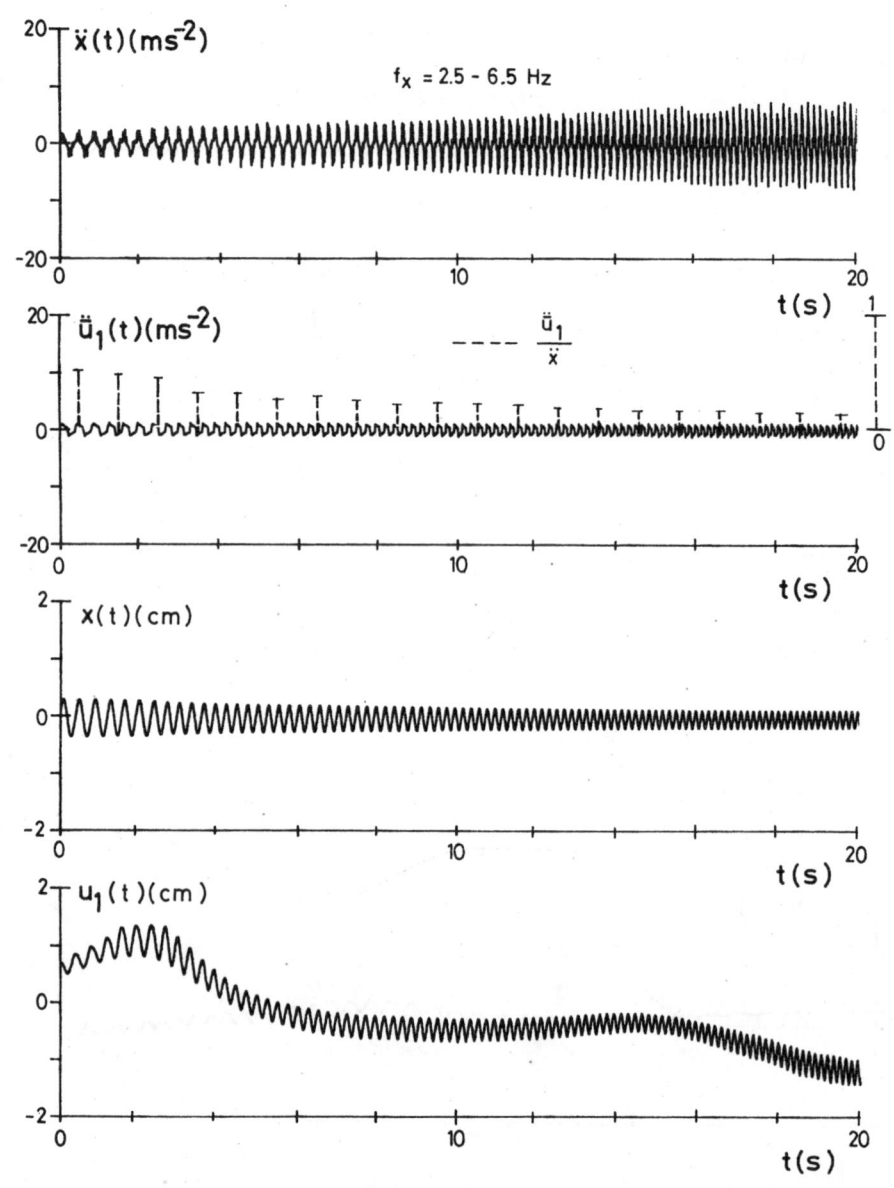

FIG. 3. HORIZONTAL HARMONIC SEISMIC LOADING AND THE RESPONSE OF RING ISOLATOR PROTOTYPE $f_x \in (2.5, 6.5)$ Hz

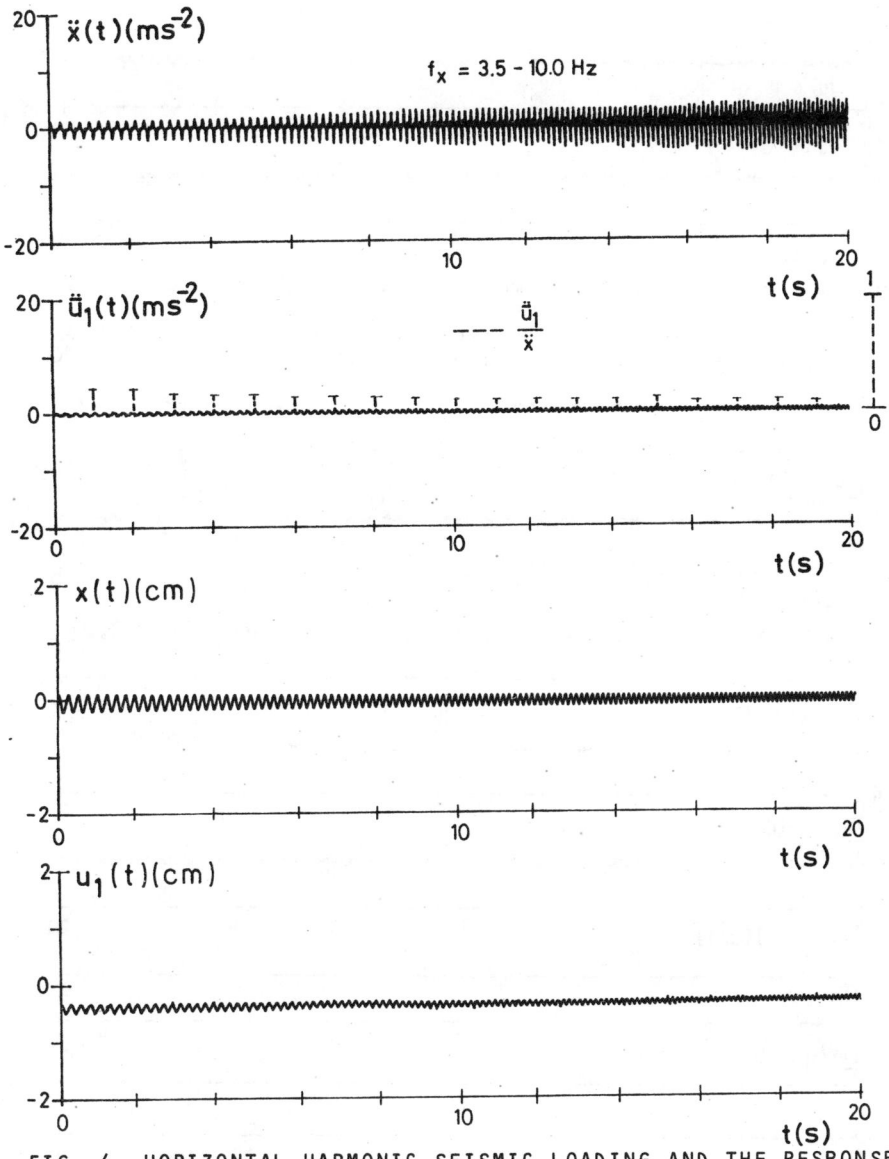

FIG. 4. HORIZONTAL HARMONIC SEISMIC LOADING AND THE RESPONSE OF RING ISOLATOR PROTOTYPE $f_x \in (3.5, 10.0)$ Hz

FIG. 5. TWO COMPONENT x-z DIRECTIONS HARMONIC SEISMIC LOADING AND THE RESPONSE OF RING ISOLATOR $f_x = f_z \in (1.5, 3.5 Hz)$

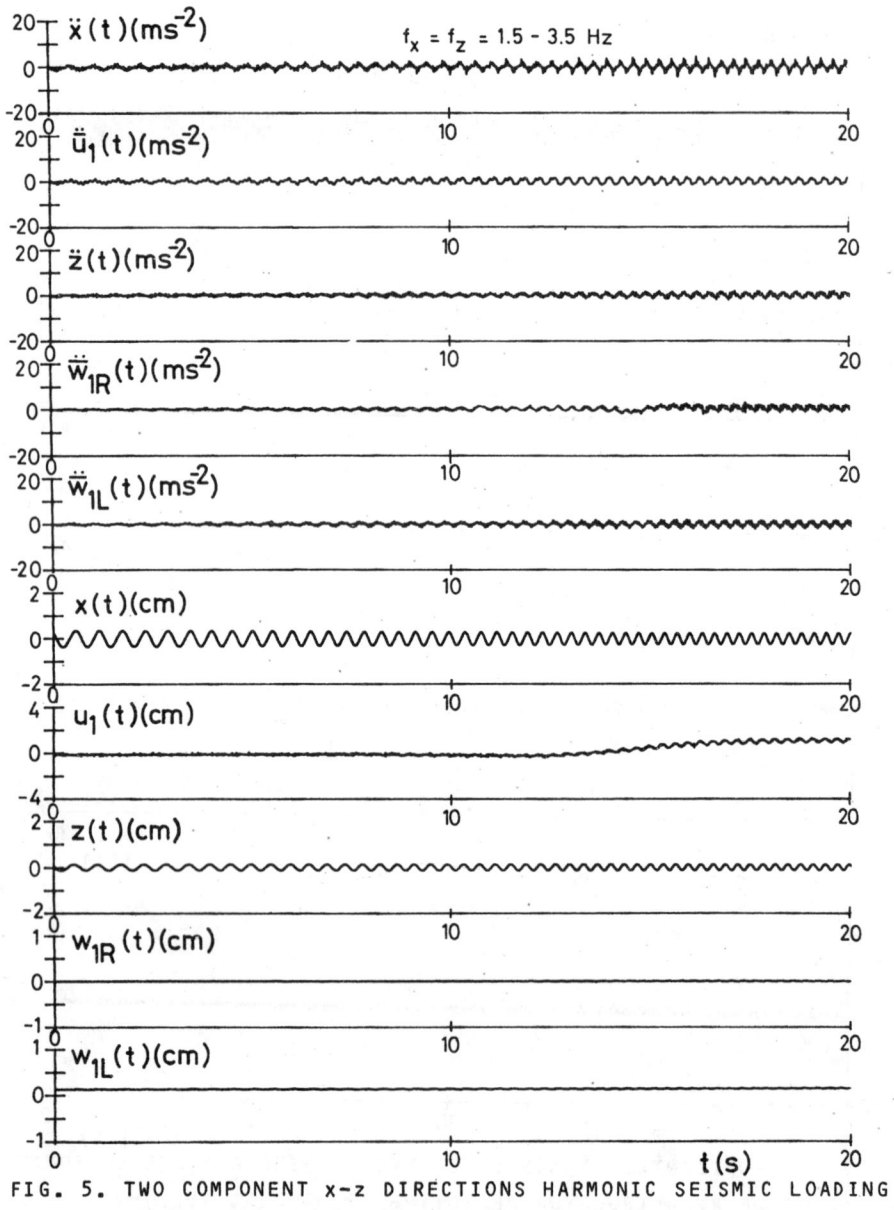

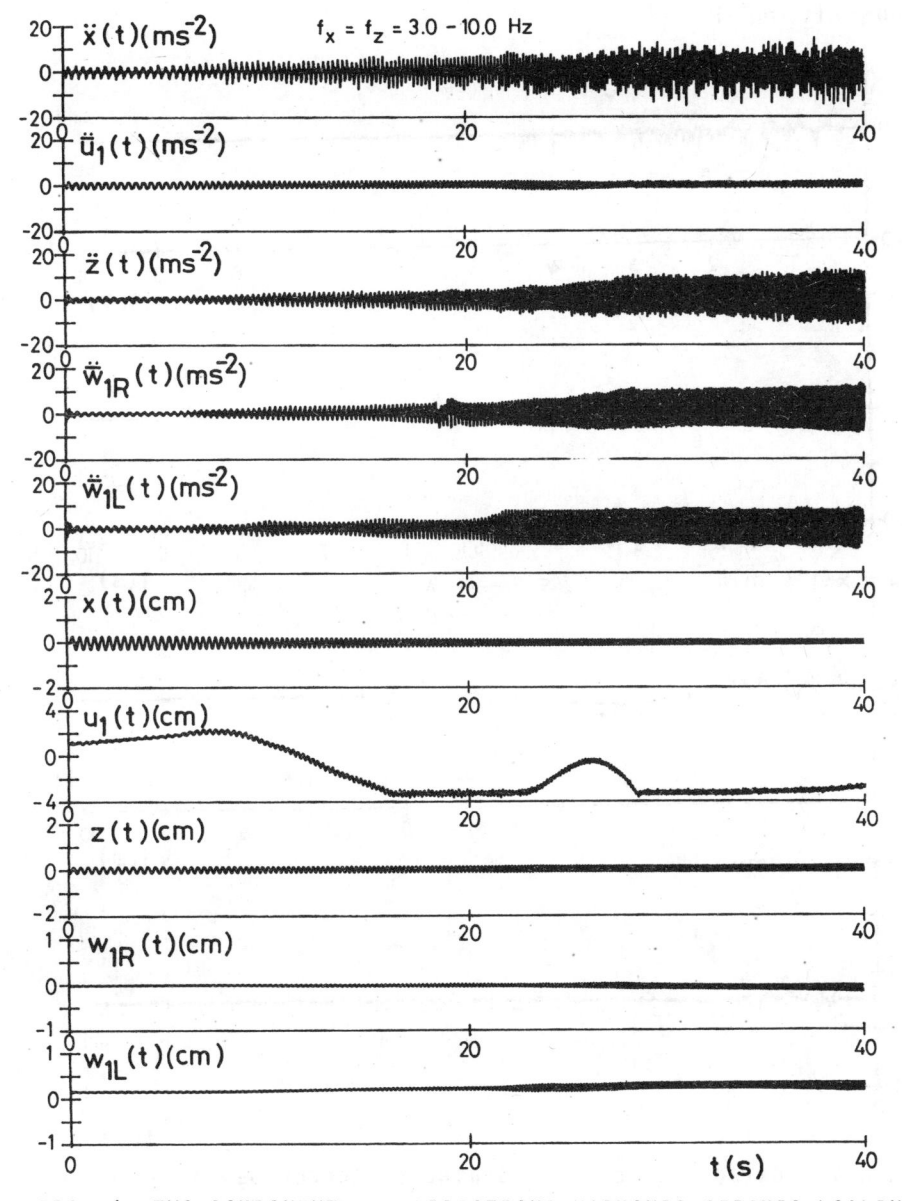

FIG. 6. TWO COMPONENT x-z DIRECTIONS HARMONIC SEISMIC LOADING AND THE RESPONSE OF RING ISOLATOR PROTOTYPE
$f_x = f_z \in (3.0, 10.0)$ Hz

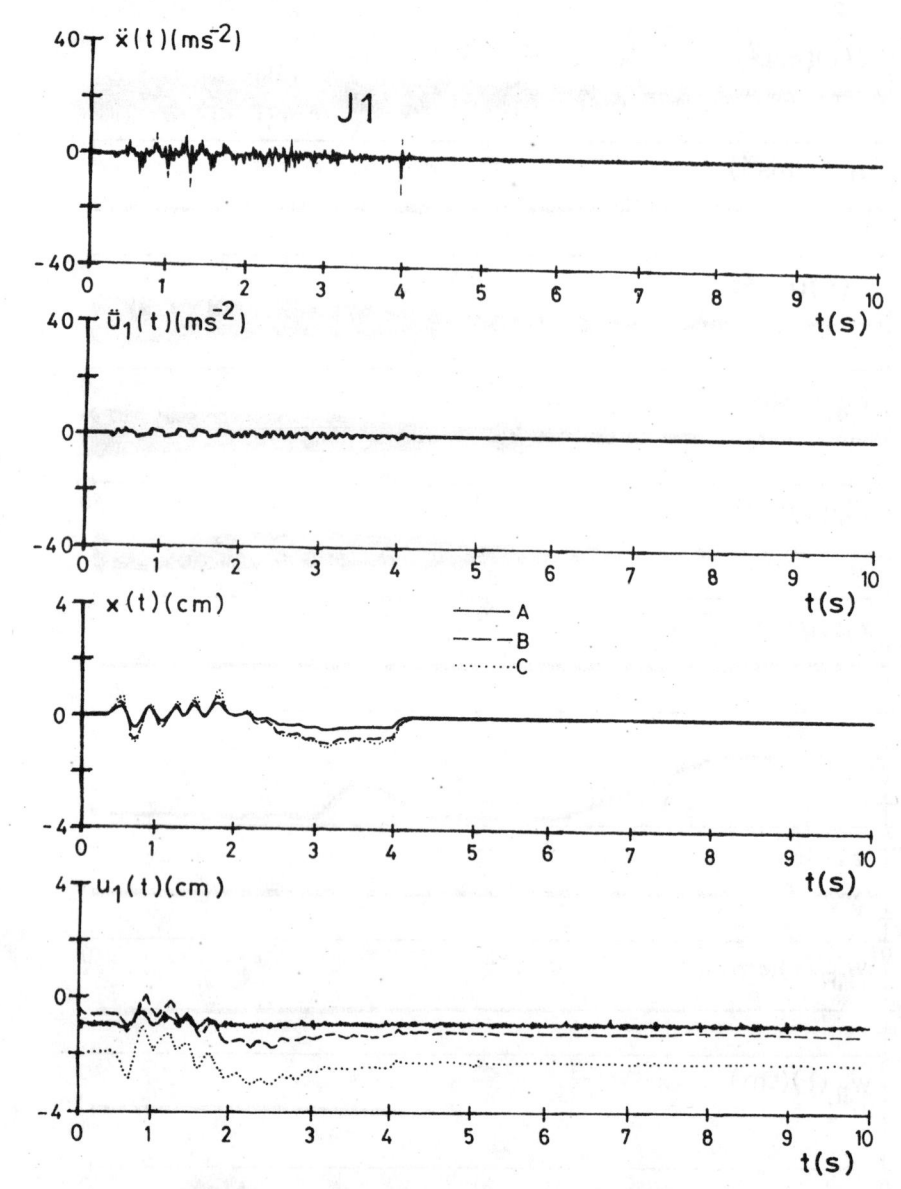

FIG. 7. HORIZONTAL SEISMIC LOADING BY ACCELEROGRAM J1 x-DIRECTION AND THE RESPONSE OF RING ISOLATOR AT DIFFERENT LOADING AMPLIFICATION A, B, C

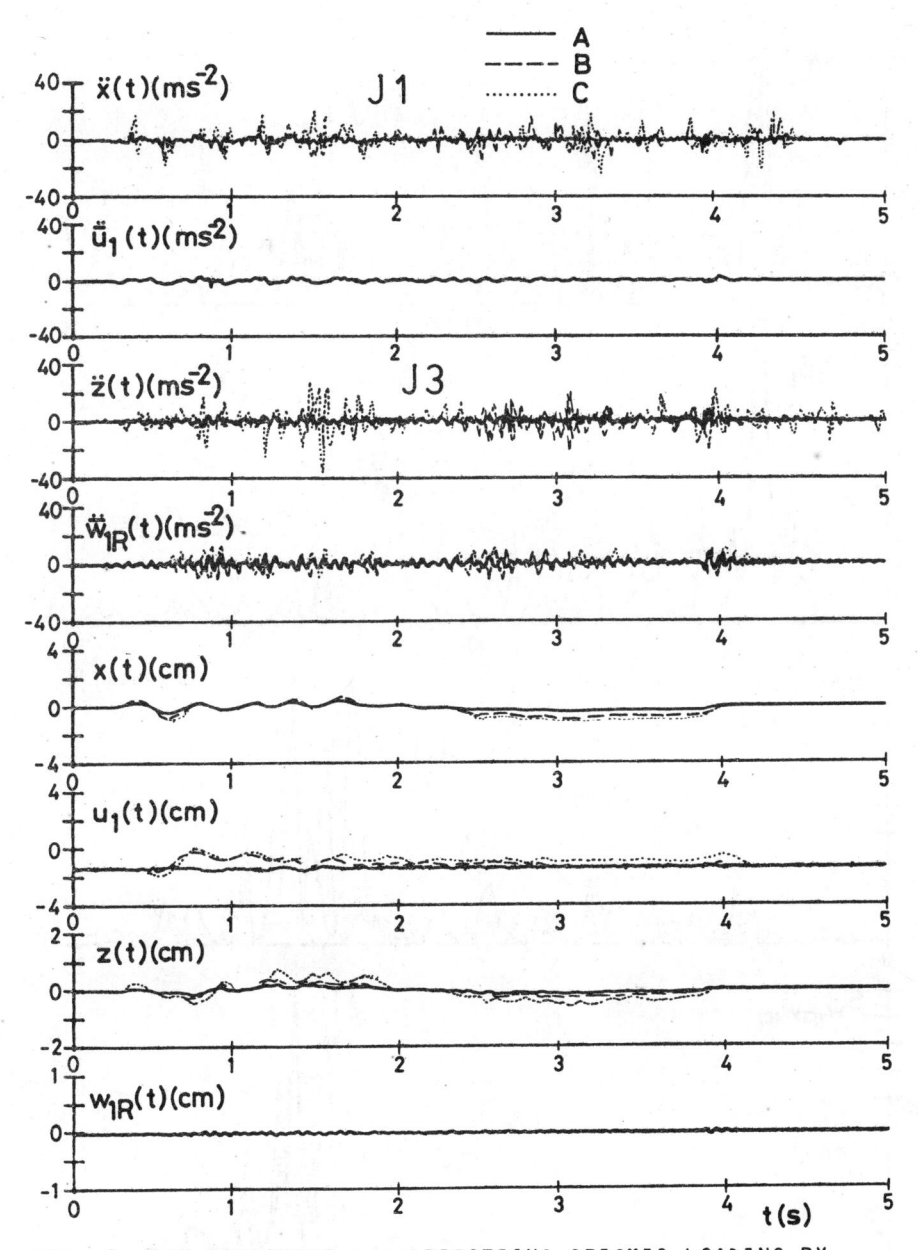

FIG. 8. TWO COMPONENT x-z DIRECTIONS SEISMIC LOADING BY ACCELEROGRAMS J1, J3 AND THE RESPONSE OF RING ISOLATOR PROTOTYPE

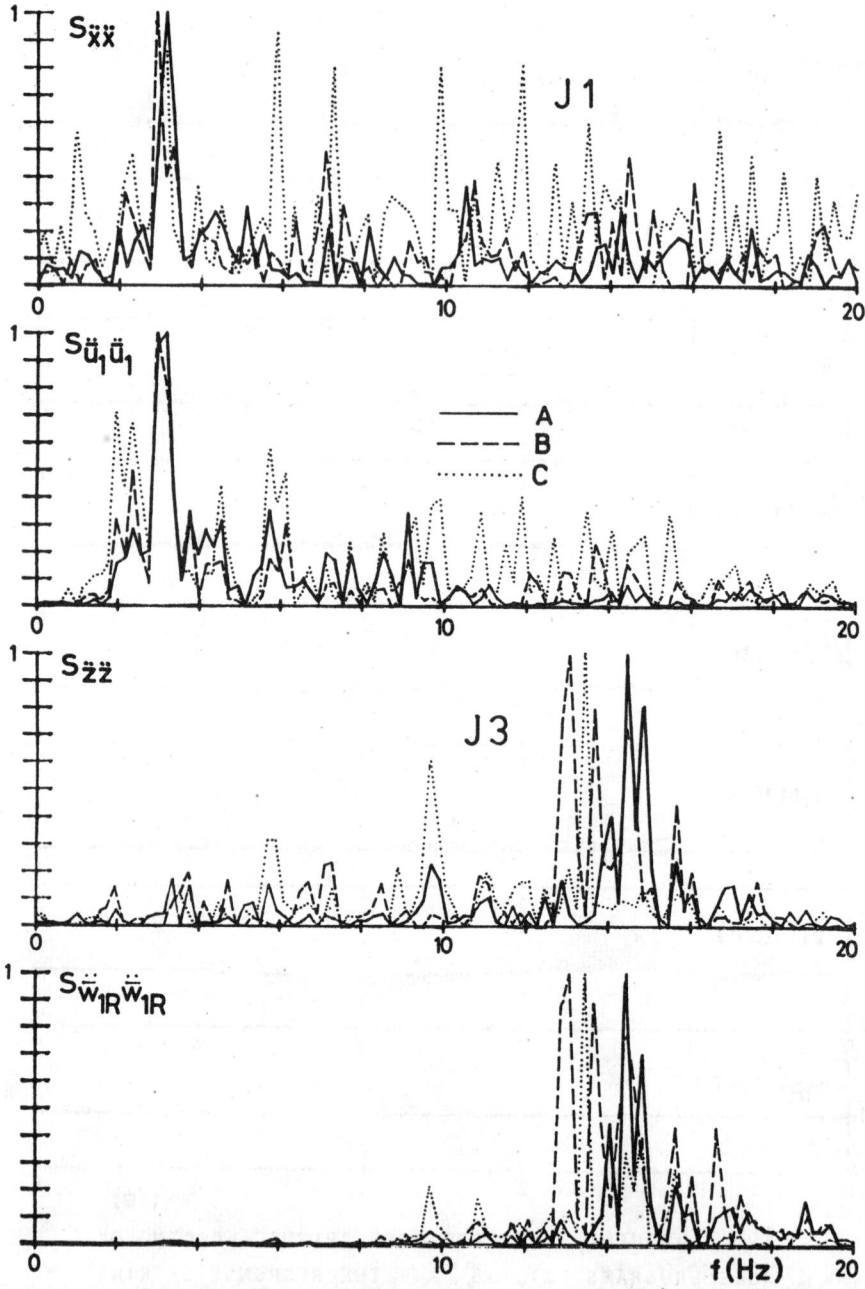

FIG. 9. POWER SPECTRAL DENSITY OF SEISMIC LOADING AND THE RESPONSE OF RING ISOLATOR. LOADING BY ACCELEROGRAMS J1, J3.

TABLE 2. EXTREMES OF SEISMIC LOADING AND RESPONSE OF RING ISOLATOR PROTOTYPES

Seismic Loading	in direction x	in direction z	max $\ddot{x}(t)$ (m s^{-2})	max $\ddot{u}_1(t)$ (m s^{-2})	max $\ddot{z}(t)$ (m s^{-2})	max $\ddot{w}_{1R}(t)$ (m s^{-2})
Harmonic:						
1.0-2.5Hz	Y	N	1.10	0.53	-	-
2.5-6.5Hz	Y	N	7.55	1.03	-	-
3.5-10.Hz	Y	N	4.53	0.55	-	-
1.5-3.5Hz	Y	Y	4.42	1.88	2.03	2.61
3.0-10.Hz	Y	Y	15.19	2.11	11.56	11.09
Seismic accelerograms:						
J1 A	Y	N	6.91	1.33	-	-
J1 B	Y	N	12.50	1.46	-	-
J1 C	Y	N	16.52	1.43	-	-
J1,J3 A	Y	Y	9.86	1.84	6.80	6.80
J1,J3 B	Y	Y	15.10	3.34	23.63	10.33
J1,J3 C	Y	Y	25.21	5.45	36.93	11.15
J5 A	Y	N	8.95	1.78	-	-
J5 B	Y	N	16.07	1.52	-	-
J5 C	Y	N	18.83	1.46	-	-
J5,J6 A	Y	Y	8.38	3.54	4.63	6.11
J5,J6 B	Y	Y	15.16	6.35	20.80	11.15
J5,J6 C	Y	Y	15.94	6.60	23.40	11.15

In Figs. 2,4 we can see that the seismic response of a stiff structure on shear ring isolators is highly influenced by the motion in a slide plane. The ratio between the response acceleration of the structure $\ddot{u}_1(t)$ and the seismic acceleration $\ddot{x}(t)$ is lower than 1 in all frequency ranges. It is falling down with increasing frequency. The result of this is that while the loading seismic acceleration is increasing, the response acceleration is nearly the same. Observing the results in Table 2, the modification of the acceleration in horizontal direction is remarkable, namely in the case of pure horizontal excitation.

In vertical direction some problems have appeared when the loading vertical seismic acceleration $\ddot{z}(t)$ was larger than 1g. In such case the contact in a part of slide area in isolators was disturbed and the behaviour of the system became irregular. But taking into account the levels of seismic accelerations as they are described by seismic scales, we see that the real seismic accelerations are much lower than those which were used during our laboratory tests. The degree of decreasing horizontal acceleration seems to be very promising.

The composition solution which has been used in ring isolator has been applied also in hat isolators. The difference is only in improvement of shock contact areas which are in action when the isolators are working in the limit positions. The seismic tests again showed the satisfactory behaviour of tested isolators.

CALCULATION APPROACHES WHEN USING THE SEISMIC SHEAR ISOLATORS

The structure which is intended to be supplied with seismic isolators should be thoroughly calculated to withstand different loading stages. The vibration of the structure system should be solved for the cases with and without sliding motion in the slide plane of isolators. Having p - number of of isolators we can describe the boundary between these two alternatives what is the equilibrium of the acting horizontal force K_a and the limit shear force K_s at the start of sliding motion.

$$K_s = K_a \, , \quad K_s = \sum_p K_{s,p} \, , \quad K_a = \sum_p K_{a,p} \, , \qquad (1)$$

$$K_{s,p} = \int_{A_z} \sigma_{z,p}(t,A) \gamma_f \, dA \, , \qquad (2)$$

$$K_{a,p} = K_{r,p}(u_{pr}) \, u_{pr}(t) \, , \qquad (3)$$

where γ_f is the coefficient of sliding friction in the slide plane, $\sigma_{z,p}$ is the changing vertical stress in the slide plane of isolator which is dependent on the axial compression force $Q_p(t)$ and on the additional bending effects caused by dynamic horizontal deflections. Taking the equivalent mass Δm_p,

$$Q_p(t) = \Delta m_p \, (g - \ddot{z}(t) - \ddot{w}_p(t)). \qquad (4)$$

A_z is the area of the sliding contact plane, $K_{r,p}(u_{pr})$ is the shear stiffness of the rubber layer in p-th isolator, $u_{pr}(t)$ is the shear relative deflection of rubber layer in p-th isolator.

The stress $\sigma_z(t,A)$ can be changed from compression to zero in a particular area of p-th isolator and this can cause the partial loss of the contact in the slide plane. Then, the integration will be done only through the contact area of the respective isolator.

In the numerical calculation there are coming the equations of motion in isolators and in the upper structure. The examples of such systems describing the motion of structure are in [2], [3], [4]. The necessity of the calculation system is to account in the every time step all changes in variables coming into calculation. Naturally the shear force , which is acting in a slide planes of isolators, should be considered in direction against the sliding motion and it is changing its sign depending on the stage of motion and vibration of the analysed structural system. The transition between stages in isolators with or without sliding motion is sudden and in this way it is considered in equations of motion.

CONCLUSION

One must admit that the progress in improving seismic isolation systems is going on. Our results show the promising behaviour of seismic isolators of different modification. The cases of their application are mostly limited on the design of important structures and equipments. Naturally the choice and the application of seismic isolation systems must be always interconnected with the actual conditions in which the structure will be situated and will be used. All influencing factors like geological, geophysical and mass-stiffness conditions should be taken into account to ensure that the resulting seismic response of the isolated structure represents a satisfactory vibration of structure.

REFERENCES

1. JUHÁSOVÁ, E. - OPRŠAL, M.: Some problems of efficiency of seismic sliding isolation systems. In: Proc. 8th ECEE, Lisbon 1986, Vol. 5, p. 8.4/17.
2. JUHÁSOVÁ, E.: Application of different damping systems in the base isolation approach. In: Proc. 14th ESEE, Ossiach 1988. OGE, Wien 1989.
3. JUHÁSOVÁ, E.: Seismic Effects on Structures. Elsevier-Veda coedition. Amsterdam - Bratislava 1991.
4. JUHĂSOVÁ, E.: New base isolation systems under space seismic excitation. For World Congress Natural Hazard Reduction, New Delhi 1992 (in print).
5. JURUKOVSKI, D.: Base isolation of structures in seismic areas. In: Proc. 9ECEE, Moscow 1990, Vol. A.
6. KELLY, J.: Aseismic base isolation: review and bibliography. Soil Dynamics and Earth. Eng., 1986, 5, 3.
7. MARTELLI, A. et al.: Research and development work in Italy on seismic isolation for industrial and power plants applications. In: Proc. 9ECEE, Moscow 1990, Vol. 3.
8. WERNER, D. - KRELL, A.: Base isolation for protection of structures against seismic influences. In: Proc. 9ECEE, Moscow 1990, Vol. 3.

SYSTEMS OF SEISMIC PROTECTION OF OLD STRUCTURES

B. Kirikov[1]

INTRODUCTION

Great experience in aseismic construction has been gained for millennia of human civilizations. Basing on this experience a great number of various building structures able to resist eartquake effects has been developed. In many opinion, it is not only interesting, but also useful to study the heritage of ancient builders. Many seismic protection methods and ideas of the ancient can be employed nowadays too. I therefore have undertaken the task difficult to fulfil that was aimed at gaining information on measures of seismic protection of buildings as used in different epochs in different parts of the globe. [6] , [9].

PRINCIPLES OF EARTHQUAKE RESISTANCE

The question is stated, that the earthquake protection methods developed by ancient masters can be treated in terms of the modern theory of earthquake resistance. The principles of earthquake resistance are based on generalized century old experience in earthquake engineering.

The principles are:
1. Principle of symmetry. Weights and rigidities should be distributed uniformly and symmetrically.
2. Principle of geometric harmony. There should be certain correlation between structural dimensions.
3. Principle of antigravity. Structures should be as more lightweight as possible with the centre of gravity located at the lowest possible level.
4. Principle of elasticity. Materials in structures should be strong, lightweight and elastic.
5. Principle of framework. Bearing members should be connected with each other to form closed contours both vertically and horizontally.
6. Principle of foundation. Deep foundations should be used to prevent non-uniform settlement.
7. Principle of earthquake isolation. Devices reducing energy transfer from oscillating soil to buildings should be used.

The basic principles of earthquake engineering as have been formulated and comments particularly those concerning the features of seismic effects, can be used to reach the more earthquake resistant structures.

STEPS IN HISTORY

We can begin in prehistoric times. Considered is the

---

1 - TsNIISK, Moscow, USSR

structure of one- and two-storey menhirs that survived during several millenia in conditions of high seismicity in the Caucasus. After the acquaintance with other ancient structures we can meet the methods of seismic protection as used in the three great river civilizations: Harrappans in the valley of the Indus; Sumerians, Babylon, Assyria and Iran in the valley of the Tigris and the Euphrates, Egyptians along the Nile.

The civilization of the Harappans that existed as far back as 3200 B.C. is interesting in terms of structures made of burnt brick on clay mortar. Brick was used to build up huge platforms ensuring sound earthquake protection of buildings placed on them. The lower floors of the buildings were made of burnt brick, the upper ones - of air brick reinforced with timber. In Babylon, Assyria and Iran various structures were made of brick on clay and bitumen. The gigantic platforms were also used here to support the whole palace complexes. The platforms were of two types, i.e. rigid and soft. Examples of the two types are given with their performance compared. The arrangement of ziggurats is described, that of the legendary Babylon tower included. The structure of vaulted floors made of brick is analysed. Especially detailed is the description of the throne-room of Tzar Khosroy. The room is roofed with a brick shell; the shell of a 27 m span has a complex outline. The structure of "The Hanging Gardens of Babylon" is also of interest. Brick supports reinforced with masonry poles bore powerful cylindric vaults above which there were the gardens. "The Hanging Gardens of Babylon" are considered to be one of the seven worders of the world. It seems particularly interesting in terms of earthquake resistance to see how the Colossus of Rhodes was constructed and destroyed.

The Egyptian Pyramids have three points of interest from the viewpoint of earthquake resistance. To begin with, it is

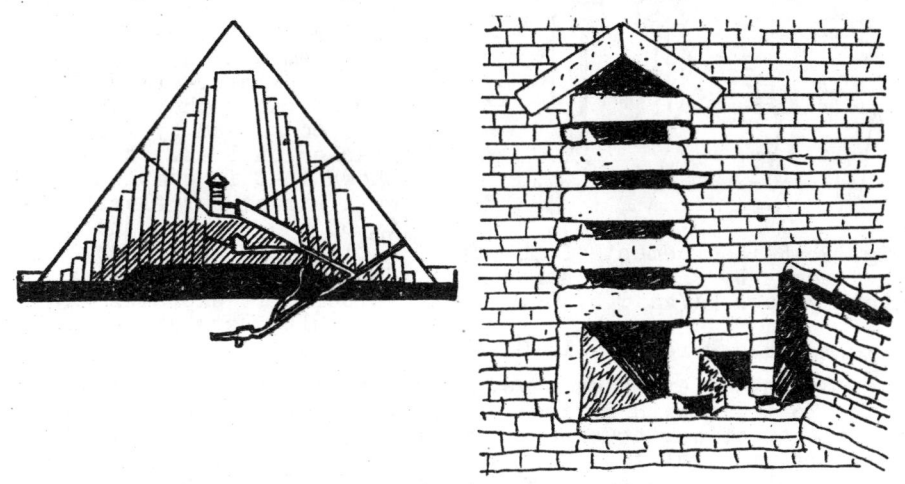

Fig. 1. Load-carrying system of the grave chamber in the Cheops pyramid.

their ideal shape. Then goes their enormous weight that influences soil-structure interaction towards the reduction of seismic effects and, finally, as is seen from the burial cell of the Cheops Pyramid, ancient builders tried to lower stress concentration in structures by arranging various unloading systems. Particular attention is paid to these systems which purpose the "Lion Gate" in Mycenae, the multilayer arches in Byzantium and the highly reliable structure of Armenian portals are considered. The structure of Egyptian temples can be analysed from the soil foundation to the masonry blocks and floor plates connected with each other by wood and stone staples. Here is Egypt shake-absorbing sand cushions appeared under separate columns and the whole structures. It seems particularly interesting to consider the evaluation of stone columns in Egyptian temples - from columns made of small elements to cast-in-place solid columns and those made of large semi-cylinders connected by wood staples.

Now we turn to studying Grecian architectural monuments. The Knossos Palace on Crete is made of stone reinforced with wood. The structure of the Palace's wood columns presents interest. To ensure reliable support for the beams the columns are made thick at the top having thin bottom parts.

The travel round the Grecian world presupposed the acquaintance with the structure of the walls of Troja VI and the domeshaped underground Tomb of Atreus. Further the structure of the famous Grecian temples was extremely simple beam-stay system with flexible inter-elements. Structurally the elements were connected by iron staples on lead. It is shown that in compliance with structural flexibility of the temples the foundations are made independent under each bearing members. It explains why no vaulted structures and lime mortars were used in Grecian constructions.

Fig. 2. Reinforcement of the stone masonry by the wood bars: Knoss palace, XVth c. b. Ch.

By way of examples several Grecian temples are dealt with. To begin with it is Erechteum in which one of the main principles of earthquake engineering, i.e. the principle of symmetry, is violated. The structure of the most famous Grecian temple, Parthenon, is also depicted. Both the temples were placed on an extremely uneven bed rock. The Greece builders understood fairly well that constructions in seismic areas should have had a homogenous base, therefore Parthenon was shifted so that it could not be placed on a rock edge. The ruins of Parthenon permit to carry on a detailed study of its structure. Here a flexible tie between the column cylinders can be observed. The roof beams made of three edgewise plates ensure a reliable performance of the whole roof.

Constructions in Grecian colonies can be described on gigantic temples in Sicily. Their great dimensions was the main cause of their destruction during earthquakes. Constructions in Asia Minor are examined furtheron, among them being the three Wonders of the world; the Temple of Artemins, the Mausoleum in Haliarnassus and Colossus of Rhodes. The study of these architectural monuments which are non-existent now permits to understand the view of old builders concerning various problems of construction. The Temple of Artemins is erected on extremely poor soil, the Mausoleum is structurally complicated whereas the Colossus has a structure unique for that time. All these problems were solvable already at that time. The Temple had a pile foundation, the Mausoleum was placed on an adobe brick cushion. Inside the Colossus there was a skeleton of metal and stone. From the Black Sea Grecian settlements can be depicted. Olbia famous for its artificial stratified foundations of clay and ash that permitted to erect heavy constructions, temples and towers on soft coastal soil. Dealt with in detail is the Tower of Zenon in Hersones. The tower presents interest because of its multi-layer structure. At different times it was enlarged and retrofitted after destruction. The central part of the construction is the tower of antique times. It is made of stone reinforced with powerful wooden bars. Let follow crypts under burial mounts in Pantikapei. It is interesting to observe the combination of the building methods of the West and the East. In one of the crypts with a roof in the form of a cylindric vaults if wedge-shaped stones-like in the East, all the stones are connected by metal staples on lead the practice of Greeks. But the most interesting structure is that of the Regal Tumulus, IVth century B.C. It seems that for the first time in the history of architecture a square structure was successfully connected with a conic dome. All subsequent history of dome-shaped buildings can be viewed upon as the struggle for the connection of the dome and the building that would exclude stress concentration anywhere in the dome. The crypt of the Regal Tumulus is a successful solution of the problem.

Now let us pass to Rome and Byzantium. Here builders began using concrete and vaulted roofs. Buildings became structurally rigid. Here in the paper I shall only remind three of them. First of all, it is Pantheon. The geometry of the construction is extremely simple as required by considerations of earthquake resistance. The building is a cylinder roofed by a spheric dome. However interesting and didactive is the structure of every member of the Pantheon. It has a deep stratified foundation with sand streaks. Thick walls are lightened by bays. There is a whole system of brick arches in the walls of cast-in-place concrete. A concrete dome is also reinforced by a double frame made of brick.

Besides, the event of building St. Paul's Cathedral in

Rome, of its dome, in particular, is described. Initially they intended to make the dome similar to that of the Pantheon. Then, however, several other variants were elaborated. The event ended in the erection of a double dome of an upraised outline designed by Michelangelo.

The central place in the analysis of earthquake protection practice in Byzantium is taken up by the analysis of Hagia Sophia. There are many errors in terms of aseismic design, i.e. a great outward thrust from the dome, non-uniform rigidity and weak foundations.

Fig. 3. Greek-Armenian temple in Garni.

We can follow also the structures of Caucasus. First of all, it is Armenian temples that have retained their earthquake resistance for millennia. The acquaintance with Armenian architecture begins from the Temple of Garni. It is shown how successfully combined are building practice of the East and the West. The solid basement is on lime mortar, columns and walls are made of stoned connected by the Greece method, using metal staples on lead. The cylindric floor is made of concrete on lime mortar with lightweight aggregates.

Particularly interesting is the structure of Zvartnotz. It consists of three cylinders put on each other. It seems remarkable how skillfully the ancient architect redistributed stresses between the members. The base of the structure is a strong lime-stone ring at the ground floor level.

All the armenian temples have three-layer walls consisting of two stone linings and a core made of untooled stone and lime mortar. It is interesting to trace back the changes these walls have undergone for the centuries. Dealt with is the structure of traditional national fighting towers.

In Azerbaijan the structure of massive Virgin Tower, formerly the temple of fire-worshippers, attracts attention. Here we mention original Mausolea of Moslem architecture.

Now we proceed to the Central Asia. The mausolea, mosques and minarets that have survived to this day testify to the advanced earthquake engineering technique of that time. The whole standard set of earthquake protection measures was actually used, i.e. the necessary clay cushions under structural foundations, the famous "ganch" mortar ensuring structural elasticity, sliding sand belts and shock-absorbers of reed spacers. There is a variety of domes in the ancient structures of the Central Asia. They are cone-shaped, spheric or elliptic. Skillfully connected with the buildings these domes were reinforced by the whole system of arches. The magnificent and cumbrous portals that appeared in the XIVth century had a diverse effect on the earthquake resistance of the above mentioned structures destroying their homogeneity and symmetry.

Fig. 4. General view of the mausoleum of Yusuf.

We continue our trip turning to the history of aseismic structures in China and Japan. Our interest here is attracted by high wooden pagodas equipped with different damoing systems. In Japan pagodas consisted of two systems of different rigidities, i.e. a flexible central shaft and rather a rigid system of floors of the whole pagoda. In China additional mass was placed at the very top of the pagoda.

We are finishing our trip in pre-Columb America. It is interesting to find out here how thoroughly fitted stone blocks in pyramids and temples were attached to each other.

I suppose, this trip in the history of earthquake engineering, will be usefull not only for the youth and the people having an inquiring mind, but also for specialist.

CONCLUSION

It is necessary to study, assess and use the valuable legacy of ancient architects. The study of ancient stonework structures in earthquake-prone areas serves general cultural purpose, helps update earthquake hazard in specific areas, and teaches us to use the experience of earthquake resistant design and construction in antiquity to build our contemporary structures.

REFERENCES

1. Proceed. Intern. Symp. on Engineering Geol. of Ancient Works, Monuments and Historical Sites. Athens, 19-23 Sept. 1988. Eds. P.G. Marinos and G.C. Konkis, VIII. Balkema, Rotterdam, 1988.

2. KARCZ, I. - KARFI, U.: Evaluation of Supposed Archaeseismic Damage in Israel//Journ. Archeological Sci. 1978. N 5. P. 237-253.

3. SOLONENKO, V.P., ed.: Paleoseismology of the Greater Caucasus. Moscow: Nauka, 1979. 188 p. (in Russian).

4. NIKONOV, A.A.: On the methodology of Archeoseismic research into historical monuments. In: The Engineering Geology of Ancient Works, Monuments and Historical Sites. P.G. Marinos and G.C. Konkis (Eds) 1988, Balkema, Rotterdam, 1988.

5. BACHINSKY, N.M.: Earthquake Resistant Design in Architectural Monuments in the Area of Soviet Central Asia. Moscow: USSR Acad. Sci. Publishing House, 1949. 48 p. (in Russian).

6. KIRIKOV, B.A.: Ancient and Contemporary Earthquake Resistant Structures. Moscow: Nauka, 1990. 72 p. (in Russian).

7. KUZNETSOV, A.V.: Tectonics and the Design of Centric Buildings. Moscow: Architektura i gradostroitel'stvo, 1951, 274 p. (in Russian).

8. PUGACHENKOVA, G.A. - REMPEL̆, L.I.: Outstanding Architectural Monuments in Uzbekistan. Tashkent. Khud. Literatura UzSSR. 1958. 292 p. (in Russian).

9. KIRIKOV, B.A.: Selected Pages of History of Earthquake Engineering, 1990. (Manuscript of the book).

THE VIBROISOLATION EFFECT OF BARRIERS IN SOIL BASES
G. Martinček[1]

INTRODUCTION

Many experimental investigations performed on real or model soil base [1-6] have shown that the barriers in soil bases may be very effective means to vibroisolation screening for objects or zones which are situated immediately behind the barriers. The barrier can be applied above all by the propagation of technical seismic efects or vibrations evoked by traffic when the majority of the frequency spectrum components has the wavelengths smaller or comparable with the dimensions of the barrier. The problem has been treated in [2] using the method of finite elements as a two-dimensional, i.e a plane strain problem.

APPLICATION OF THE METHOD OF BOUNDARY ELEMENTS

Our investigation has been performed theoretically by using the method of boundary integral equations respectively method of boundary elements [7], [8], [9], [10] assuming the soil base described by simplified dynamic model. We suppose then the depth of barrier larger or comparable with the wave length of the propagated force effect components and in this manner the influence of the shape and barrier dimensions on vibroisolating effect can be studied.

A harmonic variable dynamic force acting on the surface of the soil base generates the dilatational and shear waves, which spread into the soil base and the surface - Rayleigh waves which propagate at the surface with the velocity c_R . The amplitudes of vertical displacements w and horizontal displacements u of Rayleigh waves are concentrated at the surface (Fig. 1). It is known [11], that the maximum part of energy - 67 % is radiated by these surface waves. If the depth of the obstacle H (Fig. 1) is comparable with the wave length Λ_R, the screening effect can be expected.

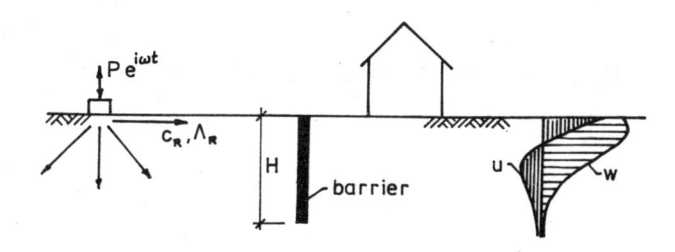

Fig. 1.

1 - ÚSTARCH SAV, Bratislava, ČSFR

THE SIMPLIFIED DYNAMIC MODEL OF SUBGRADE

Filonenko-Borodich [12], Pasternak [13], Vlasov-Leontjev [14] and others [15] have been presented the simplified models of subgrade and their applications in static analysis. The special aspects of dynamic analysis by using the coefficient of equivalent inertia and complex characteristics of elasticity have been performed in [16] and [17].

The differential equation of motion for dynamic simplified model of subgrade has the form

$$K_2^* \nabla^2 w - K_1^* w - \rho_0 K_3 \frac{\partial^2 w}{\partial t^2} = - p(r,t) , \qquad (1)$$

where $w(r,t)$ is the dynamic deflection of the subgrade, $p(r,t)$ - vertical dynamic load, ρ_0 - density of the subgrade material. The operator ∇^2 in the case of axial symmetry is given by the relationship

$$\nabla^2 = \frac{\partial^2}{\partial r^2} + \frac{1}{r} \frac{\partial}{\partial r} \qquad (2)$$

In the case of stationary dynamic tasks the vibroelastic behaviour of the subgrade may be determined by means of complex characteristics $K_1^* = K_1(1 + i\delta)$ - coefficient of uniform compression in [N/m³], $K_2^* = K_2(1 + i\delta)$ - coefficient of shear transmission in [N/m], if δ is damping parameter. The coefficient of equivalent inertia K_3 with the dimension [m] is determined from the condition, that the velocity of wave propagation approaches to the velocity of Rayleigh waves c_R for the short wave lengths. This condition is going to the relationship

$$c_R^{*2} = \frac{K_2^*}{\rho_0 K_3} \qquad (3)$$

and simultaneously the relationship is given

$$c_R^{*2} = \varkappa c_2^{*2} , \qquad (4)$$

where

$c_2^* = \sqrt{\dfrac{G_0}{\rho_0}}$ is the velocity of shear waves in the subgrade and $\varkappa$ is the function of Poisson's ratio μ according to the relationship

$$\varkappa = \left(\frac{0.87 + 1.12 \mu}{1 + \mu} \right)^2 \qquad (5)$$

$G_0^* = G_0(1+i\delta)$ is a complex shear modulus of elasticity.

The Green's function as the dynamic influence function of the deflection under harmonic unit concentrated force in the form [17]

$$G(r,t) = \frac{-i}{4K_2^*} H_0^{(2)}(\alpha_1 r) e^{i\omega t} \tag{6}$$

represents the fundamental solution. The function $H^{(2)}(\alpha_1 r)$ is Hankel's function of complex argument $\alpha_1 r$ with the positive real part and negative imaginary part. α_1 is expressed by the relationship

$$\alpha_1^2 = \frac{\omega^2}{c_R^{*2}} - \frac{K_1^*}{K_2^*} \tag{7}$$

if ω is angular frequency.

BOUNDARY INTEGRAL FORMULATION ACCORDING TO RAYLEIGH THEOREM OF RECIPROCITY

We suppose that the barrier occupies a region S_0 with the boundary Γ in the unbounded subgrade S. The radiusvector of the load point $\vec{\xi}$, of the unit force point $\vec{r}$ and boundary point $\vec{\eta}$ are denoted in Fig. 2.

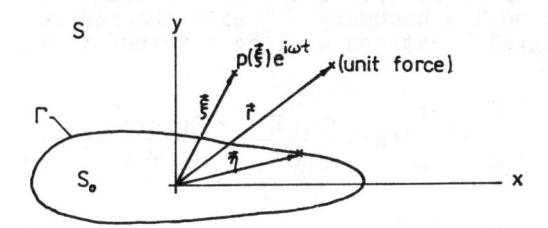

Fig. 2.

The boundary integral formulation according to the theorem of reciprocity has the form

$$\vartheta(\vec{r})w(\vec{r}) = \int_S p(\vec{\xi})G(\rho)dS + \int_\Gamma K_2^* \left[\frac{\partial w(\vec{\eta})}{\partial \vec{n}(\vec{\eta})} G(R) - \frac{\partial G(R)}{\partial \vec{n}(\vec{\eta})} w(\vec{\eta}) \right] d\Gamma +$$

$$+ \int_{\bar{\Gamma}} K_2^* \left[\frac{\partial w(\vec{\eta})}{\partial \vec{n}(\vec{\eta})} G(R) - \frac{\partial G(R)}{\partial \vec{n}(\vec{\eta})} w(\vec{\eta}) \right] d\bar{\Gamma}, \tag{8}$$

if $\vartheta(\vec{r}) = \begin{cases} 1, & \vec{r} \in S \\ 0, & \vec{r} \bar{\in} (\Gamma \cup S) \end{cases}$

and the variable distances $R = |\vec{\eta} - \vec{r}|$, $\rho = |\vec{\xi} - \vec{r}|$.
$\bar{\Gamma}$ is the fictional boundary of the region S.

BARRIER FROM DIFFERENT MATERIAL

If the barrier S_0 in the subgrade is made from different material, the integral formulation for barrier medium has the form

$$\vartheta(\vec{r})w_0(\vec{r}) = \int_\Gamma K^*_{2,0} \left[\frac{\partial w_0(\vec{\eta})}{\partial \vec{n}(\vec{\eta})} G_{ba}(R) - \frac{\partial G_{ba}(R)}{\partial \vec{n}(\vec{\eta})} w_0(\vec{\eta}) \right] d\Gamma \qquad (9)$$

where $\vartheta(\vec{r}) = \begin{cases} 1 & ; \vec{r} \in S_0 \\ 0 & ; \vec{r} \bar{\in} (\Gamma \cup S_0) \end{cases}$

$G_{ba}(R)$, $w_0(\vec{r})$, $K^*_{2,0}$ represent the fundamental solution, deflection and coefficient of shear transmission for the barrier medium.

The conditions at the boundary Γ are:

$$w(\vec{\eta}) = w_0(\vec{\eta})$$
$$K^*_2 \frac{\partial w(\vec{\eta})}{\partial \vec{n}(\vec{\eta})} = K^+_{2,0} \frac{\partial w_0(\vec{\eta})}{\partial \vec{n}(\vec{\eta})} \qquad (10)$$

By means of limit transition on the boundary Γ for $\vec{r} \to \zeta \in \Gamma$ and on the boundary $\bar{\Gamma}$ with the radius $\bar{R} \to \infty$, the boundary integral equations will be obtained from (8) and (9) in the form

$$(1 + \tfrac{1}{2}) w(\zeta) - \int_\Gamma K^*_2 \left[G(R') \frac{\partial w(\vec{\eta})}{\partial \vec{n}(\vec{\eta})} - \frac{\partial G(R')}{\partial \vec{n}(\vec{\eta})} w(\vec{\eta}) \right] d\Gamma =$$
$$= \int_S p(\vec{\xi}) G(\rho') dS , \qquad (11)$$

$$\tfrac{1}{2} w(\zeta) - \int_\Gamma \left[K^*_2 G_{ba}(R') \frac{\partial w(\vec{\eta})}{\partial \vec{n}(\vec{\eta})} - K^*_{2,0} \frac{\partial G_{ba}(R')}{\partial \vec{n}(\vec{\eta})} w(\vec{\eta}) \right] d\Gamma = 0 .$$

The distances R', ρ' are given by the relationships

$$R' = |\vec{\eta} - \vec{\zeta}|$$
$$\rho' = |\vec{\xi} - \vec{\zeta}| \qquad (12)$$

If the entire boundary is decomposed into boundary elements with N nodal points, then it is possible to transformate the system of integral equations into the system of 2N linear algebraic equations

$$1.5 \frac{w_k K_2}{\rho a^2} + \sum_{\substack{j=1 \\ j \neq k}}^N A_{kj} \frac{w_j K_2}{\rho a^2} + \sum_{j=1}^N B_{kj} \frac{\partial w_j K_2}{\partial n \, \rho a} = P_k$$

$$0.5 \frac{w_k K_2}{\rho a^2} + \sum_{\substack{j=1 \\ j \neq k}}^N C_{kj} \frac{w_j K_2}{\rho a^2} + \sum_{j=1}^N D_{kj} \frac{\partial w_j \, K_2}{\partial n \, \rho a} = 0$$

(13)

The dimensionless unknown $\dfrac{w_j K_2}{pa^2}$, $\dfrac{\partial w_j}{\partial n}\dfrac{K_2}{pa}$ at the nodal points can be determined by the solution of the system (13).

The complex coefficients of the system (13), provided constant course of subintegral function along the boundary element have the form

$$A_{kj} = K_2^* \int_{\Gamma_j} \frac{\partial G(R')}{\partial \vec{n}(\vec{\eta})} d\Gamma_j = \frac{i}{4} \alpha \frac{l_j}{a} \cos\alpha_{kj} H_1^{(2)}(\alpha \frac{R'}{a}),$$

$$B_{kj} = -K_2^* \int_{\Gamma_j} G(R') d\Gamma_j = \frac{i}{4} \frac{l_j}{a} H_0^{(2)}(\alpha \frac{R'}{a}),$$

$$C_{kj} = K_{2,0}^* \int_{\Gamma_j} \frac{\partial G_{ba}(R')}{\partial \vec{n}(\vec{\eta})} d\Gamma_j = \frac{i}{4} \alpha_0 \frac{l_j}{a} \cos\alpha_{kj}^0 H_1^{(2)}(\alpha_0 \frac{R'}{a}),$$

$$D_{kj} = K_2^* \int_{\Gamma_j} G_{ba}(R') d\Gamma_j = -\frac{i}{4} \frac{l_j}{a} \gamma_{ba} H_0^{(2)}(\alpha_0 \frac{R'}{a}),$$

(14)

where

$$\alpha^2 = (\alpha_1 a)^2 = \frac{\Omega^2}{1+i\delta} - \frac{1}{\gamma},$$

$$\alpha_0^2 = \frac{\Omega^2}{\varepsilon} - \frac{1}{\gamma_0},$$

(15)

l_j - length of boundary element and dimensionless parameters are given in the form

$$\varepsilon = \frac{C_{R,0}^{*2}}{C_R^2}, \quad \Omega = \frac{\omega a}{C_R}, \quad \gamma = \frac{K_2^*}{K_1^* a^2}, \quad \gamma_0 = \frac{K_{2,0}^*}{K_{1,0}^* a^2},$$

$$\gamma_{ba} = \frac{K_2^*}{K_{2,0}^*}.$$

(16)

In the case of uniform harmonic normal load $p\,e^{i\omega t}$ on the circular area of the subsoil with radius a, the right side of the first equation (13) is expressed by the equation

$$P_k = -\frac{i\pi}{2\alpha} J_1(\alpha) H_0^{(2)}(\alpha \frac{|\vec{\xi}-\vec{\xi}_k|}{a}),$$

(17)

where $J_1(\alpha)$ is the Bessel function of the first order.

EFFECTS OF LINEAR BARRIER

By the numerical study of the effect of linear barrier with the length $L = 9.0$ m and thickness $H = 0.6$ m and with dimensionless parameters $\gamma = 10$, $\gamma_0 = 5$ the boundary was divided into 32 boundary elements. Harmonic variable load $Pe^{i\omega t}$, uniformly distributed at the circular area with radius $a = 0.25$ m, effects the subgrade according to scheme of Fig. 3. The calculated amplitudes of dynamic deflections

$$W_A = \frac{|w|K_2}{pa^2}$$

are drawn along the boundary and along the section A-A for the cases with or without barrier.

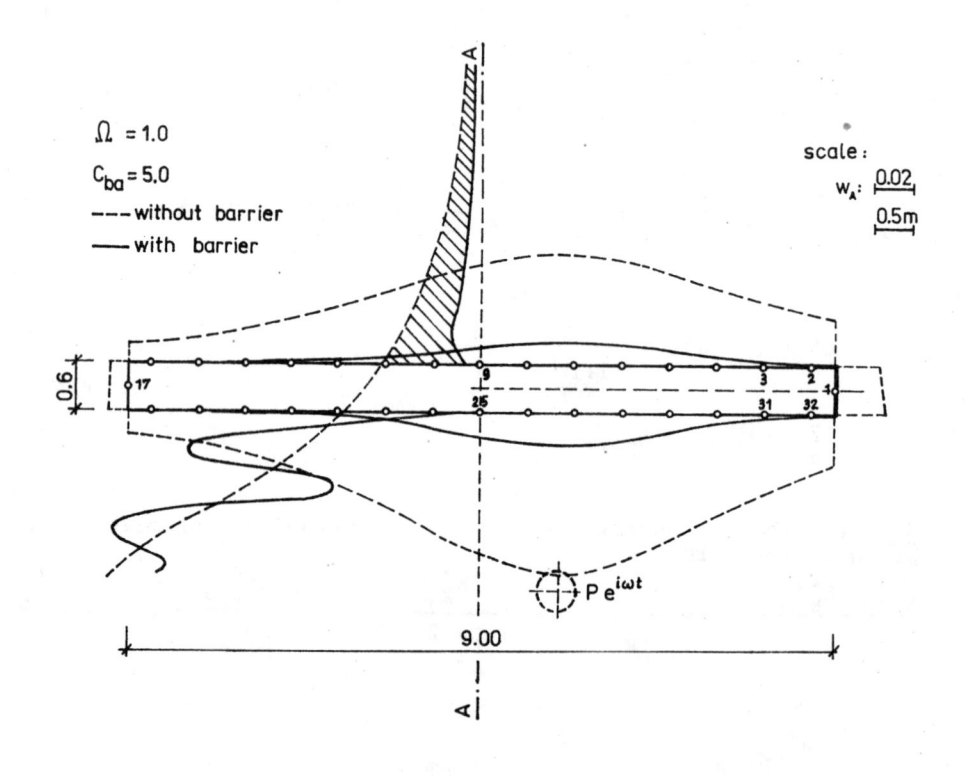

Fig. 3.

The materials of the barrier and subgrade are given by the ratio $C_{ba} = \dfrac{C_{R,0}}{C_R} = 5.0$ and dimensionless frequency of vibration $\Omega = 1.0$.

It can be seen from the Fig. 3 the radical vibroisolating effect behind the barrier and the superposition of direct and reflected waves before the barrier.

THE INFLUENCE OF THE VARIOUS BARRIER STIFFNESS ON VIBROISOLATING EFFECT

The results in Fig. 4 represent influence of parameter C_{ba} on the screening zone along the cross-section A-A at frequency $\Omega = 1.0$.

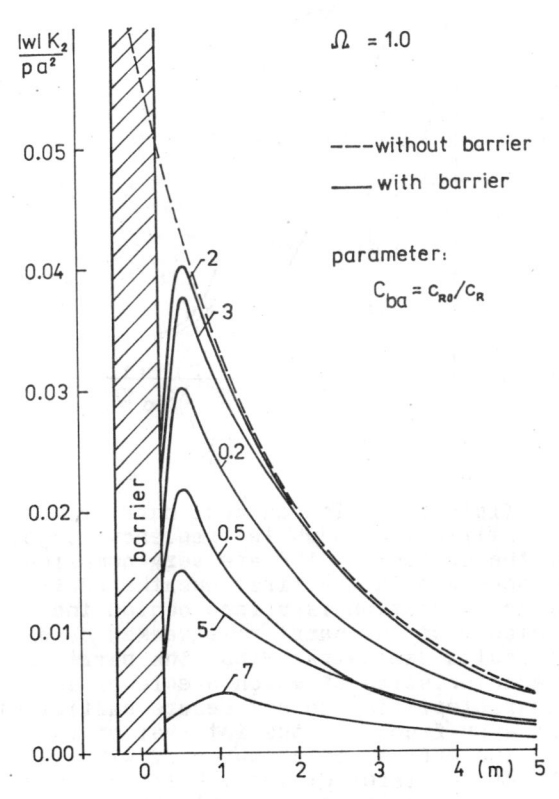

The screening effect increases with the increase of the ratio C_{ba}, if the wave characteristic of the barrier medium is greater than the wave characteristic of the subgrade.

On the other side the results are more complicated in the case, if the material of barrier has lower value of the velocity of surface waves in comparing with the subgrade.

The results of the detailed study of barrier efficiency at various frequencies Ω in interval $(0.3 \div 1.0)$ and for various values of C_{ba} are in Fig. 5. The efficiency expressed by amplitude reduction factor ARF is given in the screening zone by the ratio:

Fig. 4.

$$ARF = \frac{\int_0^{L_0} \frac{|w_{BA}|(l) K_2}{pa^2} dl}{\int_0^{L_0} \frac{|w|(l) K_2}{pa^2} dl} . \qquad (18)$$

$|w_{BA}|(l)$ is the course of the deflection amplitudes behind the barrier and $|w|(l)$ is the course of the deflection amplitudes at the same section of the subgrade without barrier. The length of section is $L_o = 5.0$ m.

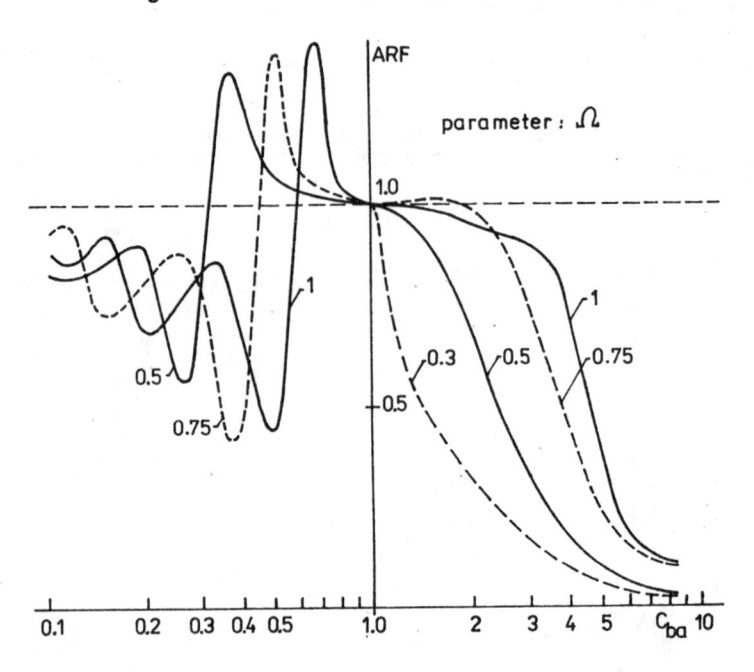

Fig. 5.

The trend of higher efficiency by increase of barrier medium rigidity is clearly certificated in whole frequency range. In the region $C_{ba} < 1.0$ the courses of ARF are very complicated. There are resonance zones and in the first one the ARF > 1, i.e. the concentration of the vibration is arised behind the barrier. However the efficiency of the barrier is weeken by other resonances too. The results demonstrate that the barriers from material the wave characteristics of which are lower in comparing with subgrade characteristics do not secure sufficient vibroisolating effect. For each frequency the interval of values C_{ba} exists, in which the barrier is a source of vibration concentration and thereby an amplifier of the vibration. These phenomena are legitimate, they are connected with the diffraction of the stress waves on the boundaries of barrier and with the resonances. The resonances arise at certain ratios of barrier thickness and wave length. It can demonstrate that the resonances take place if the wave lengths Λ_{ba} in the barrier medium have the values Λ_{ba} = 2H; H, 0.5H, ... H is the thickness of barrier.

The efficiency of the barrier falls down also in the region $C_{ba} > 1$ at very high frequencies, if the wavelength in subgrade is comparable with the thickness of barrier.

TRENCH BARRIER

In the case of trench barrier with boundary Γ the boundary condition applies

$$Q_n(\vec{\eta}) = K_2^* \frac{\partial w(\vec{\eta})}{\partial \vec{n}(\vec{\eta})} = 0 . \tag{19}$$

The boundary integral equation has the form

$$(1+\tfrac{1}{2}) w(\vec{\zeta}) - \int_\Gamma K_2^* \frac{\partial G(R')}{\partial \vec{n}(\vec{\eta})} w(\vec{\eta}) d\Gamma = \int_S p(\vec{\xi}) G(\rho) dS . \tag{20}$$

The numerical solutions of the screening effect for trench barrier of various kind have been made in [18].

SHEET PILING BARRIER

The sheet piling barriers may have arbitrary shapes, bounded or closed curves. Provided that depth of barriers is greater as the wavelength of the harmonic vibration spreaded in the subgrade, the boundary condition on Γ can be of use

$$w(\vec{\eta}) = 0 , \tag{21}$$

i.e. we suppose the vertical deflections at the barrier equal zero.

The boundary integral equation has the form

$$K_2^* \int_\Gamma G(R') \frac{\partial w(\vec{\eta})}{\partial \vec{n}(\vec{\eta})} d\Gamma = - \int_S p(\vec{\xi}) G(\rho') dS . \tag{22}$$

The efficiency of the closed sheet piling barrier in the shape of circle with radius $R = 3.0$ m was analysed numerically. The boundary curve was divided into 64 boundary elements according to Fig. 6. Dynamic harmonic force $pe^{i\omega t}$ affects the subsoil surface at the distance 1.0 m from the centre of barrier.

The courses of calculated amplitudes of dimensionless dynamic deflections $|w|K_2/(pa^2)$ along the cross-section A-A are drawn in Fig. 7 for the frequency $\Omega = 1.0$.

The courses in Fig. 7 demonstrate the superposition of direct and reflected waves inside the closed barrier and high

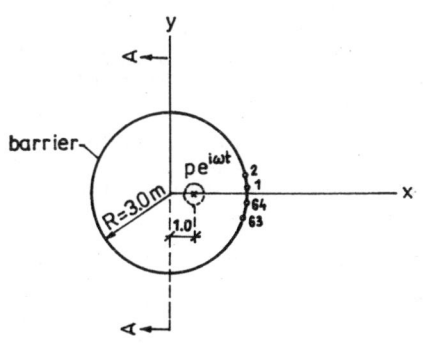

Fig. 6.

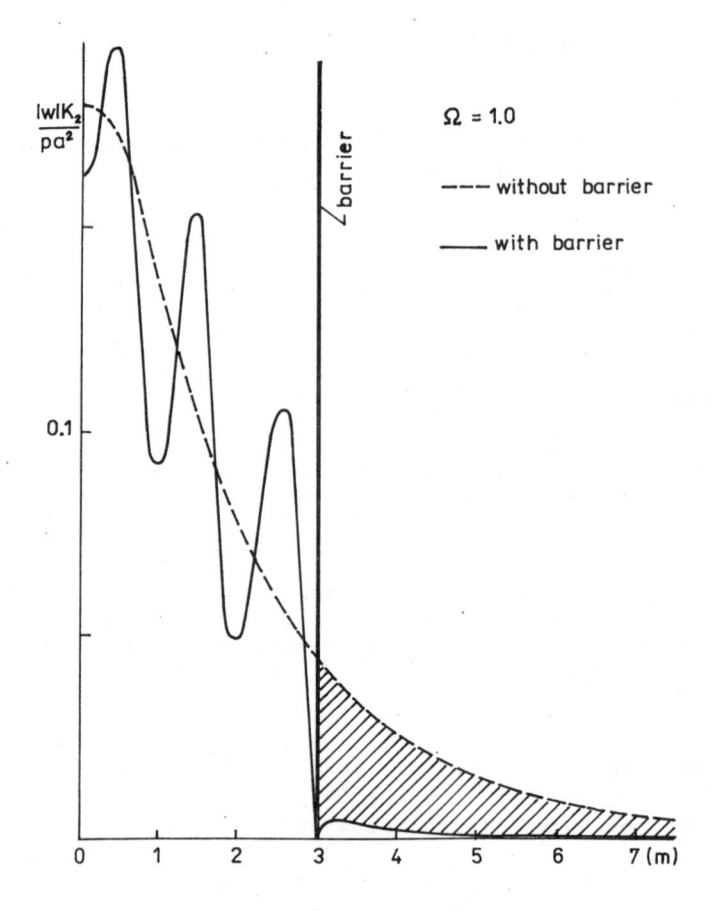

Fig. 7.

degree of screenig effect behind the sheet piling barrier. The effect of vibroisolation by linear barriers is analysed in [18].

CONCLUSION

Submited solutions refer to considerable possibilities of the method of boundary integral equations in the tasks of wave diffraction near the barriers.

In the case of barrier from different material the numerical studies proved high vibroisolating effect of barrier, if the stiffness of material medium of barrier is greater than the stiffness of material medium of soil base.

The screening effect of barrier is smaller in the case of smaller rigidity of barrier medium comparing with the subgrade medium and in resonance zones the barrier becomes the amplifier of the vibration.

The vibroisolating effect of sheet piling barrier is high.

In general the simplified dynamic model of the subgrade however limits the analysis only on the vertical deflections.

REFERENCES

[1] BARKAN, D.D.: Dynamics of Bases and Foundations. McGraw-Hill, New York, 1962.
[2] HAUPT, W.A.: Surface-waves in non-homogenous halfspace. In: Dynamic response and wave propagation in soils. A.A. Balkema, Rotterdam, 1978, pp. 335-367.
[3] WOODS, R.D.: Holografic interferometry in soil dynamic. In: Dynamic response and wave propagation in soils. A.A. Balkema, Rotterdam 1978, pp. 391-406.
[4] WOODS, R.D.: Screening of surface waves in soils. Journal of the Soil Mechanics and Foundations Division. ASCE 94 (SM4), 1968, pp. 951-979.
[5] WOODS, R.D. - BARNETT, N.E. - SAGESSER, R.: Hollography - a new tool for soil dynamics. Journal of the Geotechnical Engineering Division, ASCE 100 (GT11), 1974, pp. 1231-1247.
[6] BENČAT, J.: Otázky tienenia povrchových vĺn napätia v reálnom prostredí. In: Práce a štúdie Vysokej školy dopravy a spojov, zv. 7, Žilina 1982, 39-55.
[7] BENERDŽI, P. - BUTTERFILD, R.: Metody graničnych elementov v prikladnych naukach (Translation from English), Mir, Moskva, 1984.
[8] KRAUČ, S. - STANFILD, A.: Metody graničnych elementov v mechanike tverdogo tela (Translation from English), Mir, Moskva 1987.
[9] BREBBIJA, C.A. - TELLES, J. - WROBEL, L.: Metody graničnych elementov (Translation from English), Mir, Moskva 1987.
[10] BALAŠ, J. - SLÁDEK, J. - SLÁDEK, V.: Analýza napätí metódou hraničných integrálnych rovníc. VEDA, Bratislava, 1985.
[11] GRINČENKO, V.T. - MELEŠKO, V.V.: Garmoničeskie kolebanija i volny v uprugich telach. Naukova dumka, Kijev, 1981.
[12] FILONENKO-BORODIČ, M.M.: Nekotoryje približenye teorii uprugogo osnovanija. Učebnyje zapiski MGU, č. 46, Moskva 1940.
[13] PASTERNAK, P.L.: Osnovy novogo metoda rasčeta fundamentov na uprugom osnovanii pri pomošči dvuch koeficientov posteli. Gostrojizdat, Moskva-Leningrad 1954.
[14] VLASOV, V.Z. - LEONTJEV, N.N.: Balki, plity i oboločki na uprugom osnovanii. Gos. izd. fiz.-mat.lit., Moskva 1960.
[15] KOLÁŘ, V. - NĚMEC, I.: Energetické definice a algoritmy nového modelu podloží. Stav. čas. 26, 1978. s. 565-581.
[16] MARTINČEK, G.: Dynamický zjednodušený model podložia. Stav. čas. 32, 1984, s. 21-43.
[17] MARTINČEK, G. - ALEXIEV, V.: Varianty zotrvačných síl pri nelineárnom kmitaní podložia. Stav. čas. 35, 1987, s. 705-726.
[18] MARTINČEK, G.: Difrakcia vĺn napätia v okolí bariéry. In: Numerické metódy v mechanike. SSM-SAV, Boboty 1989, s. 21-24.

BASE SEISMIC ISOLATION FOR IMPORTANT BUILDING STRUCTURES

Rudolf Masopust [1]

1. INTRODUCTION

Base seismic isolation is a significant development in earthquake engineering that is rapidly gaining worldwide acceptance in the commercial field /1,2,3/. This approach introduces a damped flexible mechanism between the building base and the ground to decouple the structure from the most important components of seismic motion, resulting in a significant reduction of seismic loads on the structure and equipment inside. Seismically isolated building structures have been built in many countries. The actual number of structures seismically isolated against earthquake is now more than 125 /4/. This includes several hospitals, bridges, and nuclear power plants, and various similar buildings in the United States, Japan, France, Germany, New Zealand and several other countries. Crucial lifelines such as bridges and hospitals, as well as nuclear power plants and similar safety-related structures, must be operational during and after a major earthquake. In such cases the structure's utility far exceeds its construction cost, and base seismic isolation is a viable alternative. Similar reasoning may be applied to buildings housing unique artifacts, or architecturally important buildings themselves. The economic aspects of seismic isolation for important building structures are very varied /5/. Base seismic isolation of building structures is used not only to protect them against earthquakes, but also against damage from mining subsidences, subway vibrations, etc./6/.

2. THEORETICAL BACKGROUND

To gain insight into the behavior of seismically isolated building structure an elementary analysis can be developed using a simple linear two-degree-of-freedom system with linear springs and viscous dampers (Fig.1). The equation of motion has the following matrix form /7/:

$$[M]\{\ddot{q}\} + [C]\{\dot{q}\} + [K]\{q\} = -[M]\{r\}a(t) \qquad (1)$$

where

$$[M] = \begin{bmatrix} m_s+m_b & m_s \\ m_s & m_s \end{bmatrix} \quad [C] = \begin{bmatrix} c_b & 0 \\ 0 & c_s \end{bmatrix} \quad [K] = \begin{bmatrix} k_b & 0 \\ 0 & k_s \end{bmatrix} \qquad (2)$$

are the mass, damping and stiffness matrices respectively, $\{q\} = \{q_b, q_s\}^T$ is the vector of relative displacements, $\{r\} = \{1,0\}^T$, and $a(t)$ is the given time history of seismic acceleration. The displacement q_s is relative to the base, while q_b is relative to the ground. Let us assume that

- $m_b < m_s$, but is of the same order of magnitude,
- $\Omega_s = (k_s/m_s)^{1/2} \gg \Omega_b = (k_b/(m_s+m_b))^{1/2}$

---

[1] ŠKODA, Prague, ČSFR, currently at Stevenson & Assoc., Cleveland, USA

- $e = (\Omega_b/\Omega_s)^2$ is of the order of magnitude 10^{-2}
- $\beta_s = c_s/(2m_s \Omega_s)$ and $\beta_b = c_b/(2(m_s+m_b)\Omega_b)$ are of the same order as e.

The characteristic equation for natural frequencies of the system is

$$(1-æ)\,\Omega_n^4 - (\Omega_b^2 + \Omega_s^2)\,\Omega_n^2 + \Omega_b^2\,\Omega_s^2 = 0, \quad n = 1,2 \qquad (3)$$

where $æ = m_s/(m_b + m_s)$. The exact roots are given by

$$\genfrac{}{}{0pt}{}{\Omega_1^2}{\Omega_1^2} = \frac{1}{2(1-æ)}\{(\Omega_s^2 - \Omega_b^2) \pm [(\Omega_s^2 + \Omega_b^2)^2 - 4(1-æ)\,\Omega_s^2\Omega_b^2]^{1/2}\} \qquad (4)$$

Taking into account that $\Omega_b \ll \Omega_s$, and expanding the radical by a binomial series, we obtain, to the same order of e, the following results:

$$\Omega_1^2 = \Omega_b^2(1 - æ\,\Omega_b^2/\Omega_s^2) \quad \text{and} \quad \Omega_2^2 = \Omega_s^2(1-æ)^{-1}(1 + æ\,\Omega_b^2/\Omega_s^2) \qquad (5)$$

The lower of these two roots, Ω_1, represents the shifted base isolation frequency, while the higher one, Ω_2, represents the structural frequency modified by the presence of the isolation system. For practical purposes the following approximation may be accurate:

$$\Omega_1 \approx \Omega_b \quad \text{and} \quad \Omega_2 \approx \Omega_s\,(1-æ)^{-1/2} \qquad (6)$$

This means that while the isolation frequency is only slightly changed by flexibility of the structure (the change of order e), the structural frequency is significantly increased, so that the separation between the isplation frequency and the fixed-base structural frequency is increased to the same degree. Retaining only terms of order e, we get for the mode shapes $\{\phi^n\} = \{\phi_b^n, \phi_s^n\}^T$, $n = 1,2$, the following relations:

$$\{\phi^1\} = \{1, e\}^T \quad \text{and} \quad \{\phi^2\} = \{1, -(1-(1-æ)e)/æ\}^T \qquad (7)$$

As shown in Fig. 1, $\{\phi^1\}$ is approximately a rigid structure mode shape, while $\{\phi^2\}$ involves both structural and base isolation displacements. With these mode shapes we can write the relative displacements q_b and q_s in the form

$$\{q\} = [\phi]\,\{y\}, \qquad (8)$$

where

$$[\phi] = \begin{bmatrix} \phi_b^1 & \phi_b^2 \\ \phi_s^1 & \phi_s^2 \end{bmatrix} \quad \text{and} \quad \{y\} = \{y_1, y_2\}^T \qquad (9)$$

are the modal matrix and the vector of normal coordinates respectively. The basic matrix equation of motion (1) is then replaced /7/ by two equations

$$\ddot{y}_n + 2\beta_n\Omega_n\dot{y}_n + \Omega_n^2\,y_n = -\Gamma_n\,a(t), \quad n = 1,2 \qquad (10)$$

where we implicitly assume that the damping in the system is light enough to allow us to retain the orthogonality of the mode shapes. Participation factors $\{\Gamma\} = \{\Gamma_1, \Gamma_2\}^T$ are given by

$$\{\Gamma\} = \frac{[\phi]^T [M] \{r\}}{[\phi]^T [M] [\phi]} \tag{11}$$

Retaining only terms to order e, we can obtain after some manipulations

$$\Gamma_1 = 1 - \text{æe} \quad \text{and} \quad \Gamma_2 = \text{æe} \tag{12}$$

These results and those for the shifted frequencies clearly show how a base isolation system works. The participation factor of the second mode shape, which is the mode shape that involves structural deformations, is of order e and, if the original frequencies Ω_b and Ω_s are well separated, it could be very small. In addition, the frequency Ω_2 is shifted to a higher value than the original fixed-base frequency Ω_s, and if the earthquake input has large spectral accelerations at the original structural frequency, shifting it higher could shift it out of the range of dominant earthquake motion. Since the participation factor for this mode is very small, the ground motion will not be transmitted into the isolated structure. This is the real effectiveness of a seismic isolation system, which does not absopb energy, but deflects it through this property. Energy absorption is, of course, an important part of the behavior of an isolation system. The question is now how to select both the modal damping factors β_1 and β_2. Here we are able to make a very good estimation assuming approximately $\beta_1 = \beta_b$ and $\beta_2 = \beta_s$. The isolation system, if it is a laminated rubber system or a spring system with dampers, will provide damping in a range of about 5 - 15 %. The building structure will have somewhat less, about 2 - 3 %. In conventional seismic analyses it is generally assumed that the damping of the building structures is about 5%. But this higher value reflects the ductility and admissible damage factors of the structure, while in the present case of a base-isolated building structur, the aim is to reduce the seismic inertia forces transmitted into the structure. Therefore, it will probably work mostly within the elastic limits without any ductility or damage effects.

The design structure base shear coefficient C_s is usually defined as

$$C_s = k_s q_s / m_s = \Omega_s^2 q_s \tag{13}$$

When the building is unisolated (fixed-base system), then

$$C_s = \Omega_s^2 S_D(\Omega_s, \beta_s) \approx S_A(\Omega_s, \beta_s), \tag{14}$$

where S_D and S_A are the displacement and acceleration response spectra respectively. When the system is base isolated as shown in Fig. 1,

$$C_s = \{S_A^2(\Omega_1, \beta_1) + e^2(1-\text{æ})^2(1-2e) S_A^2(\Omega_2, \beta_2)\}^{1/2} \tag{15}$$

Although the second term is multiplied by e^2, it can sometimes be of the same order as the first term. This will happen if the spectrum is a constant displacement spectrum. If the spectrum is either constant velocity (like most earthquake design spectra) or constant acceleration, the second term appears to be negligible. These results indicate that the reduction in structure base shear, as compared with that of a fixed-base structure, is given by

$$S_A(\Omega_b, \beta_b) / S_A(\Omega_s, \beta_s) \qquad (16)$$

since $\Omega_1 \approx \Omega_b$ and $\beta_1 \approx \beta_b$ (approximately).
The ratio (16) is for a constant velocity spectrum of order $e^{1/2}$.

3. BENEFITS AND SOME RESTRAINTS OF BASE SEISMIC ISOLATION

The main benefits of base seismic isolation can be summarized as

- significant reduction of seismic loads on the structure and equipment
- little or no damage of structural elements and equipment during strong earthquakes
- concentration of the main deformations into the isolation bearings and dampers which are designed, constructed and tested with great care.

Real seismic base isolation systems involve resistance which is both velocity dependent (i.e. viscous) and hysteretic. Such isolation systems may be modeled for practical purposes approximately as equivalent linear viscous ones. But this equivalence is neither unique nor exact /8/. The effective period of a hysteretic system is usually defined as

$$T_{ef} = 2\pi (M/K_{ef})^{1/2}, \qquad (17)$$

where M is the mass and K_{ef} is the secant modulus either from the origin to the maximum displacement or between the extreme positive and negative displacements (Fig. 2). The difficulty with this definition is that it does not specify a unique property of the structure, since it depends on both the structural properties and the input motion. Furthermore, if the hysteretic system is made from linear spring devices in parallel with rigid-plastic energy dissipators, T_{ef} will always be shorter than the natural period of the system without dissipators because K_{ef} is inevitably higher than the elastic stiffness of the linear spring. The addition of hysteretic damping to the elastic system thus shortens its effective period, whereas viscous damping lengthens it. Effective damping may be defined as

$$\beta_{ef} = \text{Area of hysteretic loop} /(2\pi K_{ef} d_{max}^2), \qquad (18)$$

where d_{max} is the maximum displacement. However, there are again several difficulties associated with the use of Eq. 18. First, it is amplitude-dependent and so β_{ef} is not an inherent property of the structure, but depends also on the input. Second, Eq. 18 will produce reasonable results if the structure is forced through periodic oscillations. But, in a less regular motion containing reversals short enough for little hysteresis to occur, the viscous damper in the effective linear model may still dissipate significant energy. In that case the influence of damping would be overestimated. Viscous and hysteretic damping also have different effects on displacement and acceleration response. In general, the equivalent viscous damping chosen to correlate with the maximum displacements fails to predict the acceleration response. Such equivalent linear viscous models are thus useful for preliminary design and non-linear analysis is recommended for final design. Some near-field earthquake records lead to a very large displacement response in long period structures such as base isolated ones

/8/. In these cases it is essential to use realistic ground motions rather than those which have been modified to be compatible with a spectrum representative for more distant earthquakes. Pure sliding systems have no inherent natural period. The response is controlled by the coefficient of friction and by input motion. Non-linear analysis is necessary in such cases.

Protection against vertical seismic motions is regarded as less important because the vertical earthquake input is usually weaker and the structure is usually stronger in that direction. However in many cases it is necessary to incorporate both horizontal and vertical isolation systems, namely when vertical seismic excitation is the same or a similar order as horizontal.

4. BASE SEISMIC ISOLATION SYSTEMS IN COMMERCIAL USE AND NEW DEVELOPMENTS

The basic elements of a base seismic isolation system are:

- flexible bearings to achieve required tuning of the isolated system and to decouple the structure from the most harmful earthquake motion
- dampers or energy dissipators to control the relative displacements between the structure and its base on the ground
- means to provide some rigidity of the system under low-frequency loads such as wind loads, etc.

Table 1 (below) summarizes the elements in the first two groups. While flexibility is required to isolate buildings against earthquakes, it is undesibable to habve a structural system which will vibrate due to resonance with low-frequency gusting winds /9/. High flexibility of the base seismic isolation system may also detract when the isolated building is subjected to external impacts. Simple means providing some rigidity under such loads are sometimes used in the base seismic isolation systems. They are usually destroyed when strong seismic motion occurs. They must be replaced after each such event. However, specially provided elastomers and springs with dampers now have such resistance that these means are not usually necessary. In the usual type of analysis, soil-structure interaction is neglected. But when the flexibilities of the isolators and soil are comparable, the soil should be considered /10/. Structures founded on very soft soils are not suitable for base isolation.

Rubber bearings offer the simplest method of isolation, and they are easy to manufacture and relatively cheap. These bearings are made by vulcanization bonding of sheets of rubber to thin steel reinforcing plates (Fig. 3). Such bearings are relatively stiff in the vertical direction and are flexible in the horizontal direction /11,12/. Their action under seismic excitation is to isolate the building from the horizontal earthquake motion components, while the vertical components are transmitted in the building relatively unchanged. In addition, these bearings isolate the building from the high-frequency vibrations produced by underground transport, local trafic etc. Rubber pads are suitable for buildings that are rigid and for masonry or concrete ones up to eight-ten stories. Wind loads are usually dynamically unimportant for such buildings. If the main fixed-base frequency of the building is much higher than that of the isolated one, say 3 Hz compared to 0.5 Hz, the first mode of the isolated building is mainly a rigid body mode, while the second mode has a frequency about 50 to 100 % above the first fixed base frequency.

Significant progress has been made in the U.S.A. with the implemantation of rubber bearings in seismic isolation for various important buildings, bridges and items of industrial equipment /13,14/. Two such base isolated buildings are illustrated in Fig. 4 and 5. Experimental works on the rubber-isolated systems have been carried out on the shaking table at the Earthquake Engineering Research Center, University of California, Berkeley /15/. Rubber seismic isolation has been incorporated in the design of two advanced nuclear power plants with the U.S. liquid metal reactors PRISM and SAFR /16/(Fig. 6).

In New Zealand, a several isolation concepts have been applied to highway and railway bridges and buildings /17/. One of the buildings, a government office in Wellington, uses laminated natural rubber bearings each of which has a cylindrical plug of lead in a central hole (Fig. 7). The lead plug produces a substantial increase in damping of up to 15% and also increases resistance to wind loads. According to New Zealand's experience, lateral non-seismic loads (e.g. wind) should not exceed about 10% of the base isolated weight /17/. A twelve-story building has been constructed in Auckland on a base isolation concept called the "sleeved piles". This uses 12m long bearing piles within cylindrical sleeves, allowing a certain amount of lateral movement, in this 150 mm. The isolation period of this building on the piles is 4 seconds. Thus the resistance to wind loads would be inadequate with this system. In addition, the damping would be very low. To improve the behavior of the system, energy absorbing devices in the form of mild steel tapered plate beams are incorporated into the system and these lower the period to around 2 seconds. A similar principle has been applied recently to the base seismic isolation of the Wellington central police station /17/. The sleeved-pile concept is similar to the soft first story design concept often used in earthquake civil engineering /18/, but without the risk of collapse due to excessive first story lateral deflections.

Several French nuclear power plants in France and South Africa are built on a special base seismic isolation system, developed for this purpose by the company Spie Batignolles /19,20/. This system uses laminated neoprene bearings with lead bronze-stainless steel slip plates on the top of each bearing (Fig. 8). The neoprene bearings act as conventional isolators for small earthquakes, but cannot accept large displacements since they have only a few layers of elastomer. If a large earthquake should occur, sliding will take place on the slip plates. These have been designed to have a friction coefficient of 0.2 and to maintain this for the life time of the plant. The construction costs for this system are very high, but are justified in that it allows the standardization of plants to be built at any site. The Koeberg nuclear power plant in South Africa has been in commercial operation from 86. Neoprene bearings without slip plates have also been used under another nuclear power plant Cruas-Meysse in France /21/. An isolation system is used for this site because there is a probability of shallow earthquakes of low magnitude occurring close to the site, producing higher accelerations and high-frequency motions. The fixed-base frequency of this building is roughly 4.5 Hz and corresponds to the peak frequency of the ground response spectrum. With the neopren bearings, the natural frequency of the reactor building is reduced to 1 Hz, which significantly reduces inertia forces on the structure and its equipment. The maximum displacement capacity of these bearings is only 50 mm. But due to the high-frequency input, the anticipated displacements are only up to 26 mm /21/. This system really isolates only horizontally.

Interest in seismic isolation has been traditionally great in Japan (Fig.9) /22,23/, Italy, Greece, China, Soviet Union, Rumania and Yugoslavia. One of the oldest seismically isolated buildings (3-story school on rubber bearings) is in Skopje, Yugoslavia. A principally different base seismic isolation system has been developed by the company GERB, Germany /24/. They use their well-known spring units combined with viscous dampers. Helical springs achieve the required bearing capacity, stability and low tuning, while the incorporated viscous dampers provide damping. Such units (Fig. 10) may be loaded up to 1500 kN and their natural frequencies under the nominal load are about 0.5 Hz and 1 Hz in the horizontal and vertical directions respectively. Their damping may be up to 15-20%. The GERB isolators really work in all work conformably in all directions including the vertical one. Although this system is relatively expensive, it is probably the best and most reliable base seismic isolation system for very important and safety-related buildings such as nuclear power plants. Several ingenious seismic bearings have been proposed and applied for the earthquake protection of bridges. Special rubber bearings have been used on several bridges in New Zealand, Iceland and the U.S. /4,17/. GERB uses the tuned mass dampers, which are the dynamic absorbers, consisting the small weights attached to the structyure by means of the springs and dampers /6/,for the same purpose.

5. CONCLUSION

Base seismic isolation is a significant development in earthquake engineering that results in a reduced seismic response of buildings and their equipment. Many practical systems of seismic isolation have been developed in recent years and interest in the application of this technique for earthquake protection is continuously growing. The reluctance of structural engineers to use this concept is now hopefully diminishing. The research on isolation that has been carried out over the past few years and the construction of several new seismically isolated building structures enable engineers to have confidence that structures with seismic isolation can be economical. More than 20 seismically isolated structures have experienced earthquakes over the past several years /4/. Many of these events have been of rather small magnitude, but they have provided our first experience of how such systems really work. Time histories of acceleration obtained from several of these structures show clearly reduced accelerations above the isolation system, especially compared to the response of adjacent non-isolated buildings /4/.

References

/1/ Kelly, J.M., Seismic Base Isolation: Review and Bibliography. Soil Dynamics and Earthquake Engineering, Vol. 5, No. 3, 1986, pp. 202-216.
/2/ Kelly, J.M., Recent Developments in Seismic Isolation. PVP - Vol. 127. ASME, New York, 1987, pp. 381-386.
/3/ Walters, M. and Elsesser, E., Seismic Base Isolation of Existing Structures With Elastometric Bearings. PVP - Vol. 127. ASME, New York, 1987, pp. 387-398.
/4/ Buckle, I.G. and Mayes, R.L., Seismic Isolation: History, Application, and Performance. Earthquake Spectra, Vol. 6, No. 2, May 1990, pp.161-202.
/5/ Mayes, R.L. et al, The Economics of Seismic Isolation in Buildings. Earthquake Spectra, Vol. 6, No. 2, May 1990, pp. 245-264.
/6/ Elastic Support of Buildings. GERB, Berlin, 1988.

/7/ Paz, M., Structural Dynamics. 3rd edition. Van Nostrand, New York, 1991.
/8/ Stanton, J. and Roeder, Ch., Advantages and Limitations of Seismic Isolation. Earthquake Spectra, Vol. 7, No. 2, May 1991, pp. 301-323.
/9/ Henderson, P. and Novak, M. Response of Base-Isolated Buildings to Wind Loading. Earthquake Engineering and Structural Dynamics, Vol. 18, 1989, pp. 1201-1217.
/10/ Novak, M. and Henderson, P., Base-Isolated Buildings with Soil-Structure Interaction. Earthquake Engineering and Structural Dynamics, Vol. 18, 1989, pp. 751-765.
/11/ Stevenson, A., Longevity of Natural Rubber in Structural Bearings. Plastic and Rubber Processing and Applications. Vol.5, No. 3, 1985.
/12/ Derham, C.J. et al., Nonlinear Natural Rubber Bearings for Seismic Isolation. Nuclear Engineering and Design, Vol. 84, 1985, pp. 417-428.
/13/ Kelly, J.M., The Economic Feasibility of Seismic Rehabilitation of Buildings by Base Isolation. Report UCB/EERC 83/01. EERC, Berkeley,1983.
/14/ Kelly, J.M. and Tsai, H.C., Seismic Response of Light Internal Equipment in Base Isolated Structures. Report UCB/EERC 84/17. EERC, Berkeley,1984.
/15/ Kelly, J.M., The Use of Base Isolation and Energy-Absorbing Restrainers for the Seismic Protection of Large Power Plant Components. Report EPRI NP-2918. EPRI, Palo Alto, 1983.
/16/ Tajirian, F.F. et al., Seismic Isolation for Advanced Nuclear Power Stations. Earthquake Spectra, Vol. 6, No. 2, May 1990, pp. 371-401.
/17/ McKay, G.R. et al., Seismic Isolation: New Zealand Applications. Earthquake Spectra, Vol. 6, No. 2, May 1990, pp. 223-244.
/18/ Naeim, F., The Seismic Design Handbook. Van Nostrand, New York, 1989.
/19/ Jolivet, J. et al., Aseismic Foundation System for Nuclear Power Stations. Trans. 4th Int. Conf. SMiRT, paper K9/2, San Francisco, 1977.
/20/ Plichon, C. et al., Protection of Nuclear Power Plants Against Seism. Nuclear Technology, Vol. 49, 1980, pp. 295-306.
/21/ Postollec, J.C., Les Fondations Antiseismiques de la Centrale Nucleare de Cruas-Meysse. Notes du Service Etude Geni Civil d'EDF-REAM, 1983.
/22/ Fujita, T. et al., Research Development and Implementation of Rubber Bearings for Seismic Isolation. PVP - Vol. 181. ASME, New York, 1989.
/23/ Sawada, Y. et al., Seismic Isolation Test Program. Trans. 10th Int. Conf. SMiRT, Vol. K2, Anaheim, 1989, pp.691-696.
/23/ Hueffmann, G.K., Full Base Isolation for Earthquake Protection by Helical Nuclear Engineering and Design, Vol. 84, No.3, 1985.

Notes: PVP = Proceedings of the Pressure Vessel and Piping Conference
EERC = Earthquake Engineering Research Center, University of California
EPRI = Electrical Power Research Institute

Table 1 Flexible bearings and damping devices for base seismic isolation
----
Flexible bearings   unreinforced and reinforced rubber blocks (elastomers)
                    coil steel springs, air springs, roller or ball bearings
                    sliding plates, friction pendulum devices, rocking systems
                    cable suspension pendulum systems, sleeved piles
----
Damping devices   viscous dampers, plastic deformation of metal devices
                    high damping of rubber elastomers, friction
----

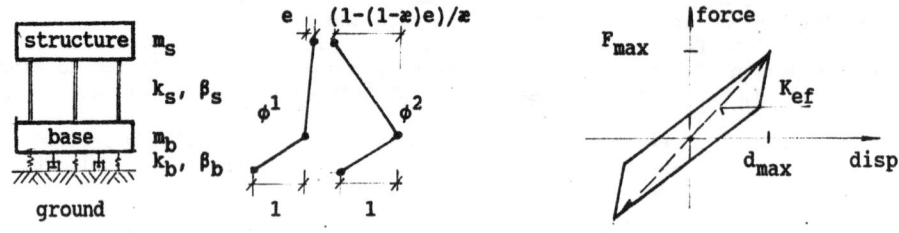

Fig. 1 Two-Degree-of-Freedom System for the Base Isolated Structure

Fig. 2 Typical Hysteresis Loop of the Rubber Bearing

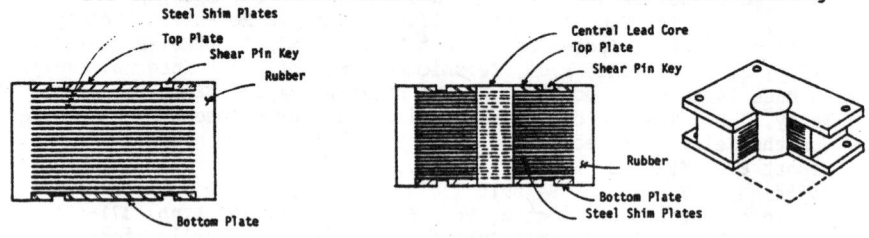

Fig. 3 Typical Rubber Bearings (Ordinary Elastometric and Lead-Rubber) for Base Seismic Isolation /3/

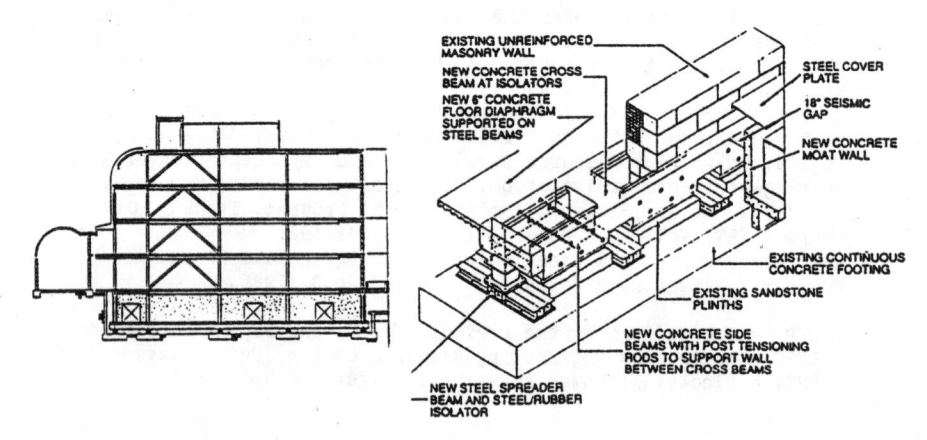

Fig. 4 Base Isolated Building of the Fuithill Communities Law and Justice Center, San Bernardino, California /4/

Fig. 5 Seismic Rehabilitation through Base Isolation - Salt Lake City and County Building, General View and Detail of Insatallation of Isolators /3,4/

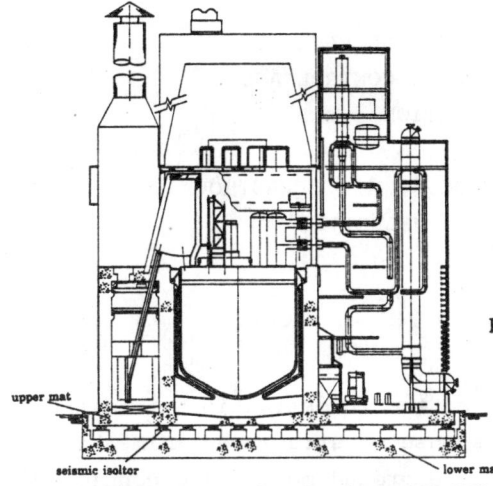

Fig. 6 Base Isolation for the SAFR NPP /16/

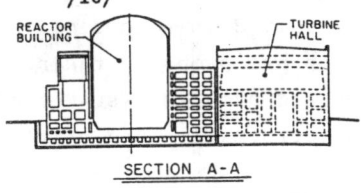

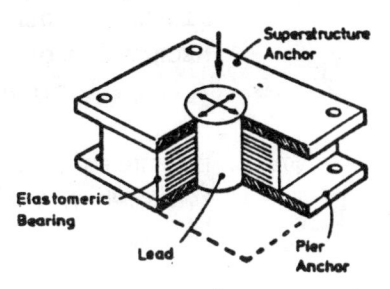

Fig. 7 Lead-Rubber Device (New Zealand) /17/

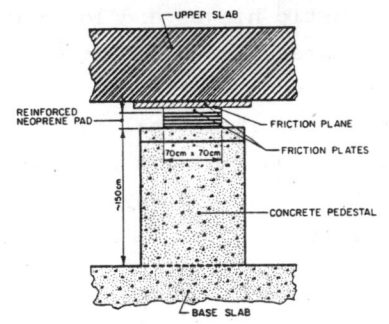

Fig. 8 French Base Seismic Isolation System for NPP, General View and Detail of Bearings /20/

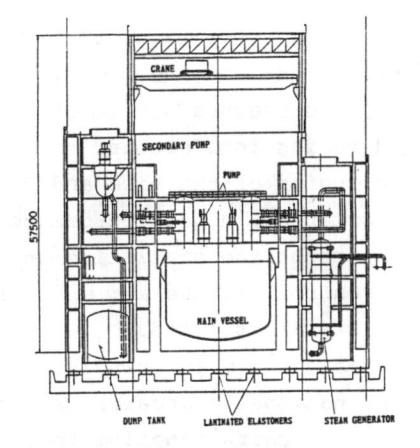

Fig. 9 Japan Approach of Seismically Isolated FBR Nuclear Power Plant /23/

Fig. 10 Prototype of the GERB Spring Unit for Base Isolation /24/

SEISMIC RESPONSE REDUCTION BY MEANS OF INELASTIC DEFORMATIONS AND FAILURE. MECHANISMS OF STRUCTURES CONTROL AND SEISMIC ISOLATION

Ja.M. EISENBERG[1], A.M. ZHAROV[1], M.T. GERASIMOVA[1], B.K. GAYAROV[2], A.M. MELENTIEV[1]

INTRODUCTION

The paper contains the results of the investigations into the methods of reducing earthquake response of structures. Time-history and spectra models of earthquake ground motion are considered taking into account seismological and geological situation in the region of the site and uncertainty involved in seismological information. Models of damage accumulation are proposed taking into account the statistics on earthquakes and aftershocks. Methods of controlling the sequence of inelastic deformation and brittle failure are studied.

Among the main principles of aseismic design as employed by many national standards is the principle of equal strength of structural members denoting that the distribution of strength in structural members is proportional to the distribution of stress that is obtained from the analysis of an elastic system for earthquake effects. Applied to practical calculations this principle brings about inpredictable distribution of plastic hinges and local brittle damage of structures during earthquakes. The sequence of these deformation depends on the random features of a given seismic motion and on the random strength spread of structural members.

Some ideas are proposed concerning rational design control of inelastic deformations and local failure; possible ways of their use in calculation procedures are depicted.

---
[1] Department of Structural Earthquake Resistance, the V.A. Kucherenko TsNIISK, Gosstroy, Moscow, USSR
[2] Scientific Research Institute for Earthquake Engineering, Gosstroy, Turkmen SSR, Ashkhabad, USSR.

IDEAS OF RATIONAL DESIGN CONTROL

Optimal design of structures should be aimed at ensuring such strength hierarchy of structural members that should be accounted for by the sequence of occurrence and development of plastic zones, disengagement (switching off) of some rigid members or ties and of local failure. This sequence creates favourable, in terms of structural behaviour during earthquakes, development of inelastic deformations, favourable energy absorption, favourable non-linearity and rigidity variation.

The sequence of plastic hinges and brittle damages should be designed as follows. At the first stage inelastic strain should develop in non-bearing vertical loads members or in special reserve members, while members resisting vertical loads remain in elastic stage. In this case the initial rigidity of primary plastified and disengaging members should be high enough as compared to that of members remaining elastic whereas the strength of the plastified and disengaging members should be lower than that of more brittle members.

A mathematical model of seismic ground motion is presented as a set of processes whose dominant frequencies belong to a definite, known by experience, frequency domain; design parameters of each of its components (spectra width, time envelope) are expressed as simple functions of the dominant frequency, the model represents a variety of dominant frequencies, spectra and other earthquake characteristics recorded in the past, it permits to predict earthquakes likely in the future. Besides, the accumulation of seismic information permits to improve the mathematical description of a seismic process in the framework of the given model modification, this being quite important for the construction of a regional model.

Mathematically, a design model is represented by a discrete set M_j of non-stationary Gaussian multiplicative processes; every j-th component of the set within the given domain $\omega_{min} \leq \omega_j \leq \omega_{max}$ can be expressed as follows:

$$F(t, \bar{\omega}_j) = \begin{cases} A(t, \bar{\omega}_j)\, \sigma(\bar{\omega}_j)\, \varphi(t, \bar{\omega}_j); & \text{at } t > 0, \\ 0 & t = 0, \end{cases} \qquad (1)$$

where

$A(t, \bar{\omega}_j)$ - standard enveloping function assigned for fixed values as Berlage impulse

$$A(t, \bar{\omega}_j) = A_0 j t e^{-\varepsilon_j t} \quad ; \quad |A_{max}| = 1 , \qquad (2)$$

where

$\sigma(\bar{\omega}_j)$ - root-mean-square acceleration,

$\bar{\omega}_j$ - carrier frequency, approximately equal to the dominant frequency of the j-th process;

$\varphi(t, \bar{\omega}_j)$ - standardized (single) stationary Gaussian process, characterized by a cosine-exponential correlation function:

$$K(\tau) = e^{-\alpha_j |\tau|} \cos \bar{\omega}_j \tau \quad ; \quad \text{or} \quad K(\tau) = e^{-\alpha_j |\tau|}(\cos \bar{\omega}_j \tau + \frac{\alpha_j}{\bar{\omega}_j} \sin \bar{\omega}_j \tau). \quad (3)$$

Basing on the regional models of seismic effects optimum parameters of non-linear and non-stationary systems of earthquake protection are analysed.

One of the varieties of non-linear and non-stationary systems is realized in the design of a large-panel building for Siberia and other regions with a severe climate and high seismicity. A distinctive feature of this structural design is the use of highly industrial horizontal and vertical joints built up at the site, which permits to save material resources owing to a considerable reduction of power and labour consumptions and, hence, of time and cost of construction.

The earthquake-adaptive properties of the building structure are represented by:
- The non-linearity due to opening of horizontal joints of wall panels;
- First and foremost breaking of mortar keys followed by energy absorption in the bottom of wall panels where vertical bars join;
- Limiting of the opening of horizontal joints and interdisplacement of wall panels in a vertical joint - the engagement of a tie with a rigid characteristic. A dry gasket in the horizontal joint increases the system's deformability and energy absorption.

Complex theoretical and experimental investigation into the elaborated structural design of the building with the adaptive seismic protection was carried out. The tests were carried out on the building's fragment, model and on a full scale installation using powerful vibration machines. The procedure of a parametric analysis of the system using single and multi-mass models had two stages.

At the first stage a non-linear statistic analysis of the building for an alternating horizontal load was performed in order to obtain the curves of design deformations of the building, to determine stress-strained state corresponding to each point of the plot and to estimate the system's limit states.

At the second stage a dynamic calculation of the equation of motion was performed for actual acceleragrams of earthquake actions; "restoring force - displacement" relations obtained at the first stage were used.

The investigation permitted to recommend for building practice reliable and efficient large-panel buildings for earthquake -prone regions with a severe climate.

Suggested are models of damage accumulation in buildings and models of losses based on the statistics of earthquakes and aftershocks.

The buildings are described by inelastic dynamic systems. The so-called damage characteristics of the system are studied (residual displacements, number of inelastic oscillation cycles, energy spent on inelastic deformation, etc.). The procedure of estimating losses due the earthquakes and choosing optimum levels of seismic strengthening of buildings operating at an elastic stage is elaborated. Losses due to earthquakes are estimated depending on the values of the damage characteristics of a corresponding inelastic system.

It is assumed that during the service life of the building a random number of earthquakes will occur, they can be divided into several classes different in the level of oscillations, their asymmetry, spectral content, durability and recurrence. Of great importance is the factor of oscillation asymmetry which noticeably influences residual displacements of the inelastic system. This factor is neglected by many researchers, in particular, insufficient attention is paid to the problem concerning the correction of a zero line in the records of earthquake motion. All this, provided residual displacements are used as the main damage criterion, leads to grave errors.

The estimation of average damage characteristics or average amount of losses corresponding to them as well as of their variances is based on a generalized model of a seismic effect which represents a set of chains of realizations of functions belonging to different classes of seismic effects. These chains are constructed with due regard for earthquake conditions in the vicinity of the site.

During the restoration of buildings after earthquakes losses get accumulated. If the buildings are not repaired after an earthquake or if they are subjected only to a cosmetic repair, damage accumulates. This, specifically, takes place after aftershocks. Practice shows that for an inelastic analysis of structures it is expedient to have four groups of relations which being independent are usually obtained by different types of experts.

The first group of the relations permits to calculate damage characteristics of an inelastic system using prescribed characteristics of a seismic effect for **different** values of the system's parameters. This group of relations is obtained by processing a set of variants of a numerical solution of the problem on the behaviour of an inelastic system under different types of seismic effects.

The set of damage characteristics determines the damage of the system or of its member. The content of the set is determined by the building's structural features, as well as by the aims and the tasks of the analysis. If the system's residual displacement is used as damage characteristic, then its asymmetry characteristics should necesserily be introduced as parameters of a dynamic system. For a generalized Prandtl curve, in particular, this characteristic is the difference between the ultimate displacements in the course of the systems movement in a positive and a negative direction. The disregard of the **above mentioned** characteristic of the system leads to calculation errors.

The other group of relations characterizes a connection between the amount of the system's damage D and the degree η of the building's damage for different values of the ultimate damage D_u. This group of the relations is obtained basing on engineering analysis of aftereffects of earthquakes, full-scale tests of the building and the use of the results of the analysis of inelastic systems for a concrete effect.

The third group of relations characterizes the connection between the amount of losses in the building L subjected to earthquakes and the degree of its damage η. The losses in the building due to the damage degree are determined by the cost of its retrofit; they depend on a proper repair. The relations can be obtained basing on the analysis of the repair cost.

Losses in buildings damaged during earthquakes can be calculated as follows:

At first, values of the system's damage are found basing on the given earthquake characteristics and specified parameters of an elasto-plastic system, simulating the perfomance of the building under consideration; the first group of the relations is used. Then, using the accepted reliability criterion and the second group of the relations, damage degree η of the building is obtained from the known values D and D_u.

The third group of the relations permits to estimate, using the calculated value of the damage degree η, losses L which correspond to the latter.

To assess the optimum level of the seismic strengthening of the building the fourth group of the relations is required. It connects the initial cost C of a seismic protection and non-dimensional load-bearing capacity of the building under a horizontal load.

The above relations are determined basing on the analysis of structural solutions of the building for different levels of its capacity under horizontal loads.

The optimum level of seismic strengthening of the building should be chosen proceeding from the principle of minimizing average total expenses over the service life of the building which consist of the cost of the initial seismic strengthening and the cost of the repair after the earthquake.

The optimum initial cost C_{opt} of structural earthquake protection and minimum total expenses C_{min} vary within a wide range. Thus in regions with earthquake intensity 7 the variation range was 5% of the building cost (from 7.1% to 11.5%), the range of amounting 2.6% of the cost of the building (from 9% to 11.6%). In regions with earthquake intensity 9 the range of C was 4.7% (from 11% to 15.7%) of the cost of the building, while the range of amounted 4.3% (from 11.6 to 15.9%). Parameters characterizing the structure varied in the certain range. If the level of structural earthquake reinforcement is specified by the accepted methods, ignoring the factors mentioned above, the total cost C_{tot} of the building's maintenance prove to be higher than the optimum cost C_{opt}, sometimes by a considerable amount of over 10% of the cost of the building.

CONCLUSION

These methods permit to take into account the type of building structures, their inelastic strain capacity, dynamic characteristics, recurrence of certain classes of earthquake and peculiarities of soil motion during earthquakes of different classes (their spectra and duration of oscillations). It was established that the influence of some of the foregoing factors on the optimum level of seismic strengthening of buildings depends on the rest of the factors and is commensurable between themselves. It is shown the change of these factors within the range embracing practically important cases tells noticeably on the optimum level of seismic strengthening of buildings.

These methods were applied to 15 regions of the USSR.

THE SEISMIC ISOLATION EFFECT OF DIFFERENT SOIL LAYERS COMBINATIONS

Y. Koleková[1] E. Juhásová[1]

INTRODUCTION

The seismic response of structure is influenced by many factors and they are playing larger or smaller part when calculating the seismic resistance of the very structure with described soil condition in the base. Usually it is assumed that the soil foundation motion influences the vibration of the structure, but the vibration of structure does not substancially affect the motion of the subsoil. This assumption can be accepted in the case of hard rock or other very stiff subsoils. When the subsoil is layered and some strata are ductile, the dynamic soil-structure interaction can affect both the vibration of the structure and the subsoil. Excluding hard rock subsoil this means the calculation of the seismic response of structure with the account of mechanical properties of individual soil layers. At the same time the more complicated calculation dynamic model should be used with the account of soil-structure interaction, seismic waves propagation in subsoil and eventual non-linear effect in dynamic behaviour either of structure or the subsoil. In this paper there is discussed the problem of non-linear region in the considered composition of subsoil and structure and the positive effect of some strata combination on the seismic response of structure.

THEORETICAL DEPENDENCES

For the linear elastic space element in a coordinate system x, y, z there are valid equations of motion

$$\rho \frac{\partial^2 u}{\partial t^2} = \frac{\partial \sigma_x}{\partial x} + \frac{\partial \tau_{xy}}{\partial y} + \frac{\partial \tau_{xz}}{\partial z} , \qquad (1)$$

$$\rho \frac{\partial^2 v}{\partial t^2} = \frac{\partial \sigma_y}{\partial y} + \frac{\partial \tau_{yz}}{\partial z} + \frac{\partial \tau_{yx}}{\partial x} , \qquad (2)$$

$$\rho \frac{\partial^2 w}{\partial t^2} = \frac{\partial \sigma_z}{\partial z} + \frac{\partial \tau_{zx}}{\partial x} + \frac{\partial \tau_{zy}}{\partial y} , \qquad (3)$$

with

$$\sigma_x = 2G \varepsilon_x + \lambda (\varepsilon_x + \varepsilon_y + \varepsilon_z) , \qquad (4)$$

---
1 - ÚSTARCH SAV, Bratislava, ČSFR

$$\tau_{xy} = G \gamma_{xy}, \tag{5}$$

$$\lambda = \frac{E\nu}{(\nu+1)(1-2\nu)}, \quad G = \frac{E}{2(1+\nu)}. \tag{6}$$

Using

$$\varepsilon_x = \frac{\partial u}{\partial x}, \quad \gamma_{xy} = \frac{\partial v}{\partial x} + \frac{\partial u}{\partial y}, \tag{7}$$

$$c_P^2 = \frac{\lambda + 2G}{\varrho}, \quad c_S^2 = \frac{G}{\varrho}, \tag{8}$$

where c_P, c_S are longitudinal and shear wave velocities, we can write

$$\frac{\partial^2 u}{\partial t^2} = (c_P^2 - c_S^2)\frac{\partial^2 u}{\partial x^2} + (c_P^2 - c_S^2)\left(\frac{\partial^2 v}{\partial x \partial y} + \frac{\partial^2 w}{\partial x \partial z}\right) + c_S^2\left(\frac{\partial^2 u}{\partial x^2} + \frac{\partial^2 u}{\partial y^2} + \frac{\partial^2 u}{\partial z^2}\right), \tag{9}$$

$$\frac{\partial^2 v}{\partial t^2} = (c_P^2 - c_S^2)\frac{\partial^2 v}{\partial y^2} + (c_P^2 - c_S^2)\left(\frac{\partial^2 w}{\partial y \partial z} + \frac{\partial^2 u}{\partial y \partial x}\right) + c_S^2\left(\frac{\partial^2 v}{\partial x^2} + \frac{\partial^2 v}{\partial y^2} + \frac{\partial^2 v}{\partial z^2}\right), \tag{10}$$

$$\frac{\partial^2 w}{\partial t^2} = (c_P^2 - c_S^2)\frac{\partial^2 w}{\partial z^2} + (c_P^2 - c_S^2)\left(\frac{\partial^2 u}{\partial z \partial x} + \frac{\partial^2 v}{\partial z \partial y}\right) + c_S^2\left(\frac{\partial^2 w}{\partial x^2} + \frac{\partial^2 w}{\partial y^2} + \frac{\partial^2 w}{\partial z^2}\right), \tag{11}$$

In the step computation let Δt be a time increment, Δs a grid increment (Fig. 1). Then for the point i,j,k in the time step r

$$u_{i,j,k,r+1} = 2u_{i,j,k,r} - u_{i,j,k,r-1} + \frac{(c_P^2 - c_S^2)\Delta t^2}{\Delta s^2}\left(u_{i+1,j,k,r} - 2u_{i,j,k,r} + u_{i-1,j,k,r}\right) + \frac{(c_P^2 - c_S^2)\Delta t^2}{4\Delta s^2}\left(v_{i+1,j+1,k,r} - v_{i+1,j-1,k,r} - v_{i-1,j+1,k,r} + v_{i-1,j-1,k,r} + w_{i+1,j,k+1,r} - w_{i+1,j,k-1,r} - w_{i-1,j,k+1,r} + w_{i-1,j,k-1,r}\right) + \frac{c_S^2 \Delta t^2}{\Delta s^2}\left(u_{i+1,j,k,r} + u_{i-1,j,k,r} + u_{i,j+1,k,r} + u_{i,j-1,k,r} + u_{i,j,k+1,r} + u_{i,j,k-1,r} - 6u_{i,j,k,r}\right), \tag{12}$$

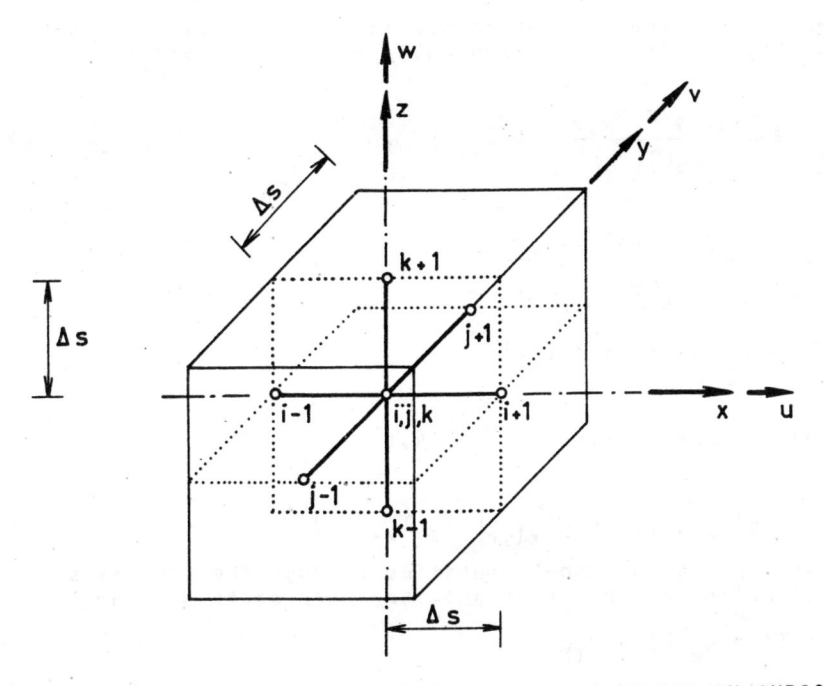

FIG. 1. SPACE GRID ELEMENT FOR THE STEP SOLUTION IN SUBSOIL

what answers to Eq. (9). Similarly we can write the discrete form of Eqs. (10), (11). At the boundaries of soil layers, which have different mechanical properties and wave velocities, the transition and the reflection of waves must be considered.

But the assumption of linear behaviour of subsoil is appropriate only for very small amplitudes of vibration. As can be seen from the laboratory and field observations for higher levels of vibration the non-linear properties of soil have to be included in the computation dynamic model. When we are solving the problem of seismic motions in subsoil with the account of non-linear properties it is convinient to include them directly into Eqs. (1 - 3). The approach is described for the case when only vertically propagated seismic shear waves are considered. Then we can write instead of Eq.(1)

$$\rho \frac{\partial^2 u}{\partial t^2} = \frac{\partial \tau_{xz}}{\partial z} , \qquad (13)$$

or in the discrete form

$$u_{k,r+1} = 2u_{k,r} - u_{k,r-1} + \frac{1}{\rho} \frac{\Delta t^2}{\Delta s} \left(\tau_{k+1,r} - \tau_{k,r} \right) . \qquad (14)$$

Supposing that for a short time interval $(t, t+\Delta t)$ over the range $\Delta t c_s(\tau)$ there is no change in the variable shear velocity $c_s(\tau)$, or in the shear modulus $G(\tau)$ we can write

$$\tau_{k,r} = \frac{u_{k,r-1} - u_{k,r}}{c_s(\tau_{k,r})\Delta t} G(\tau_{k,r}) \quad , \tag{15}$$

$$\tau_{k+1,r} = \frac{u_{k+1,r-1} - u_{k+1,r}}{c_s(\tau_{k+1,r})\Delta t} G(\tau_{k+1,r}) \quad , \tag{16}$$

or in connection with (14)

$$u_{k,r+1} = 2u_{k,r} - u_{k,r-1} + \frac{\Delta t}{\Delta s}\left((u_{k,r}-u_{k,r-1})c_s(\tau_{k,r}) + \right.$$

$$\left. +(u_{k+1,r-1} - u_{k+1,r})c_s(\tau_{k+1,r})\right) \quad . \tag{17}$$

We can express the non-linearities through the complex shear modulus with variable real and imaginary parts [1], [6]

$$G(\tau) = G_R(\tau) + G_I(\tau) \quad . \tag{18}$$

The transition and the reflection of waves have to be considered at the boundaries of adjacent soil layers which have different mechanical properties. It can be proved that at the time t for the point k, which is the point of the boundary

$$u_k(t) = (1-\alpha_0(\tau_k))u_{k-1}\left(t-\frac{\Delta s}{c_{s,k-1}(\tau)}\right) + (1+\alpha_0(\tau_k))u_{k+1}\left(t-\frac{\Delta s}{c_{s,k+1}(\tau)}\right) -$$

$$- u_k\left(t-\frac{\Delta s}{c_{s,k-1}(\tau)} - \frac{\Delta s}{c_{s,k+1}(\tau)}\right) \quad , \tag{19}$$

the transmission ratio $\alpha_0(\tau)$ is

$$\alpha_0(\tau) = \frac{1-\kappa_k(\tau)}{1+\kappa_k(\tau)} \quad , \tag{20}$$

$$\kappa_k(\tau) = \frac{\rho_{k+1}c_{s,k+1}(\tau)}{\rho_{k-1}c_{s,k-1}(\tau)} \quad , \tag{21}$$

with the point $k+1$ in the layer below the boundary, with the appropriate density ρ and variable shear wave velocity $c_s(\tau)$.

The shear wave velocity is considered variable or constant depending on the stress-strain conditions of the vibrating soil.

In this way we can analyse the soil layered system either in linear or non-linear region and follow its behaviour at the different source seismic motion. Then the question can arise whether or not there exists such a soil layers combination which for a certain group of structures has recquired seismic isolation properties. The principle is that the soil layers act as amplifiers or filters which can increase or decrease the soil surface vibration. The isolation efficiency of soil layers in optimal combinations can be remarkable.

When there exists some possibility of adjusting the thickness and material of the upper soil layers to the requirements connected with the dynamic properties of the structure and expected seismic loading we can modify the seismic motion of the soil surface. The analysis of this phenomena we have done for different number of soil layers using shear model [1],[2],[3] and considering that $x_0(t)$ is the motion of the bed-rock and $x(t)$ is the motion of soil surface. Then for the known $x_0(t)$

$$U_0(t) = x_0(t) , \qquad (22)$$

$$D_0(t) = \alpha_{10} D_1\left(t - \frac{H_1}{C_{S1}}\right) + \beta_{10} U_0(t) , \qquad (23)$$

$$U_1(t) = \alpha_{01} U_0(t) + \beta_{01} D_1\left(t - \frac{H_1}{C_{S1}}\right) , \qquad (24)$$

$$D_1(t) = \alpha_{21} D_2\left(t - \frac{H_2}{C_{S2}}\right) + \beta_{21} U_1\left(t - \frac{H_1}{C_{S1}}\right) , \qquad (25)$$

$$U_2(t) = \alpha_{12} U_1\left(t - \frac{H_1}{C_{S1}}\right) + \beta_{12} D_2\left(t - \frac{H_2}{C_{S2}}\right) , \qquad (26)$$

$$D_2(t) = \alpha_{32} D_3\left(t - \frac{H_3}{C_{S3}}\right) + \beta_{32} U_2\left(t - \frac{H_2}{C_{S2}}\right) , \qquad (27)$$

$$\vdots$$

$$U_{n-1}(t) = \alpha_{n-2,n-1} U_{n-2}\left(t - \frac{H_{n-2}}{C_{S,n-2}}\right) + \beta_{n-2,n-1} D_{n-1}\left(t - \frac{H_{n-1}}{C_{S,n-1}}\right) , \qquad (28)$$

$$D_{n-1}(t) = \alpha_{n,n-1} D_n\left(t - \frac{H_n}{C_{S,n}}\right) + \beta_{n,n-1} U_{n-1}\left(t - \frac{H_{n-1}}{C_{S,n-1}}\right) , \qquad (29)$$

$$U_n(t) = \alpha_{n-1,n} U_{n-1}\left(t - \frac{H_{n-1}}{C_{S,n-1}}\right) + \beta_{n,n-1} U_{n-1}\left(t - \frac{H_{n-1}}{C_{S,n-1}}\right) , \qquad (30)$$

$$D_n(t) = U_n\left(t - \frac{H_n}{C_{S,n}}\right) , \qquad (31)$$

$$x(t) = 2 D_n(t) . \qquad (32)$$

The coefficients α_{ik}, β_{ik} are

$$\alpha_{ik} = \frac{2}{1 + \frac{\rho_k c_{sk}}{\rho_i c_{si}}}, \qquad (33)$$

$$\beta_{ik} = \frac{\frac{\rho_k c_{sk}}{\rho_i c_{si}} - 1}{\frac{\rho_k c_{sk}}{\rho_i c_{si}} + 1}, \qquad (34)$$

where the second subscript of $\alpha_{.k}, \beta_{.k}$ denotes the layer into which the wave component is entering and the shear wave velocity c_{sk} answers to the k-th layer which has the thickness H_k [5].

The main features of obtained results can be seen in Figs. 2, 3, 4. We can see that the resulting response motion on the soil surface and secondary that of the structure is influenced by the time-frequency relation of the dynamic characteristics of layered subsoil, of the structure and input wave motion in the bed rock $\ddot{x}_o(t)$.

In some inappropriate combinations there can appear multi-resonance effects in the seismic response which are undesirable both for the structure and the subsoil. On the other hand we can see also the cases in which the motion on the soil surface is smaller than that one in the bed-rock. Naturally we can also follow the effect of layers built from artificial materials [4].

Following the eventual non-linear behaviour in subsoil we should use in shear model approach the Eqs.(14) to (21). For linear model it is sufficient to consider Eqs.(22) to (34).

CONCLUSION

On the base of obtained results it was proved the importance of considering the seismic response of layered systems and the soil-structure interaction in the solution at least for the chosen configuration of soil and structure properties. Further this approach enables to understand the linear and non-linear changes of the shear modulus and of the shear wave velocity. It seems convinient to apply the complex modulus of elasticity with different functional relationships for its real and imaginary part in the theoretical and numerical solution. The more precise analysis of layered subsoil and the soil-structure interaction enables to identify the basic unfavourable versions of structure and soil dynamic parameters. Then we can modify the structural solution in such a way as to minimize the seismic response.

REFERENCES

1. JUHÁSOVÁ, E.: Seismic Effects on Structures. Veda-Elsevier coedition. Bratislava, Amsterdam, 1991.
2. JUHÁSOVÁ, E. - KOLEKOVÁ, Y.: Behaviour of clays at harmonic and general dynamic loading. Stav. Čas., 37, 8, 1989.
3. JUHÁSOVÁ, E. - KOLEKOVÁ, Y.: Non-linear seismic motions in layered subsoil. In: Proc. 9ECEE, Moscow 1990. Vol. 4B, p. 188.
4. MAKOVIČKA, D.: Lowering of transfer of vibration into building structures from surrounding medium. Inženýrske stavby, 7-8, 1989, p. 390.
5. OKAMOTO, S.: Introduction to Earthquake Engineering. University of Tokyo Press - Strojizdat, Moskva 1980.
6. ISHIDA, K.: Dynamic characteristics of soil-foundation interaction system detected from forced vibration test and earthquake observation. J. Earth. Eng. and Struct. Dynamics, 13, 6, 1985.

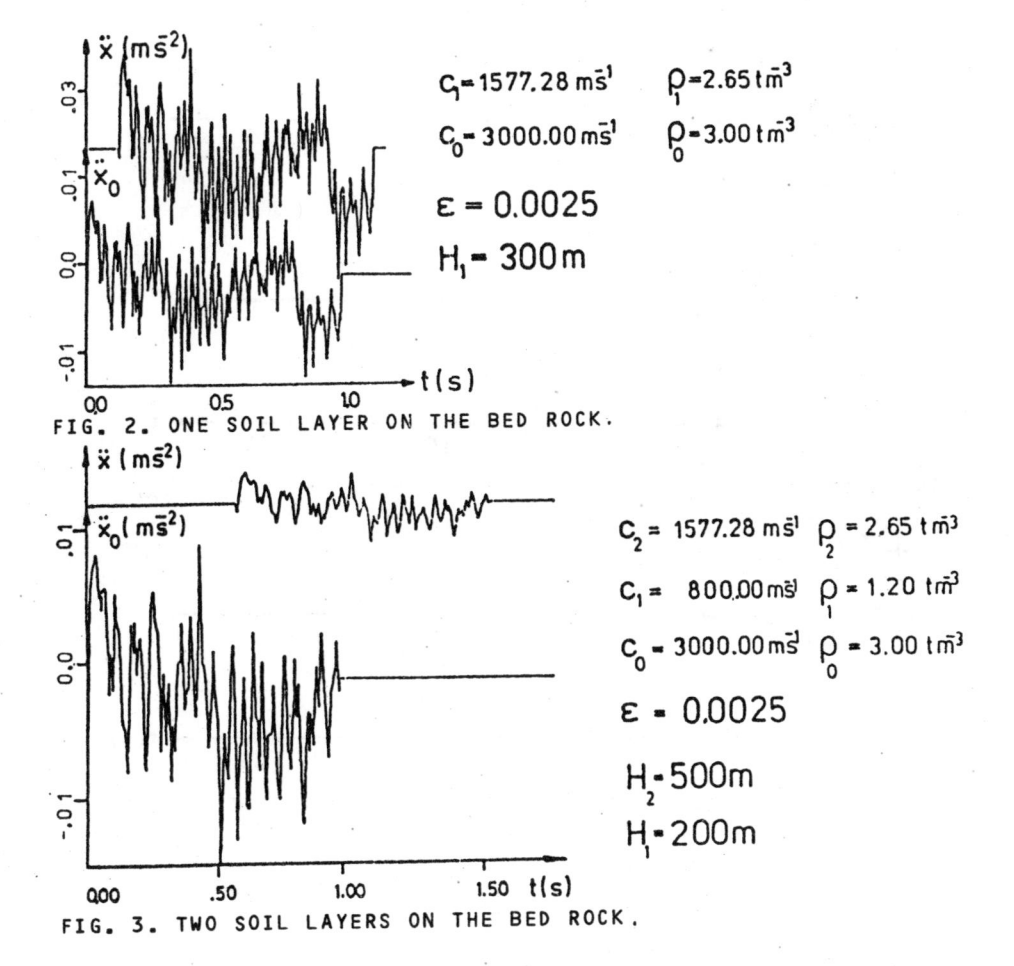

FIG. 2. ONE SOIL LAYER ON THE BED ROCK.

$C_1 = 1577.28 \, ms^{-1}$ $\rho_1 = 2.65 \, tm^{-3}$
$C_0 = 3000.00 \, ms^{-1}$ $\rho_0 = 3.00 \, tm^{-3}$
$\varepsilon = 0.0025$
$H_1 = 300 \, m$

FIG. 3. TWO SOIL LAYERS ON THE BED ROCK.

$C_2 = 1577.28 \, ms^{-1}$ $\rho_2 = 2.65 \, tm^{-3}$
$C_1 = 800.00 \, ms^{-1}$ $\rho_1 = 1.20 \, tm^{-3}$
$C_0 = 3000.00 \, ms^{-1}$ $\rho_0 = 3.00 \, tm^{-3}$
$\varepsilon = 0.0025$
$H_2 = 500 \, m$
$H_1 = 200 \, m$

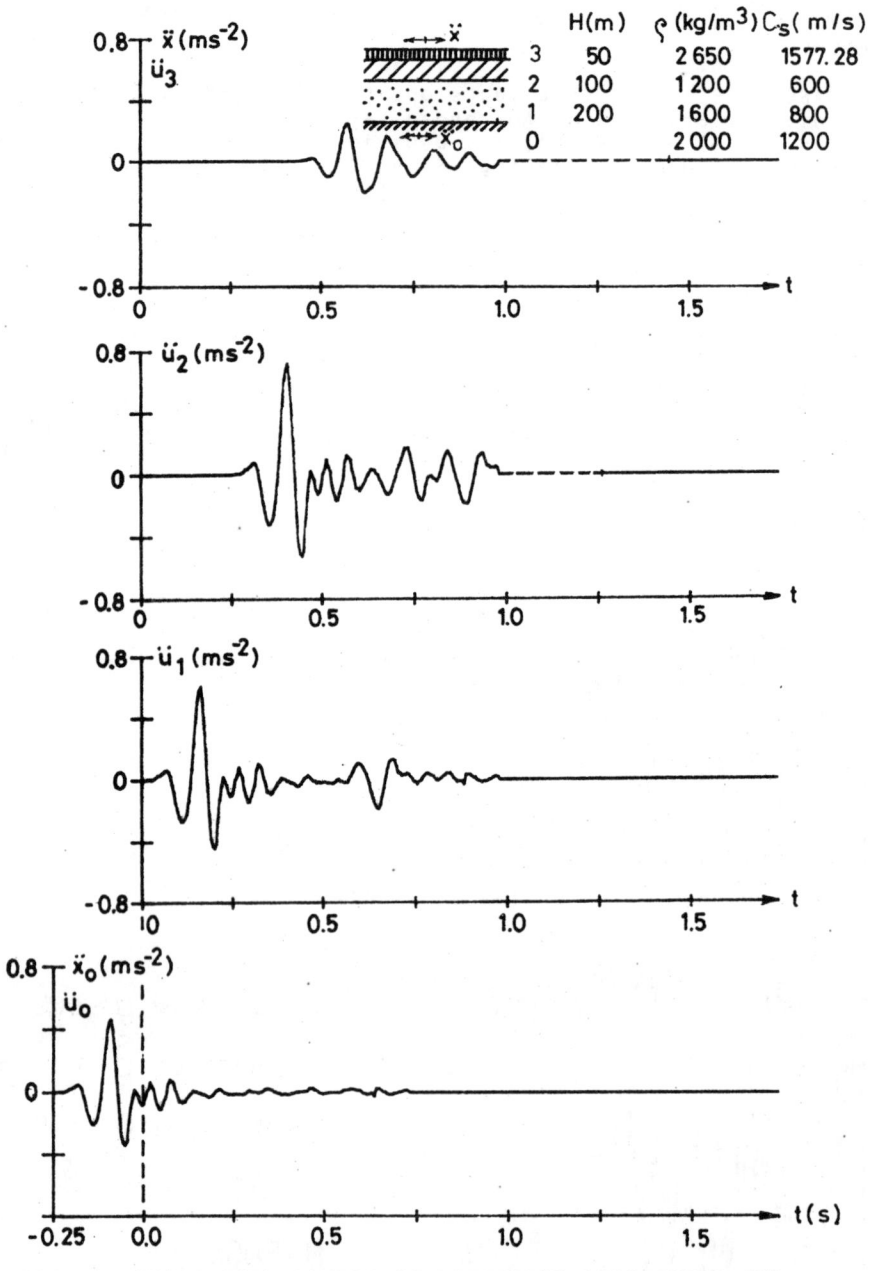

FIG. 4. THREE SOIL LAYERS OF DIFFERENT PROPERTIES AND THICKNESSES ON THE BED ROCK.

T4

SOFTWARES AVAILABLE FOR SEISMIC CALCULATION AND DESIGN.

SOFTWARE FOR ASEISMIC DESIGN OF MASONRY BUILDINGS

Dražen Aničić[1]   Dragan Radić[2]

SUMMARY

The paper focuses on main computation features of the computer program "SAMB" used on personal computers for the automatic seismic calculation of masonry structures. Seismic forces are calculated using the method of equivalent static forces, while their distribution among individual walls is based on the elastic theory. The ultimate bearing capacity of individual walls is calculated according to the limit state method. This program can also be used to make calculations for buildings whose bearing systems are composed of composite walls made of masonry and concrete.

INTRODUCTION

Despite intensive construction activities witnessed throughout the world in the period after the second world war, most currently existing buildings are in fact masonry structures. Although precise information is not available, it is estimated that traditional masonry buildings in Europe now account for 60-70% of the total number of buildings. In the developing countries of Asia and South America, the number of masonry structures is even greater. In the seismically active Mediterranean areas, there is a thousand-year-old tradition of using stone and brick as construction materials.

In the last forty years, a narrow circle of experts has been investigating the problem of seismical resistance of masonry structures. As a result of extensive experimental analyses, most questions related to the behavior of masonry structures have already been answered. The results of such analyses made on wall samples, structural assemblages, models and on full-scale buildings, enabled these experts to concieve calculation models whose behavior - when submitted to seismic action - corresponds to that of the experimentally observed phenomena.

With respect to the action of horizontal forces, masonry structures may be calculated through some general software packages used in structural design. Using such software, which is conceived to monitor elastic behavior, the walls are modelled as bars, walls or as walls with openings and, at that, the finite-element method is quite often applied. The results of such calculation are internal forces in nodes or elements, while the stress control is regularly performed in the second, separate step. The use of such software in the field of non-elastic behavior of masonry structures is not appropriate. For that reason, a specialized software adapted to the work on personal computers has been developed and is much better suited to this specific area of masonry structures calculation.

This paper focuses on the software package SAMB (Seismic Analysis of Masonry Buildings) used on personal computers, which enables seismic calculation of both new and existing masonry buildings in the process of their seismic strengthening. This software is in fact a more advanced version of an earlier program SEANZ that was developed for the use on personal computers in the course of 1985 by the second author of this paper. Some other software

---

[1] – Dražen Aničić, Professor of Civil Engineering
[2] – Dragan Radić, Assistant Professor of Civil Engineering,
   Civil Engineering Institute of Croatia, 1, Rakušina, 41000 Zagreb, Yugoslavia

packages are also used for the masonry-building calculation - they are based on calculation premises similar to those used in the software package presented in this paper /1/, /5/, /6/.

CALCULATION PREMISES

Modelling of buildings. Each building is composed of floor structures infinitely rigid in their own plane and supported by vertical plane elements - walls. The walls take on the vertical load from floors and their own weight as well as horizontal inertia forces at each floor level. Walls are modelled either as rectangular sections (without openings) or as walls with openings and, at that, the size of the opening affects the horizontal stiffness of the wall. Each wall may be one-layered (made of bricks or concrete) or two-layered (such as brick and shotcrete wall). All walls do not necessarily spread over all floors, although all walls must exist on the first floor and their number should be mentioned at the beginning of the input data list. It is not allowed to leave out a wall situated above or below a specific floor.

Position of walls is defined in the rectangular coordinate system by coordinates of initial wall points and by the angle that closes the wall with respect to the X-axis of the coordinate system. The wall geometry is established for each wall and for each floor. The geometry is defined by the floor height, wall length counting from the initial point and by the wall thickness. The size and position of wall openings is defined by the height and width of the opening as well as by its position with respect to the initial wall point.

The following material features are defined: wall density, modulus of elasticity, shear modulus, compressive strength of the wall and tensile strength (ultimate diagonal main tensile stress that causes diagonal crack formation in the column between windows). The same characteristics are also defined for the concrete, if there are concrete walls in the building. Modelling of walls composed of two materials (brick and concrete or stone and concrete) is performed by defining their material characteristics, while their joint action is ensured in the mathematical model at the floor structure levels. Properties of materials may be specified uniformly for the entire building, or for all walls on a specific floor, or separately for each wall on each floor.

The vertical load of walls is defined as an uniform load along the length of each wall at the level of each floor. These data are processed manually but the procedure is not time-consuming. The wall mass is automatically computed.

The strength of the foundation soil is taken into account through the assumed elastic springs with the defined spring constant C_z which represents the elastic constant of the vertical stiffness of the soil. The soil strength influences the rotation of the building regarded as a solid body, when submitted to the action of horizontal forces.

The calculation model presented in the form of a bar wedged in from both sides is used for calculating horizontal displacement of a wall submitted to the action of a horizontal force at its upper end. Deformations due to bending and transverse-force action are also taken into account. The floor stiffness is equal to the sum of horizontal rigidities of walls.

The torsion is taken into account in an exact manner. Eccentricity that exists between gravity of masses and gravity of stiffness is calculated for each floor. The torque equal to the product of the floor transverse force and eccentricity is confronted by bearing walls in the relationship between their horizontal stiffness and distance from the mass center.

The total displacement of a wall located on any floor is equal to the sum of displacements from the horizontal force that acts in the mass center and the displacement caused by the torque.

Modelling of seismic forces. Masonry buildings of up to 5 stories in height are characterized by short periods, i.e. their period of the first mode of oscillation amounts to T<0.4 s. The normalized spectrum of seismic acceleration of soil is constant in this period range. It is therefore allowed not to calculate oscillation periods of such buildings. Seismic forces are calculated using the method of equivalent seismic forces.

Yugoslav regulations (1981) stipulate that the total seismic force is to be calculated according to the following principle:

$$S = K \cdot W \quad (1)$$
$$K = K_o K_d K_p K_s \quad (2)$$

where

W - total mass of a building
K - total seismic coefficient
K_o - coefficient of building's importance (0.80-1.50)
K_d - coefficient of dynamic behaviour (1.0 for masonry buildings)
K_p - coefficient of plasticity and damping
K_s - coefficient of seismicity (0.025, 0.050 and 0.10 for seismic zones VII, VIII and IX)

According to the draft of the Eurocode 8, the following calculation meets better the current requirements and is more logical:

$$S = W K_s' K_d' / K_p' \quad (3)$$

where

K_s' - prescribed soil acceleration amounting to 0.10 for the seventh, 0.20 for the eigth and 0.40 for the ninth zone
K_d' - coefficient of dynamic behavior defined by the normalized design spectrum which, depending on the soil category and damping, ranges from 1.6 to 2.4. For the masonry buildings, the damping ratio normally ranges from 0.07 to 0.10.
K_p' - behavior factor varying from 1.0 to 1.5 for the non-reinforced walls, from 1.50 to 2.0 for tie-beam reinforced walls, and amounting to 2.5 for the reinforced walls. The experts have not as yet been able to provide precise values for this coefficient.

Distribution of the total seismic force along the building's height is calculated according to the following formula:

$$S_i = S \cdot \frac{Q_i H_i}{\Sigma Q_i H_i} \quad (4)$$

where

S_i - seismic force of the floor "i"
Q_i - weight of the floor "i"

H_i - height of the floor "i" counting from the top surface of foundations

Modules for calculating seismic force according to any other regulations may easily be incorporated in this software.

It is now easy to calculate internal forces affecting individual walls. Longitudinal force is automatically calculated from the defined building's geometry and the wall density, while the corresponding floor loads are defined through input parameters. Shear force for individual walls is calculated by distributing the total shear force of the S_i floor among individual walls, taking into account their horizontal stiffness. The same principle is applied for the bending moment distribution. However, the bending moment of masonry buildings is not the "cantilever moment" but rather the "frame moment", i.e. the moment we obtain by multiplying the transverse force with a half of the floor's height.

$$M_i = T_i \, 0.50 \, h \tag{5}$$

The adoption of such calculation model is based on the results of numerous experiments and observations made after seismic action (Tomažević, 1988). This method of bending moment transferring is the result of the so called "floor mechanism" of the building's response.

ULTIMATE BEARING CAPACITY OF INDIVIDUAL WALLS AND FLOORS

Ultimate bearing capacity of a wall is calculated for the defined properties of materials (wall's tensile strength), for the combined vertical and horizontal load action and for the assumed floor response mechanism. Instead of calculating the ultimate bearing capacity of the cross section (which is the usual procedure for the reinforced concrete structures) and the non-linear behavior at plastic hinges, it is the plastic behavior of the wall regarded as a whole that is modelled in the case of masonry buildings. The following formula may be applied if the failure occurs through the formation of diagonal cracks in the wall, i.e. as a result of reaching the referential tensile strength (shear failure):

$$H_u = 0.667 \, C_r \, A \, f_t \, (1 + \sigma_0/f_t)^{0.5} \tag{6}$$

The following formula is used if the failure occurs through crushing of the wall edge submitted to pressure, i.e. if it is a result of the bending action

$$H_u = 2 M_u / h = (2/h)(0.5 \, \sigma_0 \, t \, l^2)(1 - \sigma_0/f_c) \tag{7}$$

The wall failure will occur as a result of that force H_u which is weaker according to formula (6) or (7). Symbols used in formulas (6) and (7) have the following meaning:

C_r - bearing capacity reduction factor
f_t - referential tensile strength of the wall
σ_0 - normal (longitudinal) stress in the wall
M_u - ultimate bending moment
t - wall thickness

A - cross-sectional area
f_c - compressive strength
H_u - boundary horizontal force
h - floor height
l - wall length

Under the influence of the gradually increasing horizontal force, the wall attains, at a certain point, the ultimate bearing capacity. Depending on the ductility (D=1.0-1.5 for non-reinforced walls and D = 2.5 for reinforced walls), the deformation still increases under constant pressure.

The force-displacement diagram is therefore bilinear. After the wall failure, the load is distributed to other walls whose ultimate bearing capacity has not as yet been reached.

Ultimate bearing capacity of a floor is obtained by summing up ultimate bearing capacities of individual walls. A "working diagram" is defined for each wall, and the sum of all diagrams related to a single floor represents the envelope of bearing capacities of the entire floor. The smallest ultimate bearing capacity of a single floor is related to the ultimate bearing capacity of the entire building. In the case of buildings whose plan view is unsymmetrical, the floor envelope is obtained by multiple-step calculation involving 20 to 30 steps, since the new position of the stiffness center and new floor displacements (where the torsion is somewhat different when compared to the previous step) should be calculated after any single wall fails to function.

The quotient of ultimate bearing capacity and the normalized force gives us the safety coefficient for failure (of an individual wall or of an entire floor).

DEFINITION OF INPUT DATA AND ALTERNATIVE CALCULATIONS

In the SAMB program, the input data are entered in the input file. At that, a fixed format is used - as explained in the User's Guide. After the program is activated, it is ready to perform the required calculation. The read-out may be either complete or selective. At this stage, it is possible to calculate non-reinforced brick or stone walls or combined walls composed of bricks or stones and concrete (shotcrete, concrete lining etc.). It will be possible to calculate bearing capacity of reinforced walls in the second version of the program which is currently being prepared. The block diagram for calculation using the SAMB program is presented in figure 1.

When the program is used on a personal computer PC/AT 286/16, the calculation for a building up to 5 stories in height, having 12 to 15 walls, lasts approx. one minute. This enables users to perform, in a relatively short time, several alternative calculations and to compare the individual results. It can very easily be seen how some important parameters, such as tensile strength of a wall, seismic forces or wall thickness, affect (if modified) the values of safety coefficients. The elements for which an inadequate safety is proven after the first round of calculations may be strengthened in the second round of calculations or some better materials may be used. In addition, it is very easy to add new structural elements in order to increase the bearing capacity or to eliminate some existing elements so as to obtain the desired symmetry.

PROGRAM APPLICATION AND CONCLUSION

The program has already been applied in many practical situations. It is particularly useful in the preparation of seismic strengthening designs of the earhquake-damaged masonry buildings and of undamaged buildings situated in old urban centers that are due to be refurbished and strengthened.

The SAMB program is a practical and modern tool indispensable to all experts involved in calculating influence of seismic forces upon masonry structures. The authors consider that a satisfactory balance between the simplicity of entering input information, analysis method and obtained results has been achieved.

LITERATURE

1. Dražen Aničić, Miha Tomaževič, (1989), (1990), Konstruiranje i proračun zidanih zgrada, Građevinski kalendar 1990 i 1991, SGITJ, Beograd.

2. Dražen Aničić, Peter Fajfer, Boško Petrović, Antun Szavits-Nossan, Miha Tomažević, (1990), Zemljotresno inženjerstvo - visokogradnja, Građevinska knjiga, Beograd, 1-640.

3. Dragan Radić, Dražen Aničić, (1991), SAMB, kompjutorski program za seizmičku analizu zidanih zgrada, priručnik, IGH, Zagreb, 1991.

4. Miha Tomažević, (1988), Eksperimentalne osnove za proračun seizmičke otpornosti zidanih zgrada, Naše građevinarstvo, Vol. 42, 7-8, 661-672, Beograd.

5. Miha Tomažević, Polona Weiss, Tomaž Velechovsky, (1990), SREMB - računalniški program za račun potresne odpornosti zidanih zgradb, 5. seminar Računalnik v gradbeništvu, FAGG, VTOZD GG, Ljubljana, str. 268-273.

6. Antonella Pantaleo, Eugenio Rizzo, (1988), Calcolo automatico di strutture murarie, Libreria Dario Flaccovio Editrice, Palermo.

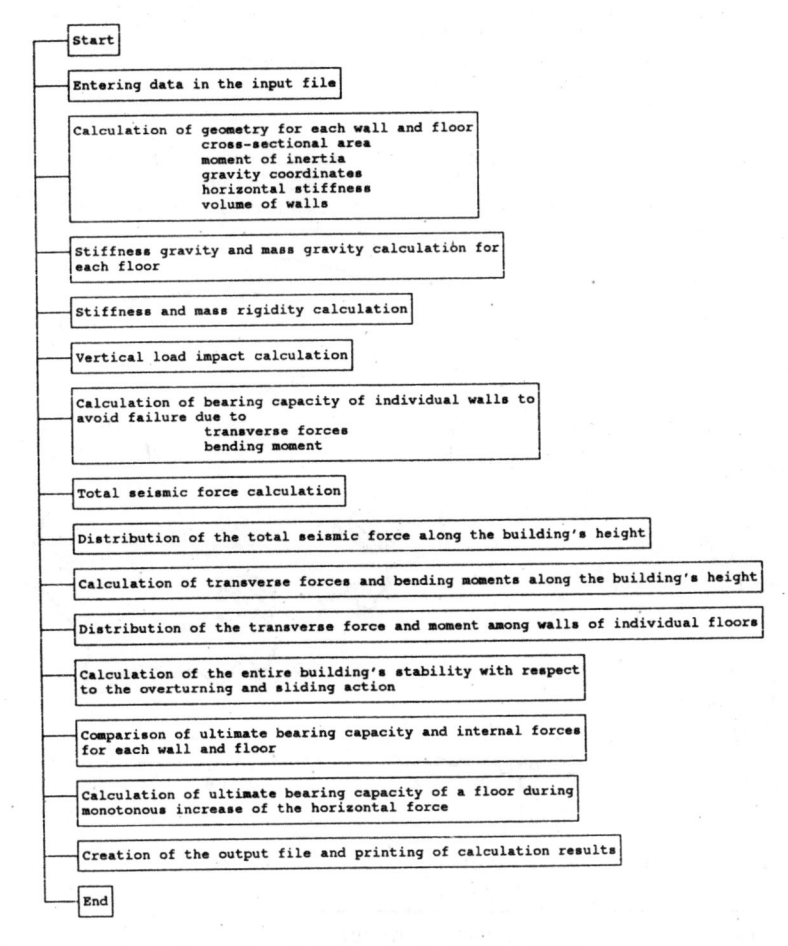

```
Start
  │
  ├─ Entering data in the input file
  │
  ├─ Calculation of geometry for each wall and floor
  │       cross-sectional area
  │       moment of inertia
  │       gravity coordinates
  │       horizontal stiffness
  │       volume of walls
  │
  ├─ Stiffness gravity and mass gravity calculation for each floor
  │
  ├─ Stiffness and mass rigidity calculation
  │
  ├─ Vertical load impact calculation
  │
  ├─ Calculation of bearing capacity of individual walls to
  │   avoid failure due to
  │           transverse forces
  │           bending moment
  │
  ├─ Total seismic force calculation
  │
  ├─ Distribution of the total seismic force along the building's height
  │
  ├─ Calculation of transverse forces and bending moments along the building's height
  │
  ├─ Distribution of the transverse force and moment among walls of individual floors
  │
  ├─ Calculation of the entire building's stability with respect
  │   to the overturning and sliding action
  │
  ├─ Comparison of ultimate bearing capacity and internal forces
  │   for each wall and floor
  │
  ├─ Calculation of ultimate bearing capacity of a floor during
  │   monotonous increase of the horizontal force
  │
  ├─ Creation of the output file and printing of calculation results
  │
  └─ End
```

Figure 1 Block diagram for calculation using the SAMB program

CORRECTION OF INTERNAL FORCES FOR COUPLED SHEAR WALLS
STRUCTURES AT POSTELASTIC STATE

A. Cholewicki[1]   J. Perzynski[2]   E. Wadecki[2]

ABSTRACT

The occurence of yielding in the lintels reinforcement and openings of the cracks in tensile zones of walls strongly influence the behaviour of loadbearing systems subjected to major earthquakes or soil tremors (the latter are frequently recorded in mining regions in Poland). A simple method of calculation of loadbearing walls with account of mentioned above nonlinearities was developed by authors. The point of the approach is the correction of the results obtained under assumption of elastic properties, this is done by increase of the lateral load (Δq method) and by increase of sectional forces and displacements (C R method).

INTRODUCTORY REMARKS

The wall structure of a building submitted to heavy horizontal loads (as it is the case in Poland with loadings originating from mining quakes, see Fig. 1) may be subject to postelastic strain/stress state.

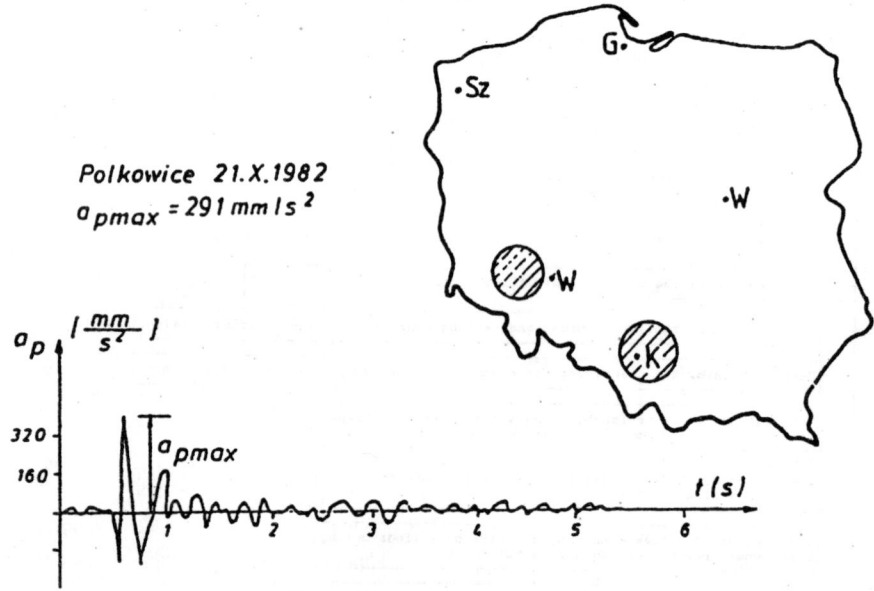

Fig. 1. Mining quakes in Poland.

1 - Building Research Institute, Warszawa
2 - Centre for Building Systems R and D, Warszawa

This is in general caused mainly by:
- lintel displacement exceeding values corresponding to lintel load capacities (most usually due to bending)
- tensile forces in the lower parts of the wall beams, mainly for prefabricated R.C. buildings affecting stresses in their compression zones (Fig. 2c).

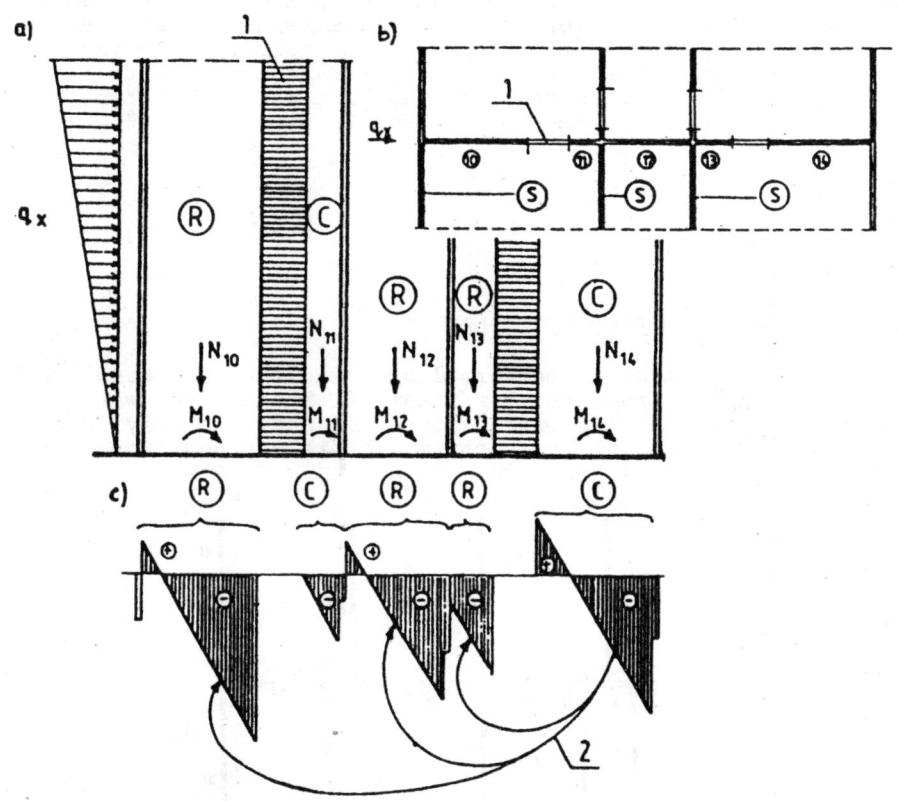

Fig. 2. Computational model (a, b). Normal stress state at the support (c).

In authors opinion while designing coupled shear walls structures one should not avoid the admittance of those above mentioned states of postelastic efforts. On the contrary, their occurrence with incidental, strongest shocks should be accepted as fact. In Poland this concerns mining shocks characterized by subsoil acceleration $a_p = 400$ mm/s^2.
Important postelastic efforts existing in the structure demand taking into consideration in practical design approach. Therefore, relatively to the basic calculations one should undertake the following supplementary steps:
- correcting the distribution of internal forces determined by means of linear calculation (for a wholly elastic structure),

- for this new internal forces distribution - create an appropriate design of the structure members and foremost its reinforcements.

The paper discusses an approximate method in executing correction procedure worked out by the authors. This method constitutes an alternative problem formulation in relation to rigorous methodics postelastic state analysis by means of finite elements method (often work consuming and expensive).

Considering the appearance of postelastic efforts in lintels the method of correction has been named "Δq" method, considering the wall beam tensile forces state - the "CR" method.

COMPUTATIONAL MODEL OF STRUCTURE

The correction is performed by using the wall beam (wall strip) model of structure. For practical reasons the wall beams should be adopted with rectangular sections only (see Fig. 3). In this model the vertical joints are positioned at wall crossing both sides.

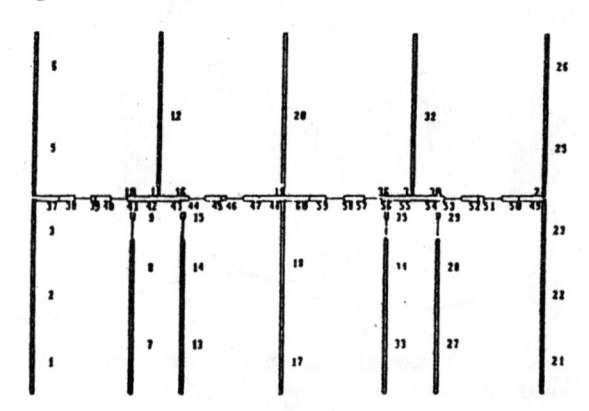

Fig. 3. Subdivision of the computational model into wall beams.

Moreover, it is assumed that:
- the structure is submitted to simultaneous action of vertical and horizontal loading,
- the design distribution of horizontal loading caused by mining shocks is triangular, with zero value at the base of structure,
- horizontal loads acting paralelly to both main axes of the building (X and Y) are analyzed in turn (separately),
- the section of the structure fixed support is placed at the floor level immediately above the underground (basement)

part of the building.

Contrary to appearances shifting fixed level downwards (e.g. to foundation level) and consequently greater effective height of the building leads to decreased efforts in the fir first above ground storey.

CORRECTION OF INTERNAL FORCES IN REGARDS TO POSTELASTIC STRESS/STRAIN CONDITIONS - THE "Δq" METHOD

The principle of the method is based on the changing of the "excess" of the wall beams coupling moment (that cannot be supported by lintels) into an additional external loading "Δq", (see Fig. 4).

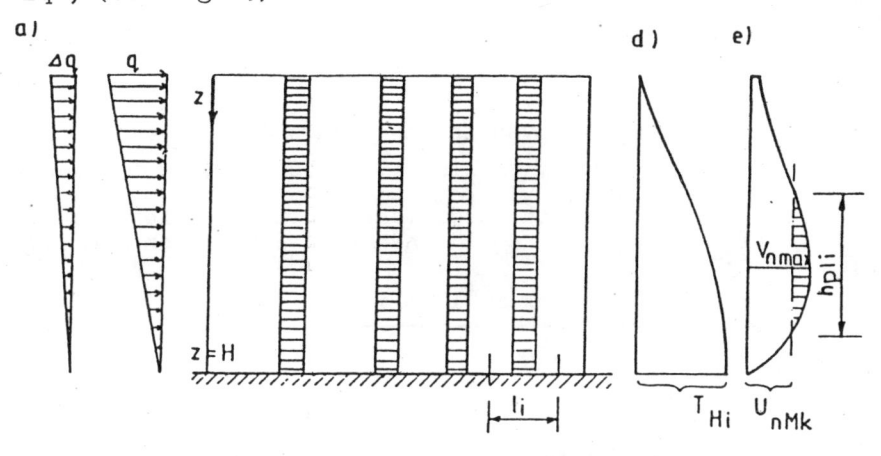

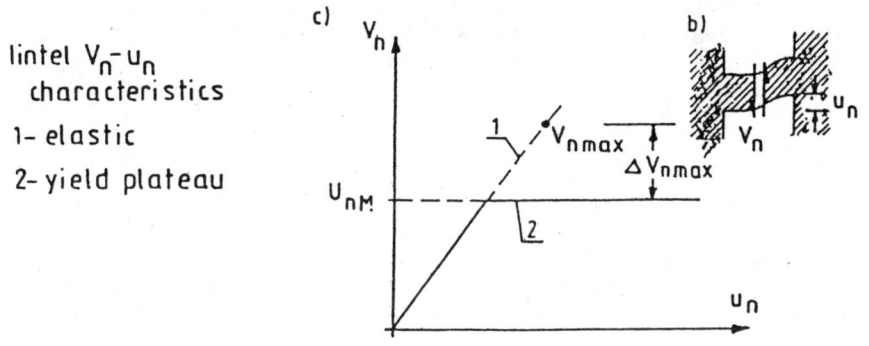

Fig. 4. Δq method.

The wall beams coupling moment is total sum of partial products of lintels shearing forces "T_{Hi}" and the distances between adjacent wall beams axes "l_i".

For each vertical row of lintels "i" where following a linear elastic calculation an excess over lintel load capacity has been stated, the increment of shearing force "$\Delta V_{n\ max\ i}$" is being calculated:

$$\Delta V_{n\ max\ i} = V_{n\ max\ i} - U_{n\ Mi}, \quad (1)$$

and then the excess of the wall beams coupling moment equals:

$$\sum_i \Delta M_{ni} = \sum_i \Delta V_n \frac{h_{pli}}{h} l_i, \quad (2)$$

where:

$V_{n\ max\ i}$ – max. shearing force a lintel of a row "i" in linear elastic model of structure,

h – storey height,

h_{pli}, l_i – dimensions given in the Figure 4a) and e)

U_{nMi} – load capacity of lintel, in bending, in the row "i" (Fig. 4c).

The top value of the additional horizontal load (of triangular distribution) could then be derived from the formula:

$$\Delta q = \frac{3 \sum_i \Delta V_{n\ max\ i}\ l_i\ h_{pli}}{h\ H^2}, \quad (3)$$

For buildings not exceeding 5 storeys formula (3) may be simplified to:

$$\Delta q = \frac{3 \sum_i \Delta V_{n\ max\ i}\ l_i}{h\ H}. \quad (4)$$

The corrected values of internal forces in the shear walls are now calculated by multiplying the proper values obtained through the linear elastic calculation by the coefficient:

$$\mathcal{H}_n = 1 + \frac{\Delta q}{q}, \quad (5)$$

where:

q – max. value of horizontal load of triangular distribution taken for linear elastic calculation.

Presently, the authors are investigating the possibility of application of $\mathcal{H}_n$ coefficient functions to determine reduced structure rigidity (due to postelastic behaviour).

CORRECTION OF INTERNAL FORCES IN REGARDS TO WALL BEAMS TENSILE FORCES STATE – THE "CR" METHOD

The principle of the method is an appropriate change of the wall beams bending moments, determined from the elastic analysis, as the result of the appeareance of "uncontrolled" tensile stress state at least in one wall beam parallel the direction of structure horizontal loading (see Fig. 2c). There are two types of wall beams specified for given horizontal loading direction (Fig. 5) i.e:

" C " type - supporting compression only, unable to resist tension ("uncontrolled" tensile zone),

" R " type - supporting both compression and tension, of generally linear elastic characteristics ("controlled" tensile zone).

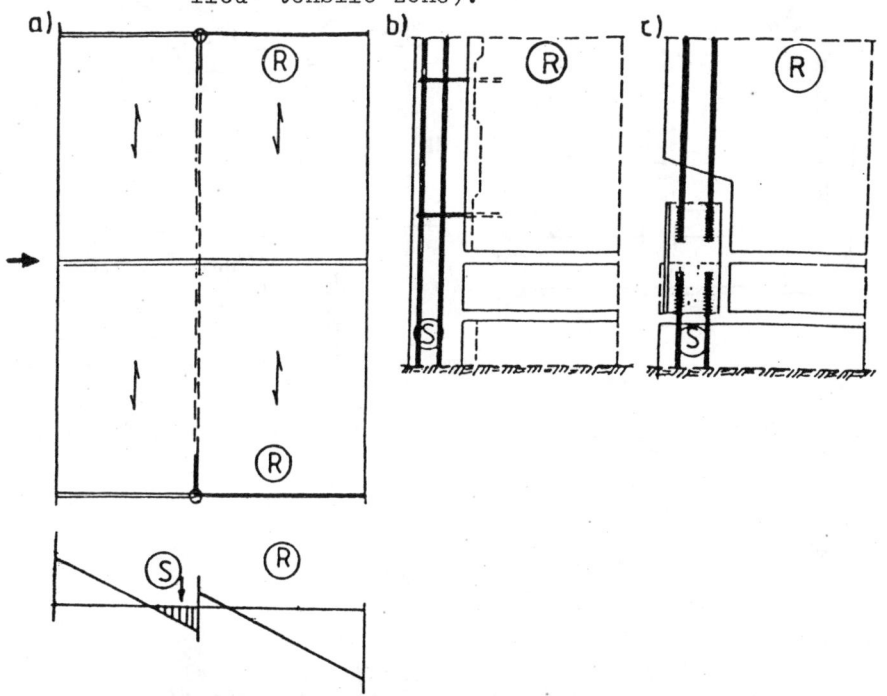

Fig. 5. R type wall beams.

The principles of R type wall structure (and consequently their design) could be various, for instance commonly used solutions are:
- connection by rigid vertical joint (there are universally used keyed joints) to a part wall under compression, (Fig. 5a) or to a perpendicular wall submitted to vertical loading balancing tensile forces in R wall beam (see wall denoted by symbol "S" in Fig. 2b),
- connection by a rigid joint to the continuous vertical tie beam (properly reinforced) (Fig. 5b),
- sufficient vertical reinforcement in the presumed tension zone (Fig. 5c).

The CR method may be considered as an approximate, qualitative and quantitative, description of the phenomenon of bending moments redistribution for coupled shear walls structure computed as linear elastic model structure.

The original bending moments are being redistributed from C type to R type wall (Fig. 2c). It is assumed that the tension zone of R beams, or their connections with the adjacent compressed walls are rigid enough that one may ignore the influence of R beams crackings upon the increase of maximum stresses, designated as $\sigma_{iM\ max}$ (Fig. 6d).

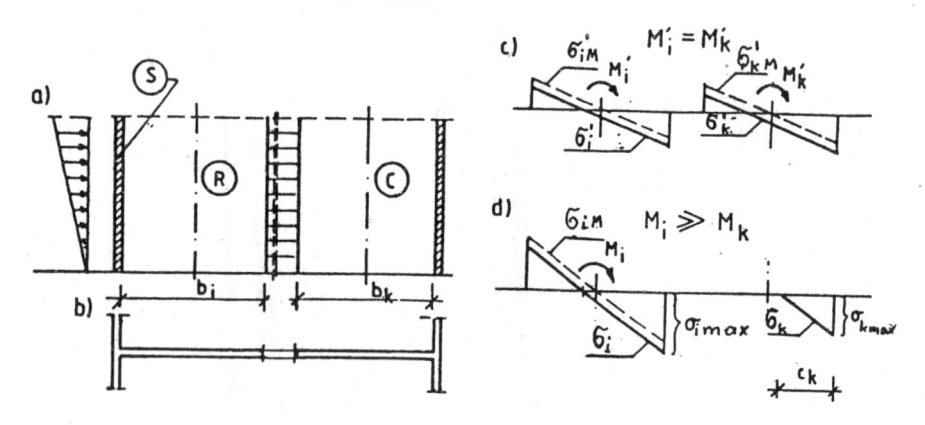

Fig. 6. Normal stresses state c) before correction, d) after correction.

The basic equation of CR method has the form:

$$\sum_i M_i' + \sum_k M_k' = \sum_i M_i + \sum_k M_k, \qquad (6)$$

where:

M_i', M_k' - values of wall beams moments, respectively R and C type, determined for the linear elastic structure model,

M_i, M_k - as above but after redistribution.

The equation (6) expresses an equality of total of wall beams bending moments before and after redistribution. The final moments distributions are being determined with regard to the following assumptions:

M_i, M_k - bending moments M_i and M_k (respectively in R and C type wall beams) are mutually proportional to beams sections stiffnesses, i.e.:
original section stiffnesses for R beams and reduced to compression zone stiffnesses in case of C beams. In the latter case the reduction of wall beam rigidity corresponds to ratio, (see Fig. 6d):

$$\left| \frac{c_k}{b_k} \right|^3,$$

- the distribution of stresses within compression zone of C type beam is triangular, (Fig. 6a)

and then

$$c_k = \frac{2 N_k}{h_{vk} \sigma_k}, \qquad (7)$$

where:

N - axial normal force in C type wall beam,

h_{vk} - wall beam thickness,

σ_k - stress at C beam's compressed edge (Fig. 6d)

- axial normal forces in wall beams, before and after redistribution remain constant, thus:

$$N_k = N_k' \qquad (8)$$

The sum of moments appearing on the left side of the equation (6) is directly determined through the initial calculations. In case of a structure with uniformly planned walls disposition this sum must be derived exclusively from horizontal loads (the vertical loadings have no prectical influence on the wall beams coupling moments). On contrary, the axial forces "N_k" within C type beams must be determined taking into consideration both vertical and horizontal loading.

The values of moments on the right side of the equation (6) may be conveniently expressed as a function of unknown "σ^*". Edge compression stress "σ^*" means the corrected value (after redistribution) for a comparative wall beam "R" (of width "b^*") chosen from among R type wall beams (the largest of the R beams is recommended).

The corrected bending moments in R and C type wall beams may then be expressed as follows:

$$M_i = \frac{h_{vi} b_i^3}{6 b^*} |\sigma^*|, \qquad (9)$$

$$M_k = \frac{1}{6} \sqrt{\frac{N_k^3 b^*}{h_{vk} \sigma^*}}, \qquad (10)$$

where:

h_v - wall beam thickness, indexes "i" correspond as previously to R beams and indexes "k" to C beams.

Using of these relationships and applying a substitute unknown $\alpha = \sqrt{|\sigma^*|}$, one may bring the equation (6) to the regular form:

$$\alpha^3 + p_1 \alpha + p_2 = 0. \qquad (11)$$

This equation (11) is being solved by habitual methods contained in all mathematic handbooks. Thus, it is coveniently

to solve it by the trigonometric method.
After finding value G^*, one calculates in turn:
- values of compressed zones widths in beams C:

$$c_k = \sqrt{\frac{N_k\, b^*}{h_{vk}\, G^*}}, \qquad (12)$$

- eccentricity of normal forces actions in relation to compressed zones of beams C axes (Fig. 5a)

$$e_k = \frac{b_k - c_k}{2}, \qquad (13)$$

and
- a coefficient expressing the influence of the vertical loading acting on the eccentricities (13) within the C wall beams subjected to cracking

$$\mathcal{X}_e = 1 - \frac{3\sum_k N'_k\, e_k}{q\, H^2}. \qquad (14)$$

All these operation executed, one may proceed to calculate the stress distribution pattern within the beams. Stress values at the compressed edges of R wall beams amount to:

$$G_i = G^* \frac{b_i}{b^*}\, \mathcal{X}_n\, \mathcal{X}_e + \frac{N'_i}{h_{vi}\, b_i}, \qquad (15)$$

Stress values at compressed edge of the C wall beams amount to:

$$G_k = -2\sqrt{\frac{G^*\, N'_k}{h_{vk}\, b^*}}\, \mathcal{X}_n\, \mathcal{X}_e, \qquad (16)$$

where:

$\mathcal{X}_n$ - a coefficient calculated according to formula (5).

If the total width of a given C beam is compressed one should in place of N_k put into the formula (10) the value

$$N'^o_k = \frac{M'_k}{W_k}\, h_{vk}\, b_k, \qquad (17)$$

where:
W_k - indicator of section modulus

h_{vk}, b_k - appropriate wall beam thicknesses and widths.

A condition to keep the structure in balance whole, with wall beams partially under tension is the transmitting R wall beams tension forces to adjacent walls. These are mainly stabilizing walls perpendicular to mentioned R wall beams.

COMPARISON OF APPROXIMATE AND PRECISE CALCULATION RESULTS

The results obtained by means of the discussed correction method have been verified by calculations performed by means of nonlinear FEM method (program ROZA 2). The object of investigation was prefabricated R.C. wall building (Fig. 7a) of technology W70 SG system, being submitted to the design horizontal static loading along the longitudinal direction.

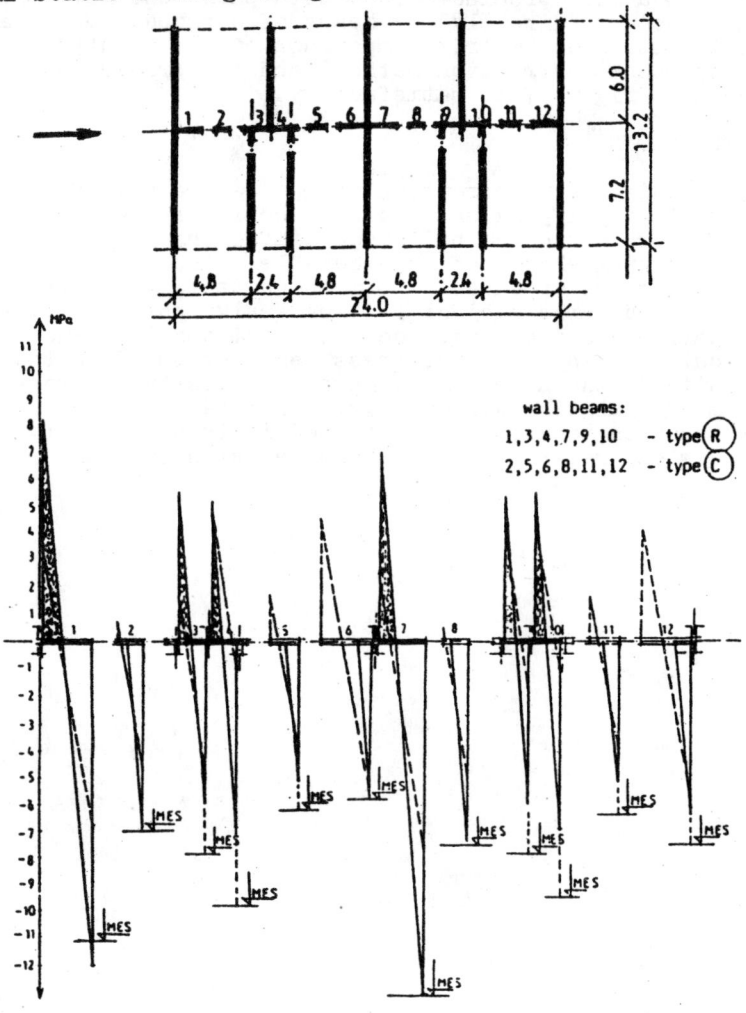

Fig. 7. Example of the comparison of exact results and of the approximate ones (----- results acc. to elastic analysis before correction, ——— results acc. to elastic analysis after correction).

In a approximate method a correction of normal stress distribution was performed taking into account two phenomena:
- yelding of lintels in the main longitudinal wall (method Δq),
- appearance of tension state within C wall beams of this wall (method C R).

In a more precise method both phenomena have been taken into account "directly" by means of a proper modifications of the stiffness matrix comparison of the results according to approach (correction method) and the precise one (FEM) shown in Figure 7 is satisfactory.

CONCLUSIONS

1) The discussed correction method can be a useful design tool simplifying and reducing costs of the computation of the prefabricated buildings stiffening structures exposed to mining shocks of considerable intensity.
2) Making use of CR and Δq methods it is possible to compute structures subject to progressive postelastic efforts wall beams cracking, their connections (lintels) plastical yielding, and stiffness degradation of vertical joints without the necessity of applying costly and work consuming analytic methods (e.g. non-linear FEM).
3) Based on results thus obtained it is possible to take decisions regarding the reinforcement arranging.

REVIEW AND COMPARISON OF METHODS FOR EVALUATING INELASTIC DESIGN SPECTRA

L. Galano[1]   A. Vignoli[1]

INTRODUCTION

In seismic design, the response spectrum method is one which is often employed. For structures having plastic resources, inelastic design response spectra (IDRS), derived from linear elastic design response spectra (LEDRS) through certain criteria, are used. In the literature there are many proposals: in this work we refer to the methods briefly illustrated here as follows. For an elastic, perfectly-plastic system (EPP) in the first method (fig. 1), known as the linear equivalent system method, the equivalent damping ν_{eq} and the natual period T_{eq} parameters are evaluated with reference to a linear system /1/, /2/; the equivalent system then replaces the real system, and the seismic actions are derived directly from LEDRS. The equivalent parameters are given by:

$$\nu_{eq} = \nu + \left[\mu_d - 1 - \ln \mu_d \right] \frac{2\,r}{\pi\,\mu_d} \qquad [1]$$

$$T_{eq} = T \left[1 + \frac{2}{3} \sqrt{\mu_d} - \frac{2}{3\mu_d} \right], \quad \mu_d = u_{max}/u_y \qquad [2, 3]$$

where ν is the viscous damping of the real system, r is an ad hoc reduction factor to take into account the non-ideality of the hysteresis loops ($r \leq 1$), and μ_d is the design ductility factor. This method is based upon the equality of the hysteretic energy dispersed through the EPP system with the viscous damping energy from the equivalent system. In this work we also consider the ductility factor method and the Newmark-Hall method. In the former, it is assumed that the maximum displacements for the elastic structures and for the EPP structures are the same for a given earthquake (fig. 2); in this way, the reduction factor of the design actions, and therefore of the spectral values, is equal to the pre-established ductility μ_d, namely:

$$f_y = f_{max} / \mu_d, \qquad \mu_d = u_{max} / u_y \qquad [4]$$

TABLE I: Newmark-Hall method for IDRS

T	S_{dp}	S_{vp}	S_{ap}
$T \geq T_2$	S_d	$S_v / \sqrt{\mu}$	S_a / μ
$T_1 \leq T < T_2$	$S_d\,\mu/\sqrt{2\mu-1}$	S_v	$S_a / \sqrt{2\mu-1}$
$T < T_1$	$\mu\,S_d$	$S_v\,\sqrt{\mu}$	S_a

---

1 - Department of Civil Engineering, University of Florence, Italy

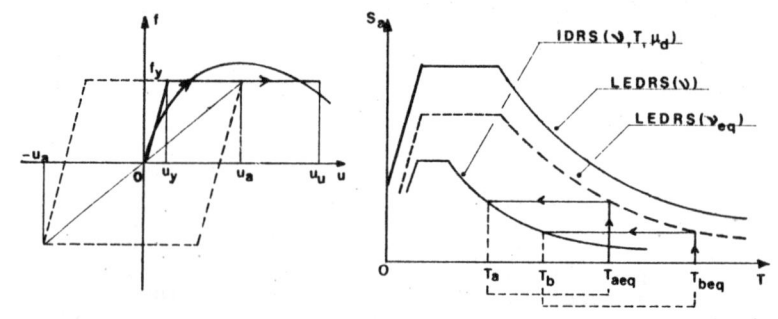

Fig. 1 - EPP model used in the equivalent linear system method. Deduction of IDRS from LEDRS

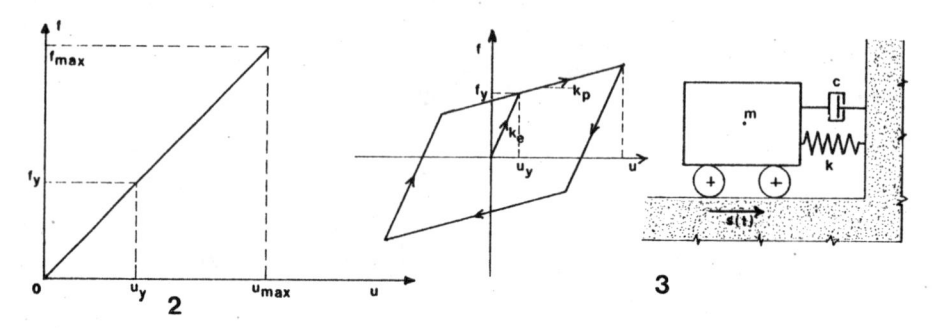

Fig. 2 - Ductility factor method
Fig. 3 - Model employed in the analysis: displacement-force relation

In the Newmark-Hall method /3/, the elastic spectral values S_a are reduced by the ductility factor μ_d into the three divisions in which the spectrum is divided, as can be seen in Table I. These relations are derived from the criteria of displacement equality ($T \geq T_2$), of velocities ($T_1 \leq T < T_2$) and of accelerations ($T < T_1$) for the EPP real system and for the linear equivalent system.

NUMERICAL COMPUTATIONS

Our work is limited to the case of a single-degree-of-freedom system with linear viscous damping and hysteretic behaviour, as can be seen in fig. 3. The normalized elastic design spectra to which we refer are the Eurocode 8 spectra /4/; they are analytically described by the follow relations (fig. 4):

$$S_a = 1 + (\beta_0 - 1)s\, T/T_1 \quad \text{for } 0 \leq T < T_1 \qquad [5]$$

$$S_a = \beta_0\, s \quad \text{for } T_1 \leq T < T_2 \qquad [6]$$

$$S_a = \beta_0\, s\, (T_2/T)^k \quad \text{for } T \geq T_2 \qquad [7]$$

The values of T_1, T_2, β_0, k and s for 5 % viscous damping are shown in Table II; they are provided in the aforesaid code as first-indication values.

TABLE II: Eurocode 8 elastic response spectra parameters

Geotechnical soil profile	s	T_1 [s]	T_2 [s]	k	β_0	d_0 [cm]
A	1	0.2	0.4	1	2.5	60
B	1	0.2	0.6	1	2.5	90
C	0.8	0.3	0.8	1	2.5	120

In fig. 5 are shown the inelastic design spectral accelerations for geotechnical soil profile A and for design ductility μ_d variables from one (elastic case) to ten. From this comparison it can be seen that, for low natural frequencies, the first method is unsafe with respect to the others, while for high frequencies it is the most secure. To investigate the performance of the three classical methods proposed, fifteen earthquakes, simulated by means of SIMQKE, a computer generation program /5/, were generated; these accounts have a duration of 20 s, and are digitalized with a constant sampling time of 0.01 s. The peak ground accelerations are equal to 0.35 g. In the numerical investigations the following design constants were pre-established (fig. 3):

K_e = 100 KN/cm   elastic stiffness of system
K_p = 25 KN/cm   plastic stiffness of system
ξ = 0.05   viscous damping ratio

Many numerical step-by-step dynamic integration analyses, using the explicit Newmark algorithm, were developed; the integration step employed was 0.01 s, and the natural period of the systems was variable from the bottom value T_{inf} = 0.1 s to the top value T_{sup} = 3 s, with discretization step of 0.1 s. In this way, the maximum displacement error in some critical cases was no greater than 1 % of the same value. For a pre-established natural period T and then for a given mass m of the system, the yielding threshold f_y was derived from inelastic spectra for the following increasing ductility values:

μ_d = 1, 2, 4, 6, 8 e 10

Fig. 4 - Eurocode 8 normalized elastic spectra

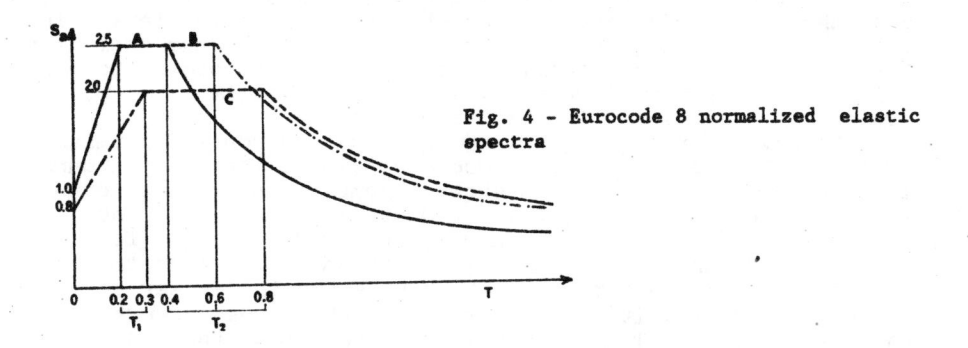

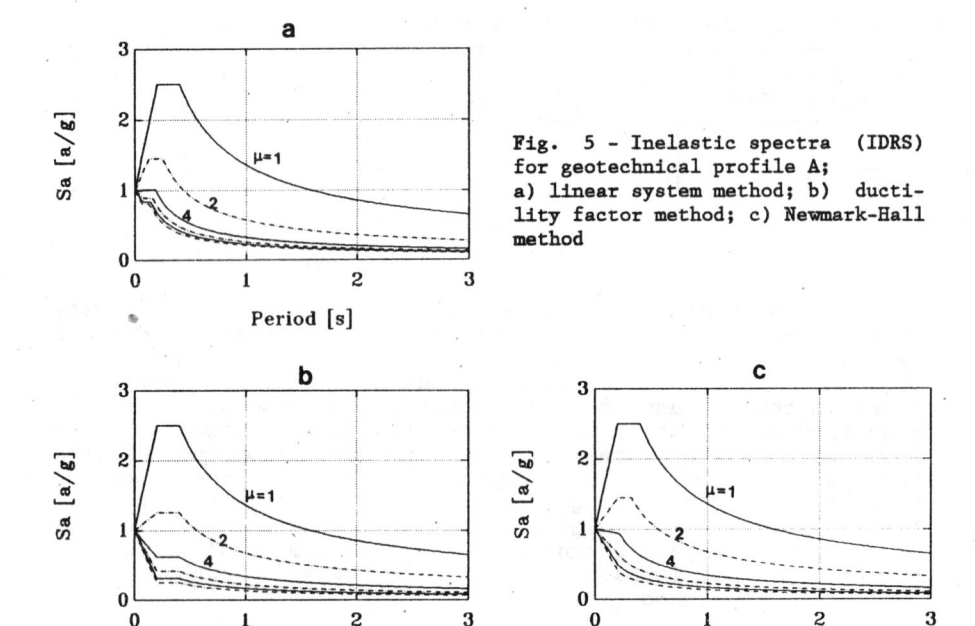

Fig. 5 - Inelastic spectra (IDRS) for geotechnical profile A; a) linear system method; b) ductility factor method; c) Newmark-Hall method

A first evaluation of the performance of the three methods under consideration can be developed from a comparison of the design ductility μ_d values with the ductility μ_r values, required by the system subject to seismic input.

DISCUSSION OF THE RESULTS

Influence of system frequency

At first a great variability is noted, especially in the required ductilities μ_r with the natural period T of the system. The design ductility factor μ_d is sometimes considerably exceeded; this trend is particulary evident for higher values of μ_d (fig. 6). The ductility factor and the Newmark-Hall methods provide better performances compared to the former due to the lower natural frequencies of the system (T $\geq\approx$ 1 ÷ 1.5 s), and the differences are reduced when the ductility level increases. For the higher frequencies (T $\leq\approx$ 0.4 s), the ductility factor method systematically shows a peak corresponding to the value of the T_1 period. The Newmark-Hall method is shown to be preferable, especially for high frequencies.

Moreover, the variability of the responses obtained, compared to the 5 simulated accelerograms of the same groups, is very great; in fig. 7 is shown a comparison of the events for spectrum A for the levels of ductility $\mu_d = 8$ and for the second method. The results show a greater dispersion for low frequencies, and tend to increase with an increase in the values of μ_d; the Newmark-Hall method presents the lesser overall dispersion of required ductility μ_r, with values in very good agreement for T $\leq T_1$ and not very different in the subsequent

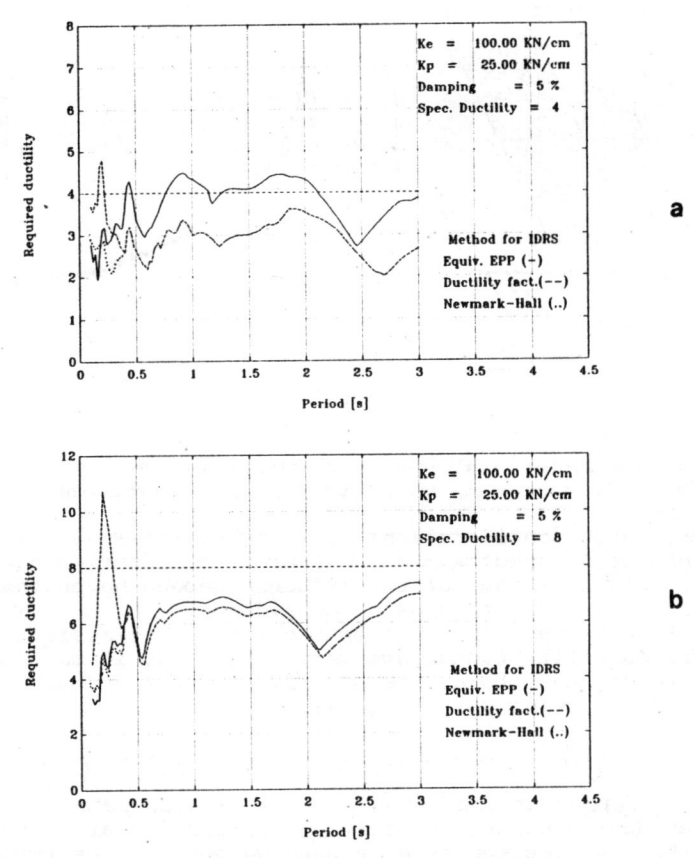

Fig. 6 - Comparison of the results obtained with the three methods: a) μ_d = 4, accelerogram E5, b) μ_d = 8, accelerogram E5

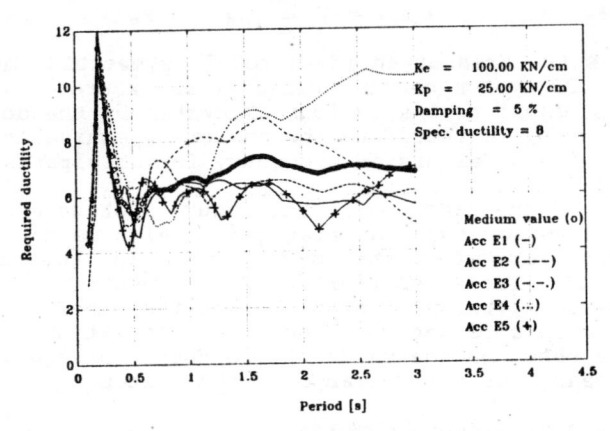

Fig. 7 - Comparison of the results computed for the 5 A-type spectrum accelerograms: ductility factor method

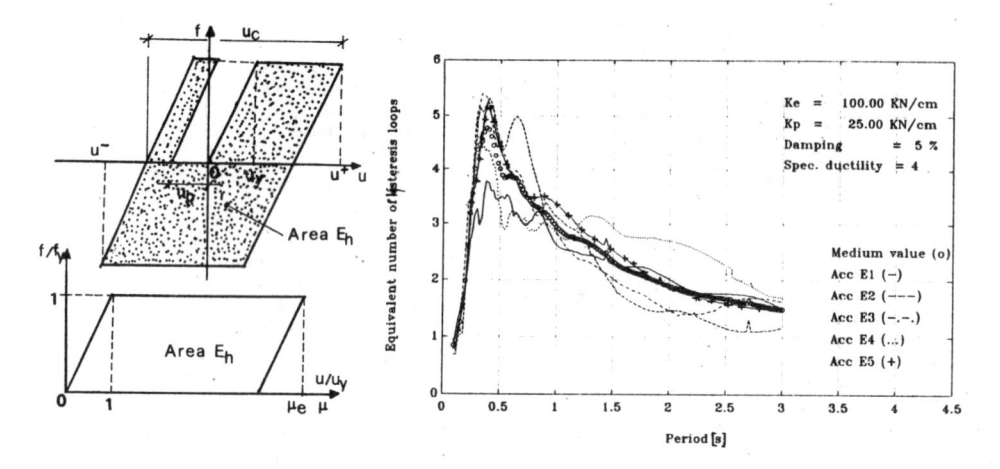

Fig. 8 - Definition of alternative ductility factors
Fig. 9 - Comparison of n_e values for the 5 accelerograms of group A

range. These considerations can also be repeated for the other two groups of spectra B and C (Eurocode), confirming therefore the random behaviour of the response depending on the particular seismic excitation being considered.

Because the results computed with the classical definition of the ductility factor are very scattered, it may be appropriate to define alternative ductility factors. Some definitions are the following /6/, (fig. 8):

$$\mu_c = u_c/u_y, \qquad \mu_p = u_p/u_y, \qquad \mu_e = 1 + E_h/(f_y u_y) \qquad [8]$$

E_h represent the total hysteretic dissipated energy of the system during seismic excitation. In particular, with reference to the devices employed in passive energy absorption design /7/, it is possible to consider the following equivalent ductility factor parameter, in connection with E_h and μ_d:

$$n_e = E_h/A_{ci}, \qquad (A_{ci} = 4 u_y (\mu_d - 1) f_y) \qquad [9]$$

where A_{ci} is the area of an ideal total symmetric hysteresis loop of the EPP system characterized by an excursion in a plastic field μ_r equal to the maximum allowable by the device. The defined factor n_e is therefore the number of equivalent hysteresis loops with μ_d excursion necessary for dissipating a total energy equal to E_h.

For one of the cases analyzed, figure 9 shows a comparison of the n_e values for the 5 accelerograms relative to the A-type spectra, together with the mean curve. In the cases illustrated, the lesser scattering of the results is evident. This circumstance, which is also true of the other cases analyzed, suggests an alternative method for calculation of the plastic spectra based on the value of the number of equivalent n_e cycles necessary for the collapse of the system.

Influence of viscous damping

We consider this influence by modifying the normalized elastic spectra for 5 % damping, with the following empirical factor (present in many seismic codes):

$$\eta = \sqrt{5/\nu}$$ [10]

By calculating in this manner, we consider that viscous damping influences the response in the non-linear field in the same way as it does in the linear elastic field. An examination of the diagrams in fig. 10, relative to the ductility $\mu_d = 4$, reveals that the ductility-factor method tends to underrate the design structure actions when ν increases, thus confirming that the importance of viscous damping is overrated. With the linear equivalent system method, the differences are lesser, especially for high frequencies; and, contrary to the previous case, the required ductilities decrease as the damping coefficient increases. This method demonstrates, therefore, its greater adaptability for design uses involving cases in which structural damping factors greater than 5 % must be taken into account for safety's sake.

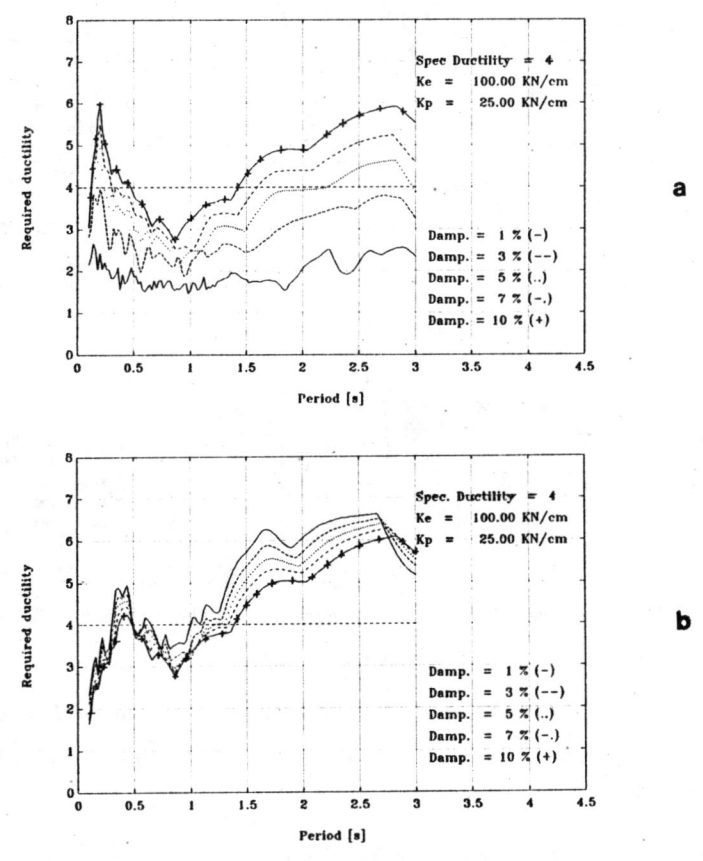

Fig. 10 - Required ductility for different damping ratios; a) ductility factor method; b) equivalent linear system method

Influence of the structural constants K_e e K_p (fig. 3)

The results of this part of the numerical analysis show a considerable dispersion of the response: e.g. fig. 11 shows that strain hardening is beneficial for certain frequencies while being actually unfavourable for others. Similar considerations can be made for the stiffness K_e; from figure 11, it is also evident that, for every natural period of the system, there is a different K_e value to minimize the required μ_r ductility. It is not possible, therefore, to compute optimal values for the K_e constant and for the s.h. constant K_p/K_e; but their choice is dependent on the seismic excitation and on the natural T period.

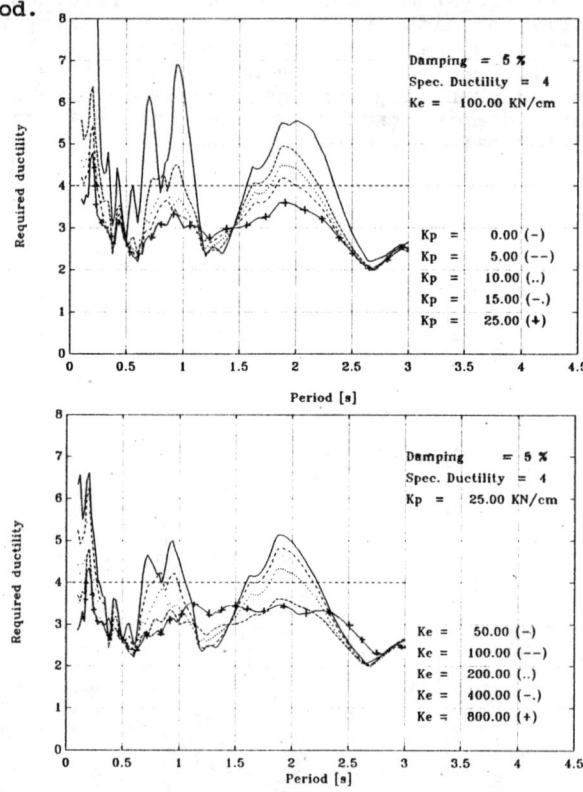

Fig. 11 - Required ductilities according to the different K_e and K_p constants

EVALUATION OF THE q BEHAVIOUR FACTOR

An alternate method for evaluating the efficiency of the criteria for inelastic spectra consists of a computation of the behaviour factor. According to the Eurocode 8 definition, this computation is made by:

$$q = a_u/a_y \qquad [11]$$

where a_u is the peak ground acceleration for a pre-established seismic event corresponding to the collapse of the system, and a_y is the same value for yielding. The q factor was computed by assuming attainment of the μ_d value to represent the collapse criterion for the system. The optimal criterion for IDRS is the one for which $q = \mu_r$ results, so that all the plastic resources of the system are exploited. In Table III, the mean values of ratio q/μ_r for the five accelerograms derived from spectrum A are visible as functions of period T and for ductility factor $\mu_d = 4$. This comparison shows that the equivalent system method is the one which presents the lowest dispersion and the best q/μ_r ratio values; in most cases, this ratio is q > 1, which means that the methods examined are sufficiently secure, at least at the mean level, in a group of simulated earthquakes. In fig. 12, the $q - \mu_r$ diagram in a typical case is shown.

TABLE III: medium values of q/μ_r, spectrum A, $\mu_d = 4$.

Method T [s]	Equiv. system q/μ_r	var.	Duct. factor q/μ_r	var.	Newmark-Hall q/μ_r	var.
0.10	1.33	0.13	0.94	0.10	1.27	0.13
0.20	1.52	0.13	0.95	0.07	1.55	0.13
0.30	1.39	0.04	1.39	0.05	2.00	0.04
0.40	1.34	0.26	1.87	0.50	1.87	0.50
0.50	1.96	0.31	2.48	0.43	2.48	0.43
0.60	1.55	0.19	2.11	0.40	2.11	0.40
0.70	1.42	0.11	1.87	0.21	1.87	0.21
0.80	1.43	0.22	1.91	0.28	1.91	0.28
0.90	1.39	0.15	1.84	0.16	1.84	0.16
1.00	1.35	0.25	1.77	0.31	1.77	0.31
1.10	1.20	0.10	1.65	0.13	1.65	0.13
1.20	1.25	0.11	1.73	0.22	1.73	0.22
1.30	1.23	0.14	1.69	0.24	1.69	0.24
1.40	1.09	0.11	1.51	0.23	1.51	0.23
1.50	1.12	0.13	1.56	0.24	1.56	0.24

CONCLUSION

The numerical analyses developed are suited to systems provided with special seismic-isolation devices and with energy dispersion devices having elastoplastic-type F-u response. Generally, the three methods under consideration (linear equivalent system, ductility factor and Newmark-Hall) are unsatisfactory only for high values of design ductility μ_d. Another evident result is the considerable dispersion of the analytical ductilities required of the system with the different simulated earthquakes (seismic excitations with the same duration, peak values and frequency spectrum). The q behaviour-factor computations show that the linear equivalent system method furnishes the lowest dispersions and the best ratio q/μ_r values between the reduction factor of the spectral ordinates q and the required ductility μ_r. These results confirm the great difficulties encountered in seismic non-linear structure design. As is well known, the use of the response spectrum method as the mean spectrum assumed in the time domain step-by-step dynamic analysis gives rise to results which are random according to the variations of the particular seismic excitation employed. From this work it has been found that the best criterion for an evaluation of inelastic spectra (IDRS), and therefore for the simple system considered here, consists of the use of an alternative ductility reduction factor of seismic action, so that it is possible to reduce the dispersions: in this case, the number of equivalent n_e cycles provides good results.

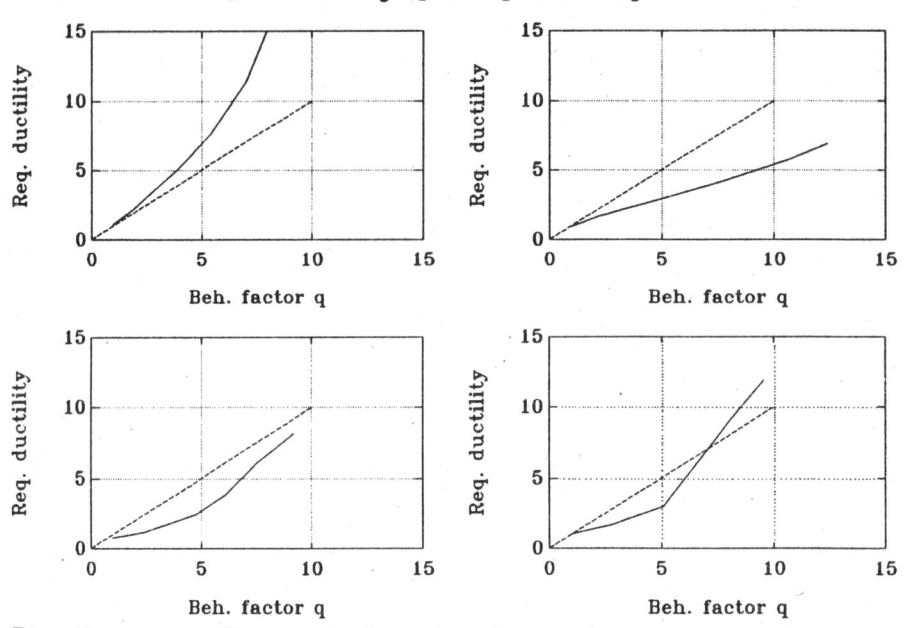

Fig. 12 - q - μ_r diagrams in four tipical cases

REFERENCES

/1/ IWAN W.D., <<The Response of Simple Degrading Structures>>, Proceedings, Sixth World Conference on Earthquake Engineering, Vol. II, January 1977, pp. 1094-1099.

/2/ IWAN W.D., GATES N.C., <<The Effective Period and Damping of a Class of Hysteretic Structures>>, Journal of the International Association of Earthquake Engineering, Vol. 7, n° 3, May-June 1979, pp. 199-221.

/3/ NEWMARK N.M., HALL W.J., <<Procedures and Criteria for Earthquake Resistant Design>>, Building Practice for Disaster Mitigation, Building Science Series 45, National Bureau of Standards, Washington, D.C., February 1973, pp 209-236.

/4/ EUROCODE 8, <<Seismic Actions for Structures>>, Part 1, Geneneral Part and Buildings.

/5/ SIMQKE: <<A Program for Artificial Motion Generation>>, Users's Manual, Department of Civil Engineering, Massachussets Institute of Technology, November 1976.

/6/ MAIHN S.A., BERTERO V.V., <<An Evaluation of Inelastic Seismic Design Spectra>>, Journal of the Structural Division, ASCE, n° ST9, September 1981, pp. 1777-1795.

/7/ SKINNER R.I., TYLER R.G., HEINE A.J., ROBINSON W.H., <<Hysteretic Dampers for the Protection of Structures from Earthquakes>>, Bulletin of the New Zealand Seismic Nat. Society for Earthquake Engineering, Vol 13, n° 1, March 1980.

An Equivalent Column in Lateral Load Analysis of Pile-Structure Interaction

J.Gluck[*]   M. Sternik[†]

1 Introduction

A very popular foundation method for structures are in-situ casted reinforced concrete piles fixed at the bottom in the rock subgrade. Vertical loads are transmited by the pile by friction on the mantle of the lateral contact area between the pile and the rock. In transmiting horizontal forces the piles are laterally restrained columns fixed at their bottom and conected at the top to horizontal beams and columns. For structural analysis purposes the lateral restraint of the column may be assumed as a Winkler elastic medium.

In lateral load analysis of the structure commonly used practice is a uniform distribution of the horizontal load between the piles and then the individual piles are analysed for this horizontal load assuming at the top a horizontally sliding fixed or articulated end. This method of lateral loads distribution is acceptable for piles having approximately equal length. In case of piles having different lengths as in sloped rock subgrade a more refined analysis is required. The horizontal stiffness at the top of a horizontally sliding

---

[*]Professor in Civil Engineering, Technion–Israel Institute of Technology, Haifa, Israel.

[†]Graduate student, Faculty of Civil Engineering, Technion–Israel Institute of Technology, Haifa, Israel.

fixed or articulated end may be calculated [1] and the horizontal load is distributed proportional to this stiffnesses. This distribution do not take into account the interaction between the piles and structure. To include such a interaction in the analysis a standard computer code for structural analysis may be used, but this kind of programs do not include as a standard element the beam on elastic subgrade. The laterally restrained column may be modeled in this case by subdividing the column and introducing lateral spring at the joints. The stiffnesses of the springs represent the adequate Winkler elastic medium. This kind of modeling may be time consuming when a structure is modeled as three dimensional.

In this paper a simple method is presented in which the laterally restrained column is modeled as a laterally unrestrained column having equivalent geometric properties; thus a standard computer code with no beam on elastic foundation may be used since the pile is modeled as a column with no lateral restraints.

2   Formulation of the equivalent pile model

The stiffness matrix for a two dimensional finite beam on Winkler elastic foundation was developed in [1]. For a laterally restrained column, having only two degrees of freedom at the top and fixed at the bottom, the stiffness matrix is given by (see fig. 1):

$$\overline{S} = \begin{bmatrix} \overline{S}_{11} & \overline{S}_{12} \\ \overline{S}_{21} & \overline{S}_{22} \end{bmatrix} \quad (1)$$

where the stiffness coefficients $\overline{S}_{ij}$ can be expressed as

$$\overline{S}_{11} = \frac{\sqrt{2}E_s T A_1(\alpha)}{C(\alpha)} \quad (2)$$

$$\overline{S}_{12} = \frac{E_s T^2 A_2(\alpha)}{C(\alpha)} \quad (3)$$

$$\overline{S}_{22} = \frac{\sqrt{2}E_s T^3 A_3(\alpha)}{C(\alpha)} \quad (4)$$

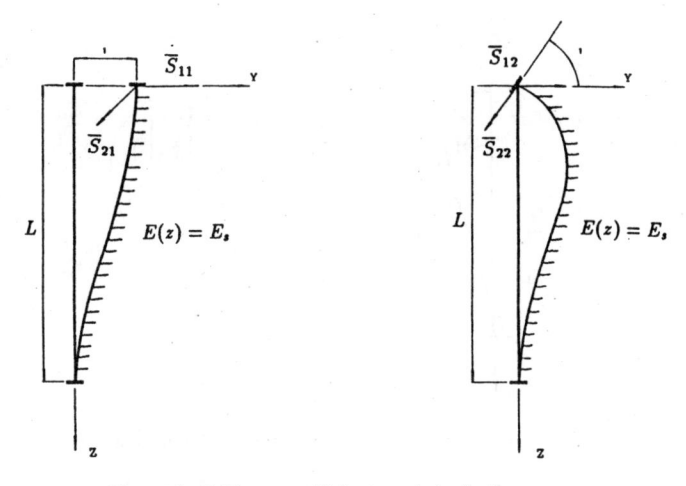

Figure 1: Stiffness coefficients – Actual pile.

in which

E_s = foundation modulus. Units: $\dfrac{F}{L^2}$

T = relative stiffness factor = $\sqrt[4]{\dfrac{E_p I}{E_s}}$

$\alpha = \dfrac{L}{T}$

L =real length of the pile

$A_1(\alpha) = \cosh(\tfrac{\sqrt{2}}{2}\alpha)\sinh(\tfrac{\sqrt{2}}{2}\alpha) + \cos(\tfrac{\sqrt{2}}{2}\alpha)\sin(\tfrac{\sqrt{2}}{2}\alpha)$

$A_2(\alpha) = \sinh^2(\tfrac{\sqrt{2}}{2}\alpha) + \sin^2(\tfrac{\sqrt{2}}{2}\alpha)$

$A_3(\alpha) = \cosh(\tfrac{\sqrt{2}}{2}\alpha)\sinh(\tfrac{\sqrt{2}}{2}\alpha) - \cos(\tfrac{\sqrt{2}}{2}\alpha)\sin(\tfrac{\sqrt{2}}{2}\alpha)$

$C(\alpha) = \sinh^2(\tfrac{\sqrt{2}}{2}\alpha) - \sin^2(\tfrac{\sqrt{2}}{2}\alpha)$

These equations will be applied to a pile embedded in an ideal homogeneous elastic soil, having a subgrade coefficient E_s which does not vary with depth.

The stiffness matrix for the equivalent laterally non-restrained column having only two degrees of freedom at the top (see fig. 2) is given by [2]:

$$S = \begin{bmatrix} S_{11} & S_{12} \\ S_{21} & S_{22} \end{bmatrix} \qquad (5)$$

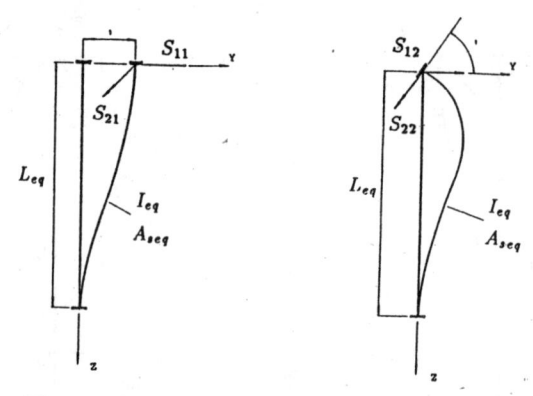

Figure 2: Stiffness coefficients – Equivalent column.

where

$$S_{11} = \frac{12 E_p I_{eq}}{(1+\phi) L_{eq}^3} \tag{6}$$

$$S_{12} = \frac{6 E_p I_{eq}}{(1+\phi) L_{eq}^2} \tag{7}$$

$$S_{22} = \frac{(4+\phi) E_p I_{eq}}{(1+\phi) L_{eq}} \tag{8}$$

in which

$$\phi = \frac{12 E_p I}{G A_{sheq} L_{eq}^2} \tag{9}$$

L_{eq} = equivalent length

I_{eq} = equivalent moment of inertia

A_{seq} = effective shear area of the equivalent column

E_p = Young's modulus of the pile

G = shear modulus of the pile

The equivalence between the two stiffness matrices (1) and (5) requires that

$$\overline{S}_{11} = S_{11} \tag{10}$$

$$\overline{S}_{12} = S_{12} \qquad (11)$$

$$\overline{S}_{22} = S_{22} \qquad (12)$$

Substituting in eqs. (10), (11), and (12) the relations given in (6), (7), and (8)

$$\overline{S}_{11} = \frac{12 E_p I_{eq}}{(1 + \frac{12 E_p I_{eq}}{G A_{seq} L_{eq}^2}) L_{eq}^3} \qquad (13)$$

$$\overline{S}_{12} = \frac{6 E_p I_{eq}}{(1 + \frac{12 E_p I_{eq}}{G A_{seq} L_{eq}^2}) L_{eq}^2} \qquad (14)$$

$$\overline{S}_{22} = \frac{(4 + \frac{12 E_p I_{eq}}{G A_{seq} L_{eq}^2}) E_p I_{eq}}{(1 + \frac{12 E_p I_{eq}}{G A_{seq} L_{eq}^2}) L_{eq}} \qquad (15)$$

Solving equations (13), (14), and (15) for L_{eq}, I_{eq} and A_{seq} we obtain

$$\frac{L_{eq}}{L} = \frac{\sqrt{2} A_2(\alpha)}{\alpha A_1(\alpha)} \qquad (16)$$

$$\frac{I_{eq}}{I} = \frac{A_2(\alpha)(2 A_1(\alpha) A_3(\alpha) - A_2^2(\alpha))}{C(\alpha) A_1^2(\alpha)} \qquad (17)$$

$$\frac{G L^2}{E_p I} A_{seq} = \frac{6 \alpha^2 A_2(\alpha)(2 A_1(\alpha) A_3(\alpha) - A_2^2(\alpha))}{C(\alpha)(6 A_1(\alpha) A_3(\alpha) - 4 A_2^2(\alpha))} \qquad (18)$$

Therefore, for a certain length and relative stiffness factor T, parameters L_{eq}, I_{eq}, and A_{seq}, corresponding to the equivalent laterally non-restrained column can be found. Variation of the equivalent length L_{eq}, equivalent moment of inertia I_{eq}, and equivalent shear area A_{seq} as a function of $\alpha = \frac{L}{T}$ are given in figures 3, 4, and 5, respectively. After analysing the structure with piles replaced by equivalent columns, the shear forces Q_0, bending moments M_0, and deformations at the top of the equivalent columns are obtained. With these end forces and moments which are equal to those acting at the top of the real piles, it is possible to calculate the distribution of the internal forces and displacements along the piles embedded in elastic medium.

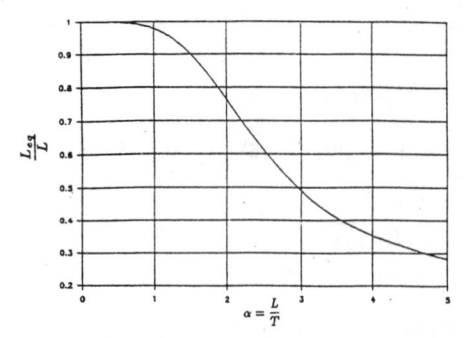

Figure 3: Equivalent lengths. Constant soil stiffness

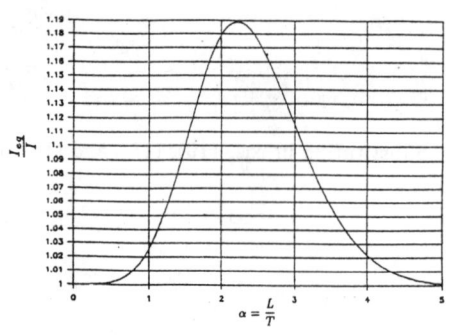

Figure 4: Equivalent inertias. Constant soil stiffness

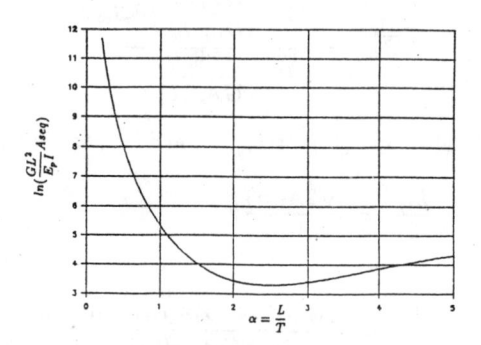

Figure 5: Equivalent shear areas. Constant soil stiffness

The general expressions for the bending moment $M(z)$, shear forces $Q(z)$, and displacements $Y(z)$ along the pile, after introducing the parameters T and α are [3]:

$$M(z) = 2E_s T^2 y_0 f_3(\tfrac{z}{T}) + 2\sqrt{2} E_s T^3 \theta_0 f_4(\tfrac{z}{T}) + M_0 f_1(\tfrac{z}{T}) - \sqrt{2} T Q_0 f_2(\tfrac{z}{T}) \quad (19)$$

$$Q(z) = -\sqrt{2} E_s T y_0 f_2(\tfrac{z}{T}) - 2E_s T^2 \theta_0 f_3(\tfrac{z}{T}) + 2\sqrt{2}\frac{M_0}{T} f_4(\tfrac{z}{T}) + Q_0 f_1(\tfrac{z}{T}) \quad (20)$$

$$Y(z) = y_0 f_1(\tfrac{z}{T}) + \sqrt{2} T \theta_0 f_2(\tfrac{z}{T}) + 2\frac{M_0 T^2}{E_p I} f_3(\tfrac{z}{T}) - 2\sqrt{2}\frac{Q_0 T^3}{E_p I} f_4(\tfrac{z}{T}) \quad (21)$$

where

$$f_1(\tfrac{z}{T}) = \cosh(\tfrac{\sqrt{2}}{2}\tfrac{z}{T})\cos(\tfrac{\sqrt{2}}{2}\tfrac{z}{T}) \tag{22}$$

$$f_2(\tfrac{z}{T}) = \tfrac{1}{2}(\cosh(\tfrac{\sqrt{2}}{2}\tfrac{z}{T})\sin(\tfrac{\sqrt{2}}{2}\tfrac{z}{T}) + \sinh(\tfrac{\sqrt{2}}{2}\tfrac{z}{T})\cos(\tfrac{\sqrt{2}}{2}\tfrac{z}{T})) \tag{23}$$

$$f_3(\tfrac{z}{T}) = \tfrac{1}{2}\sinh(\tfrac{\sqrt{2}}{2}\tfrac{z}{T})\sin(\tfrac{\sqrt{2}}{2}\tfrac{z}{T}) \tag{24}$$

$$f_4(\tfrac{z}{T}) = \tfrac{1}{4}(\cosh(\tfrac{\sqrt{2}}{2}\tfrac{z}{T})\sin(\tfrac{\sqrt{2}}{2}\tfrac{z}{T}) - \sinh(\tfrac{\sqrt{2}}{2}\tfrac{z}{T})\cos(\tfrac{\sqrt{2}}{2}\tfrac{z}{T})) \tag{25}$$

The translation y_0 and rotation θ_0 at the top of the pile can be expressed in terms of the flexibility coefficients F_{11}, F_{12}, and F_{22}, and the end moment M_0 and shear force Q_0 as follows

$$y_0 = F_{11}Q_0 + F_{12}M_0 \tag{26}$$

$$\theta_0 = F_{21}Q_0 + F_{22}M_0 \tag{27}$$

The flexibility coefficients are obtained by inversion of the stiffness matrix (1), leading to:

$$F_{11} = \frac{\sqrt{2}C(\alpha)A_3(\alpha)}{E_sTB(\alpha)} \tag{28}$$

$$F_{12} = -\frac{C(\alpha)A_2(\alpha)}{E_sT^2B(\alpha)} \tag{29}$$

$$F_{22} = \frac{\sqrt{2}C(\alpha)A_1(\alpha)}{E_sT^3B(\alpha)} \tag{30}$$

where

$$B(\alpha) = 2A_1(\alpha)A_3(\alpha) - A_2^2(\alpha) \tag{31}$$

Replacing eqs. (26) and (27) in (19) leads to

$$M(z) = 2E_sT^2(F_{11}Q_0 + F_{12}M_0)f_3(\tfrac{z}{T}) + 2\sqrt{2}E_sT^3(F_{21}Q_0 + F_{22}M_0)f_4(\tfrac{z}{T}) + M_0f_1(\tfrac{z}{T}) - \sqrt{2}Q_0Tf_2(\tfrac{z}{T}) \tag{32}$$

or in a more consistent form

$$M(z) = A_m TQ_0 + B_m M_0 \tag{33}$$

in which

$$A_m = \sqrt{2}\frac{C(\alpha)}{B(\alpha)}\left(2A_3(\alpha)f_3(\tfrac{z}{T}) - 2A_2(\alpha)f_4(\tfrac{z}{T}) - \frac{B(\alpha)}{C(\alpha)}f_2(\tfrac{z}{T})\right) \quad (34)$$

$$B_m = 2\frac{C(\alpha)}{B(\alpha)}\left(-A_2(\alpha)f_3(\tfrac{z}{T}) + 2A_1(\alpha)f_4(\tfrac{z}{T}) + \frac{B(\alpha)}{2C(\alpha)}f_1(\tfrac{z}{T})\right) \quad (35)$$

The non-dimensional expression (34) can be considered as the moment influence function due to a unit shear force ($Q_0 = 1$) acting at the top (see fig. 6). In the same manner, B_m can be considered as the moment influence function due a unit bending moment ($M_0 = 1$) acting at the top (see fig. 7). Replacing eqs. (26) and (27) in (20) leads to

$$Q(z) = -\sqrt{2}E_s T(F_{11}Q_0 + F_{12}M_0)f_2(\tfrac{z}{T}) - 2E_s T^2(F_{21}Q_0 + F_{22}M_0)f_3(\tfrac{z}{T}) +$$
$$2\sqrt{2}\frac{M_0}{T}f_4(\tfrac{z}{T}) + Q_0 f_1(\tfrac{z}{T}) \quad (36)$$

Separating the terms in Q_0 and M_0

$$Q(z) = A_v Q_0 + \frac{B_m}{T}M_0 \quad (37)$$

where

$$A_v = 2\frac{C(\alpha)}{B(\alpha)}\left(-A_3(\alpha)f_2(\tfrac{z}{T}) + A_2(\alpha)f_3(\tfrac{z}{T}) + \frac{B(\alpha)}{2C(\alpha)}f_1(\tfrac{z}{T})\right) \quad (38)$$

$$B_v = \sqrt{2}\frac{C(\alpha)}{B(\alpha)}\left(A_2(\alpha)f_2(\tfrac{z}{T}) - 2A_1(\alpha)f_3(\tfrac{z}{T}) + \frac{2B(\alpha)}{C(\alpha)}f_4(\tfrac{z}{T})\right) \quad (39)$$

As before, A_v is the shear influence function along the pile due to a unit shear force acting at the top (see fig. 8), and B_v is the shear influence function due a unit bending moment acting at the top (see fig. 9).

Replacing now eqs. (26) and (27) in (21)

$$Y(z) = (F_{11}Q_0 + F_{12}M_0)f_1(\tfrac{z}{T}) + \sqrt{2}T(F_{21}Q_0 + F_{22}M_0)f_2(\tfrac{z}{T}) -$$
$$2\frac{M_0 T^2}{E_p I}f_3(\tfrac{z}{T}) + 2\sqrt{2}\frac{Q_0 T}{E_p I}f_4(\tfrac{z}{T}) \quad (40)$$

Separating the terms in Q_0 and M_0 we obtain

$$Y(z) = \frac{T^3}{E_p I}A_y Q_0 + \frac{T^2}{E_p I}B_y M_0 \quad (41)$$

in which

$$A_y = \sqrt{2}\frac{C(\alpha)}{B(\alpha)}\left(A_3(\alpha)f_1(\tfrac{z}{T}) - A_2(\alpha)f_2(\tfrac{z}{T}) + \frac{2B(\alpha)}{C(\alpha)}f_4(\tfrac{z}{T})\right) \quad (42)$$

$$B_y = \frac{C(\alpha)}{B(\alpha)}\left(-A_2(\alpha)f_1(\tfrac{z}{T}) + 2A_1(\alpha)f_2(\tfrac{z}{T}) - \frac{2B(\alpha)}{C(\alpha)}f_3(\tfrac{z}{T})\right) \quad (43)$$

where A_y is the displacement influence function along the pile due to a unit shear force at the top (see fig. 10), and B_y is the displacement influence function due a unit bending moment at the top (see fig. 11).

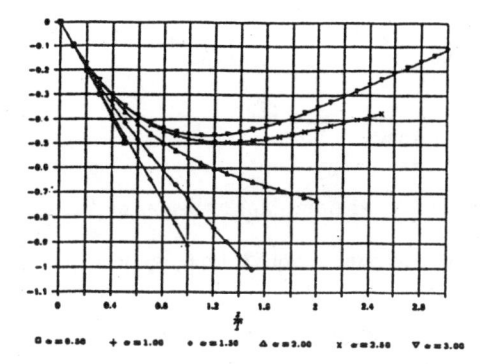

Figure 6: Moment function due to $Q_0=1$. Constant soil stiffness

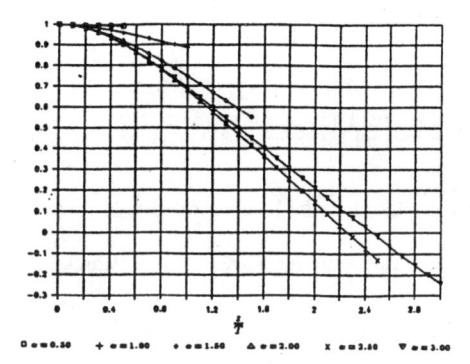

Figure 7: Moment function due to $M_0=1$. Constant soil stiffness

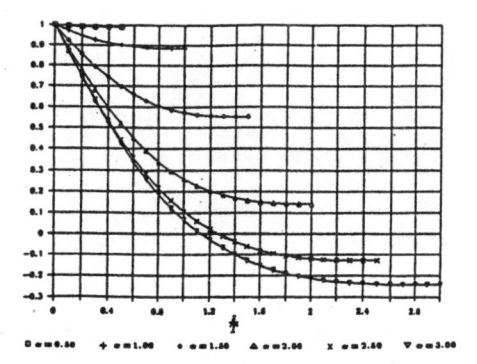

Figure 8: Shear function due to $Q_0=1$. Constant soil stiffness

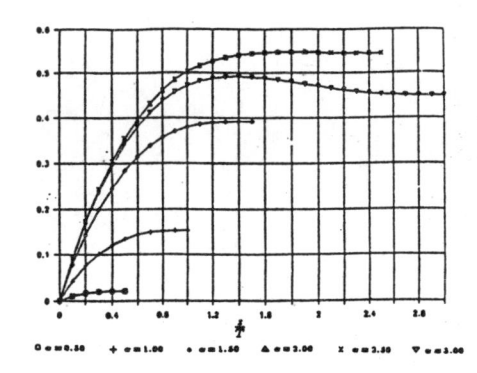

Figure 9: Shear function due to $M_0=1$. Constant soil stiffness

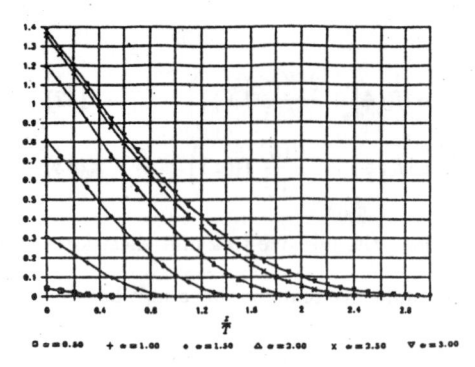

Figure 10: Displacement function due to $Q_0=1$. Constant soil stiffness

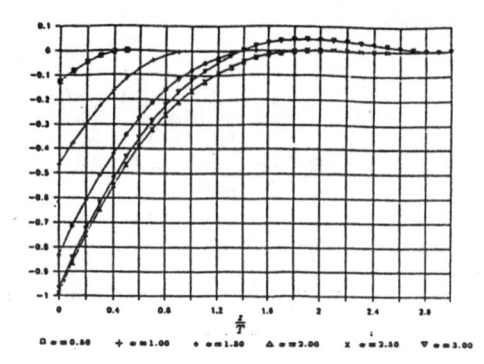

Figure 11: Displacement function due to $M_0=1$. Constant soil stiffness

References

[1] Clastornik, J., Eisenberger, M., Yankelevsky, D.Z., and Adin, M.A., "Beams on Variable Winkler Elastic Foundation", J. of Applied Mechanics, Vol.53, 1986, pp. 925-928.

[2] Przemieniecki, J.S., "Theory of Matrix Structural Analysis", McGraw Hill, 1968.

[3] Heteny, M., "Beams on Elastic Foundation", U. of Michigan, Ann Arbour, 1964.

T5

RECOMMENDED STRUCTURAL SOLUTIONS FOR MORE SEISMIC RESISTANT STRUCTURES. SEISMIC CODES.

EUROCODE 8 IN RELATION TO ITALIAN SEISMIC STANDARD

Alberto CASTELLANI [1]

INTRODUCTION

The origin of Eurocode 8 may be dated in 1979, when CEB established an international committe, including representatives of all european countries, USA, and New Zealand, to produce a Model Code on Seismic Design. The aim was to plan well recognized technical developments already established but more a less spread in the researches of recent years. A guide to this could have been a similar experience developed in USA, see the tentative code ATC-3-06, in the earthquake engineering.

A few years later, the Commission of European Communities promoted the project of Eurocodes, in view of activating a Common Marked in the field of Civil Engineering, and removing obstacles given by differences in the design procedure. This project was intended to establish a set of common rules as an alternative to those in force in the various Member States, and also to serve as a guide to the development of national rules.

The Eurocodes plan comprises nine documents, the first one devoted to basic principles and actions, six others to structures of different materials, one to foundations and, the EC8, to structures in seismic regions.

Eurocode 8 in its turn is subdivided in 5 different sections:

Part. 1 - GENERAL AND BUILDINGS
Part.1.1 - Seismic actions and generals requirements and rules for design
Part.1.2 - Buildings in seismic regions. General rules for design
Part.1.3 - Buildings in seismic regions. Specific rules for different materials and elements
Part.1.4 - Buildings in seismic regions. Strengthening and repair
Part. 2 - BRIDGES
Part. 3 - TOWERS, MASTS, CHIMNEYS
Part. 4 - SILOS, TANKS
Part. 5 - FOUNDATIONS, RETAINING STRUCTURES AND GEOTECHNICAL ASPECTS.

Presently only part 1, 1.1, 1.2 and 1.3 are published and distributed for trial use and comments within the Community while other parts are ready as first drafts.

It must be said, first of all, that since the original idea to have Eurocodes as alternative codes to those in force by the various national codes, a further agreament was reached at a political level: no new national law, within the technical codes, can be established in contrast with some Eurocode. Without going into deeper discussion, suffice it to say that the concept of alternative codes is no more valid. By this agreement

---

1 - Politecnico di Milano, Italy

therefore, a redefinition of the legal value of the Eurocodes is needed. It will be a political decision.

The difficulty of this decision is given both by the detailing and by the extention of the Eurocodes. With respect to the existing national codes, for ordinary structures the Eurocodes go into a deep detail. In a minority opinion,the cogency of the code should be restricted to a small corpus of basic requirements, leaving the rest for manuals or commentaries. In fact a manual, or a code of practice, needs not to be rigorous as a law is supposed to be, moreover, manuals are much easier to update than laws.

A part from ordinary structures, Eurocodes also cover unusual or special structures (ex. towers, or pipelines), for which general requirements and guidelines could be more apt and practicable than detailed rules. A separation between principles and advices is again suitable, according to some observers, accompanied by a relaxation of the mandatory nature of the applications rules.

Whatever the political decision, and the subsequent legal value of them, there is a lot a expectation around Eurocodes. In view of activating a Common market in the field of Civil Engineering we need to establish first a common basis for comparison of different design procedure, and then a reference text.

From an historical point of view,it can be reminded that, since 1970, the United Nations Economic Commission for Europe have studied the possibilities of harmonizing the technical content of regulations in force in member countries. The Working Party on the Industry of the ECE Committee on Housing, Building and Planning has produced a review of the building regulations in the ECE in 1974. The related standardization work is laid down in a Policy Statement during an ECE Seminar held in London in October 1973.

Furthermore, a number of work has been accomplished by United Nations, in particular by UNDRO and by the European Association of Earthquake Engineering, [4,5,6,7,8,9].

In recent times some activity has been devoted to the specification of the seismic input, see [10,11,12]. In fact the seismic ground motion is likely to occur in a variety of conditions, not yet uniquely recognized. Besides the way the seismic activity is included in the earthquake regulations still requires some work of harmonization [13,14].

FORMAL POSITION OF THE ITALIAN RESEARCH

With a few exceptions, earthquake engeneering began to grow in the european countries at the mid of 60', when a few small but well qualified centers of research arised in Italy, Portugal, and Greece. These same centers provided part of the staff engaged in the drafting of Eurocode 8. Still now Italy is largely present in the editorial group of it. Not suprising therefore, there is a strong tendency in favor of this Eurocode among researchers in Italy.

In the present paper a few comments collected during the national congress of earthquake engeneering will be discussed.

THE BEHAVIOUR FACTOR

In the Eurocode the earthquake ground motion is defined throught a response spectrum. Fig.1 shows a typical diagram of it for a seismically active zone of the mediterranean countries. This quantity is a seismo-logical information, and is referred to the site.

As to the constructions, the response spectrum to be referred to in the design stage, i.e. the design response spectrum, is obtained by reducing the ordinates of the site response spectrum according to a "behaviour coefficient" q, which is a function of the ductility and the regularity of the building

$$q = q\ (\text{ductility, regularity of the building}).$$

The codes then provides some categories to define the ductility of the buildings, and some categories to define the regularity.

First of all let discusses the numbers suggested by the Code for the behaviour coefficient and figures observed during historical earthquakes. The highest number for q in the Code is 4.5, assigned to buildings that satisfy all ductility and regularity requirements.

It will be shown however that a vast category of structures in different countries benefits of an "experimental" behaviour coefficient even greater than 20. See Fig.2.

The picture, due to R.Meli, shows four response spectra of recent strong motions and design spectra specified by the codes for the same area where the records were obtained. All response spectra assume linear behavior and five percent damping. Design spectra have been reduced as allowed by the codes for well-detailed ductile concrete frames. In Fig. 2a, the spectrum for the record of the Loma Prieta earthquake (1989) obtained at Corralitos is compared with the design spectra specified by the UBC code for California. It must be pointed out that the magnitude of the Loma Prieta earthquake is well below the maximum that can be expected in the area. Nevertheless the maximum peak of the response spectrum of that record is more than 20 times greater than the design ordinate. In fig. 2b, the SCT-EW record obtained in the lake-bed area of Mexico City, for the 1985 earthaquake, is compared with the design spectrum of the 1987 code. For the critical period, a tenfold difference is shown. The Loleo record of the 1985 Chilean earthquake is extraordinarily severe, its response spectrum reaching a peak of 2.4 g for a period of about 0.3 sec. The maximum ordinate of the design spectrum for this country is 0.1 g, as can be seen in Fig. 2c. Finally, the 1986 San Salvador earthquake was a local event of moderate magnitude (Ms = 5.4); nevertheless the response spectrum, shown in fig. 2d, has a peak of about 2g for a 0.3 sec period. The maximum base shear coefficient for the area is 0.12.

Common to all the afore-mentioned cases is the extremely large difference between the ordinates of the response spectra of events recorded at the sites and those of the corresponding design spectra. However there is experimental evidence that a large number of buildings built according to the relevant code did not suffere severe damage, or even not damage at all, during these seismic events. This suggests two considerations. One is that we still ignore how severe the earthquakes can be. And it is surprising in particular that the amplitude of the response spectra shown in the picture was not evident in the events recorded in the previous decades. It could be motivated by the spreading of the recording instruments in the seismically active zones of the World, or by the higher sensitivity of the instruments of the new generation or by something else. At any rate the Community is not prepared to specify, as response spectra for the mediterranean sites, requirements as severe as represented in the previous picture.

The second note is concerning the fact that buildings intentionally designed to withstand a given amount of lateral forces did not suffer damage in the actuality of forces several times larger than that. A few times ago it was assumed that ductility alone could explain the whole difference. More recently, at least two other factors have been identified: additional damping and overstrength. See commentaries to UBC code, ref.3, and to the Mexico City Code, ref.15. For a deeper discussion, see Meli, ref. 16.

As to the effect of energy dissipation during severe earthquake shaking, an experimental evidence has been offered by Decanini, ref. 18. In the literature a number of records are available of the acceleration at the basement of buildings and that at some elevation. Let then compute the ratio FA of the maximum acceleration recorded at the top of the building and that at his basement, (which is different from that at the soil free field, as is known). Fig. 3 reports this ratio for a given set of earthquakes and of buildings. The average value of this ratio is 2.82 (and the standard deviation 0.89), which could be recognized in a good agreement with the expected amplification resulting from the current numerical models of dynamic analysis. However the picture shows with experimental evidence, that this ratio depends on the intensity of the ground shaking. This supports an intuitive concept that all linear elastic models are to be biased toward higher damping values (than the usual 5 to 10 percent), when analysing intense shaking.

REGULARITY

Ductility is resistence beyond the elastic limit. It is only required in points where a high concentration of stress occurs. If stable hysteresis loop can develop, in these points a large amount of energy may be dissipated. However the engagement beyond the elastic limit may be too severe if only a few ,isolated,plastic hinges occur. On the other hand, when a number of such points are available, more or less distributed in a "regular" fashion, then the mechanical energy dissipated in each one is limited. The building is called in this case "regular".

Regularity ,in general ,is a question of uniform distribution of masses, stifnesses, resistences, and ductility capacities. Eurocode 8 recognizes three classes of regularity: high, medium and low regularity. Only for the upper class the semplified calculations of the dynamic behaviour are allowed. In fact only for "regular" buildings the deformation patterns are well established and under control.

Moreover Eurocode 8 recognizes a higher "behaviour factor", i.e. lower horizontal design forces, to the buildings pertaining to the higher regularity classes. But it singles up the above mentioned classes only on the basis of geometry, see for example Fig.5 and 6.

As to the horizontal distribution of masses and rigidities, it is well known that an irregular distribution is likely to induce torsional vibrations. Not surprising therefore that simplified methods of dynamic analysis are no more valid in these cases. However, as soon as the first plastic hinges appear, torsional vibrations blow up. An asymmetric progression in the plastic range occurs, in which the first plastic vertical elments are required to finally dissipate the entire kinetic energy. Thus Eurocode prescribes for buildings in these conditions an increase of the design seismic forces. G.W.Housner pointed out the fact already thirty years ago. A general agreament is therefore among Italian researchers as to the suitability of increased design forces in presence of irregular plants.

As to the vertical distributions of structural elements, Eurocode prescribes larger design forces for buildings characterised by "setbacks", see again Fig.5 and 6. This however is not justified, on the basis of several reseaches. Mazzolani et Al. for instances, working numerically with some hundreds buildings, with more or less pronounced vertical irregularities, have shown that it is possible to obtain a more favourable response by buildings with setbaks than by regular buildings. Kato and Akiyama have even shown that by a well prescribed tapering, buildings with setbacks can be optimized from the point of view of their seismic behaviour. A comparable result was obtained by Truman and Chen, and separately by Guerra et Al. in Italy.

According to similar considerations, a prevailing opininion is that the classification of regularity based only on the geometrical dimensions of the building, is likely to offer unjustified design loads, in particular when the vertical organization of the structural elements is concerned.

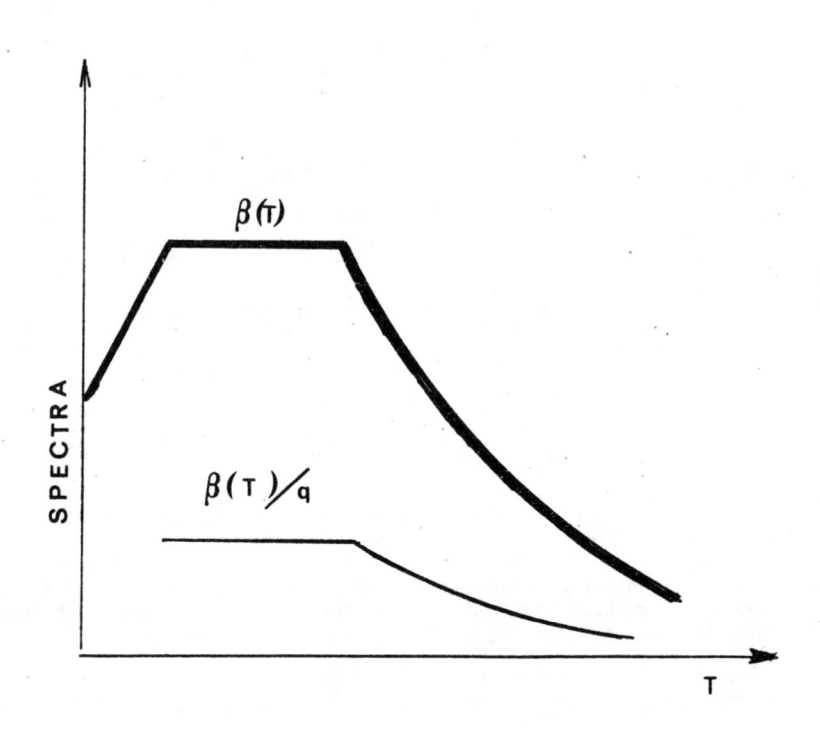

FIG 1

REFERENCE

(1) CEB "Model Code for Seismic Design of Concrete Structures", Final Draft, October 1983. Comitè Eurointernational du Beton.

(2) Eurocode n.8 - "Structures in Seismic Regions - Design. Part 1: General and Buildings". May 1988 Commission of the European Communities.

(3) Structural Engineering Association of California. "Recommended Lateral Forces. Requirements and Commentary", Seismology Committee, San Francisco, California, Fifth Edition, 1987.

(4) Cibula, Evelyn,"International Comparison of Building Regulations", Building Research Station, Garston, Watford, Current Paper 38/70.

(5) Branckov G., Sachanaski B., Tsenov L.,"Comparison of Seismic Coefficients and Forces Recording to the Norms of European and Mediterranean Countries." Earthaquake Engineering, Proceedings of the Third European Symposium on Earthquake Engineering. Sofia-Bulgaria, September 14-17, 1970,Sofia 1971.

(6) Policy Statment-Economic Commission for Europe, Committee on Housing. Building and Planning, Fourth Seminar on the Building Industry-London 1973.

(7) Sandi H., "Harmonization of the Technical Content of Building Regulations. Projet 4.1 Part II, Seismic Conditions and Seismic Loading"- United Nations-Working Party on the Building Industry, June 1975.

(8) Règlementation de la construction dans la règion de la CEE, New York 1975.

(9) United Nations-Economical and Social Council "Committe on Housing Building and Planning", ECE Programme of work on the Harmonization of Building Regulations. Sept. 1975.

(10) Hosser D., Heintzel E., Schneider G., "Proposal for Harmonized Rules for the Determination of Seismic Input Data", Universität Karlsrube, 1989.

(11) Castellani A. and Boffi G., "Rotational components of the surface ground motion during an earthquake", Earthquake eng.struc.dyn. 14, 751-767 (1986).

(12) Castellani A., Boffi G., "On the Rotational Components of Seismic Motion", Earthquake eng.Struct.Syn., 18, 785-797 (1989).

(13) Ravara A., "Harmonizations of Seismic Codes", 14th Regional Seminar of the EAEE, sept.1988 Carinthia, Austria.

(14) EAEE Working Group 1 "Calibration and harmonization of seismic codes".

(15) Rosenblueth, E. and R.Gómez, "Comentarios a las Normas Tècnicas Complementatias de Diseño pos Sismo del D.F", Instituto de Ingenieria, Internal Report, 1990.

(16) Meli R., "Code-Prescribed Seismic Actions and Actual Seismic Forces in Buildings". Italian Nat. Seminar on Earth.Eng., Palermo 1991.

(17) Dolce M., Evangelista L., "Calibrazione della Resistenza dei Pilastri per una Corretta Risposta Sismica degli Edifici". Italian Nat.Seminar on Earth.Eng., Palermo 1991.

(18) Decanini L., "Azioni Sisniche su Elementi non Strutturali e Appendici in Edifici con Struttura in ca". Italian National Seminar on Earth. Eng., Palermo 1991.

(19) Pinto, P.E., "Trends and Developments in the European Seismic Code for the New Renforced Concrete Constructions" International Meeting Earthquake Protection of Buildings, Ancona, June 6/8,1991.

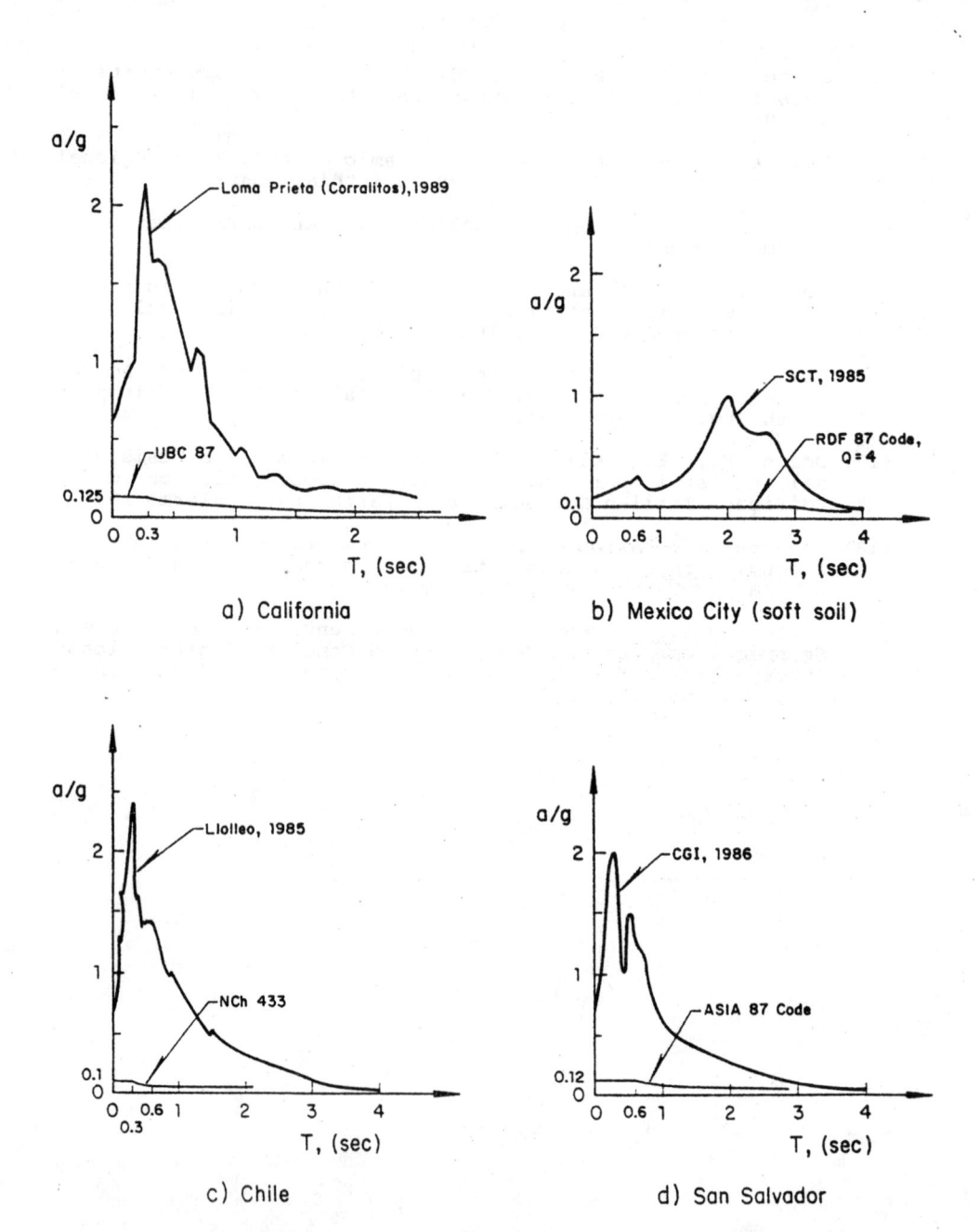

Fig 2. Comparison of response spectra of actual strong motions with design spectra specified by codes for ductile structures
(from R. Meli)

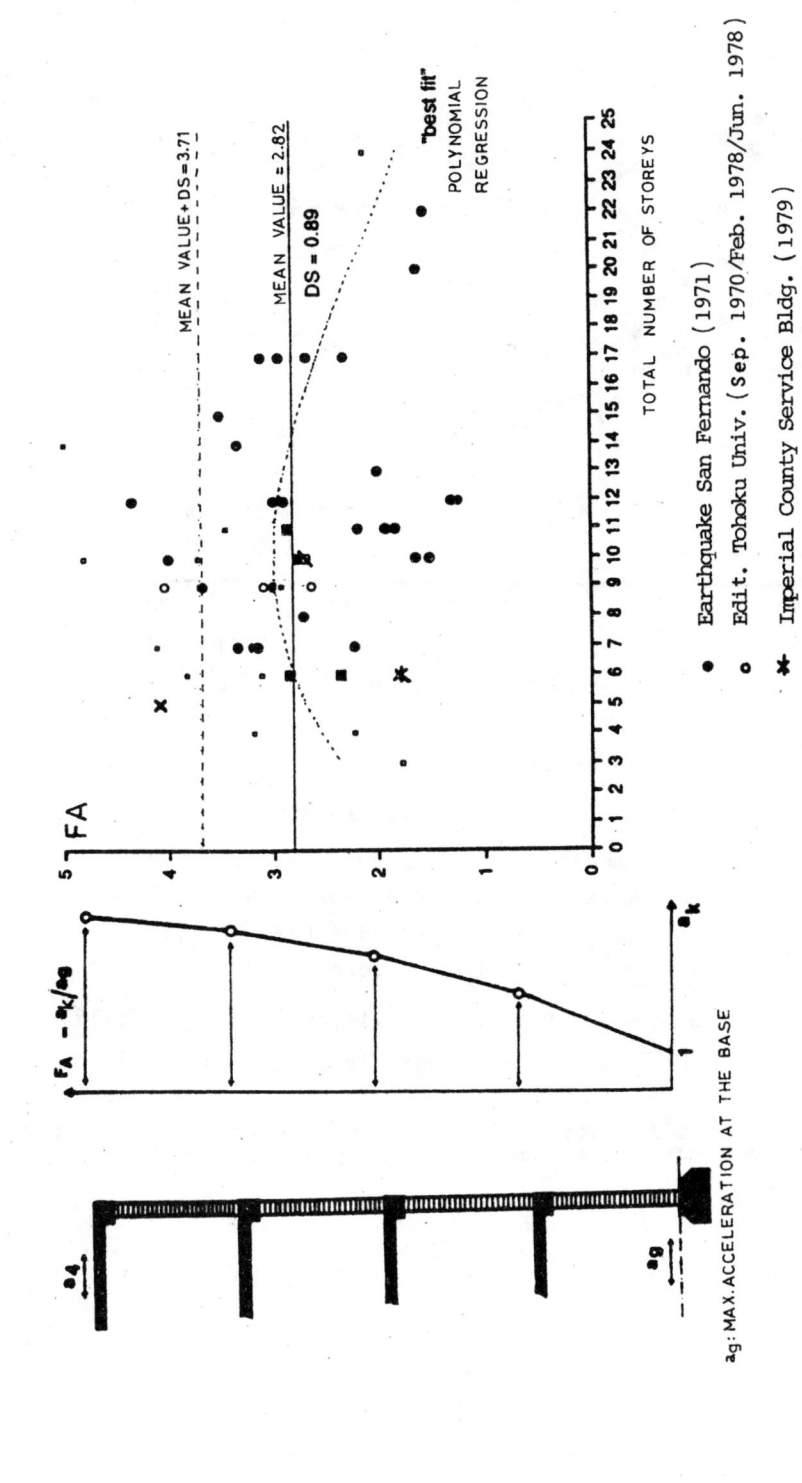

Fig. 3. Amplification factors FA of the horizontal acceleration, i.e. ratio between the maximum horizontal acceleration at the top of the building and the basement maximum acceleration (from L. Decanini).

• Earthquake San Fernando (1971)
○ Edit. Tohoku Univ. (Sep. 1970/Feb. 1978/Jun. 1978)
✱ Imperial County Service Bldg. (1979)
■ Earthquake Izu-Hanto-Oki (1980)
◢ Earthquake S. Salvador (1986)
· Earthquake Loma Prieta (1989)
× Edit. di Maiano, Friuli (replic. Jun. 1976)

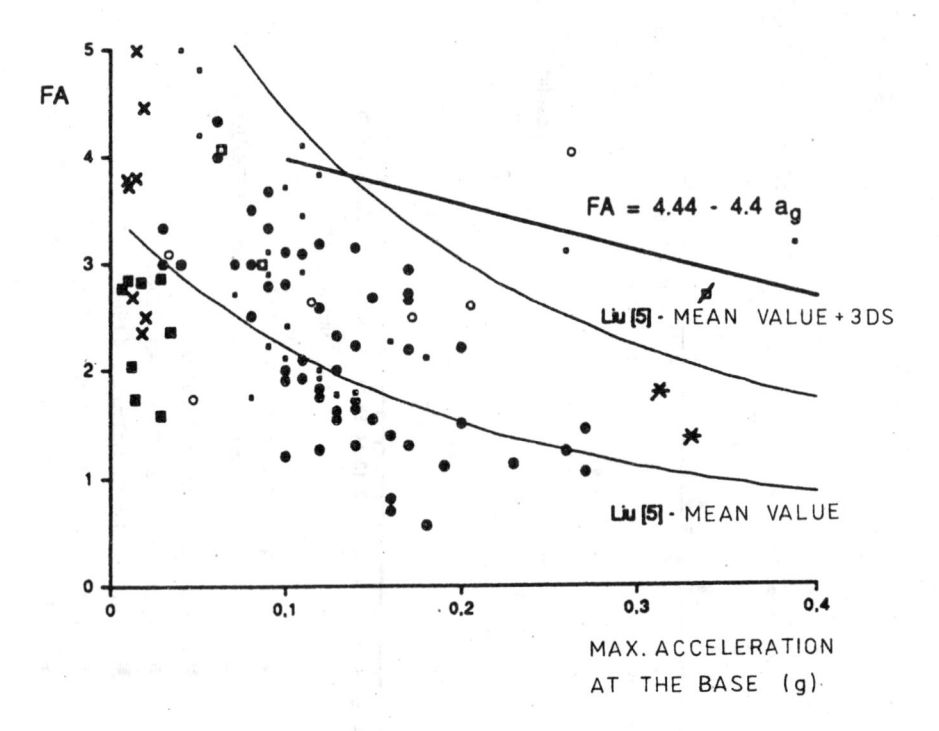

- Earthquake San Fernando (1971)
- Edit. Tohoku Univ. (1970/Feb. 1978/Jun. 1978)
- Imperial County Service Bldg. (1979)
- Earthquake Izu-Hanto-Oki (1980)
- Earthquake S. Salvador (1986)
- Earthquake Loma Prieta (1989)
- Edit. Maiano, Friuli (replic. Jun. 1976)
- Edit. Braila, Romania (Aug. 1986)

Fig.4. Amplification factors FA in function of the basement maximum horizontal acceleration, (from L. Decanini).

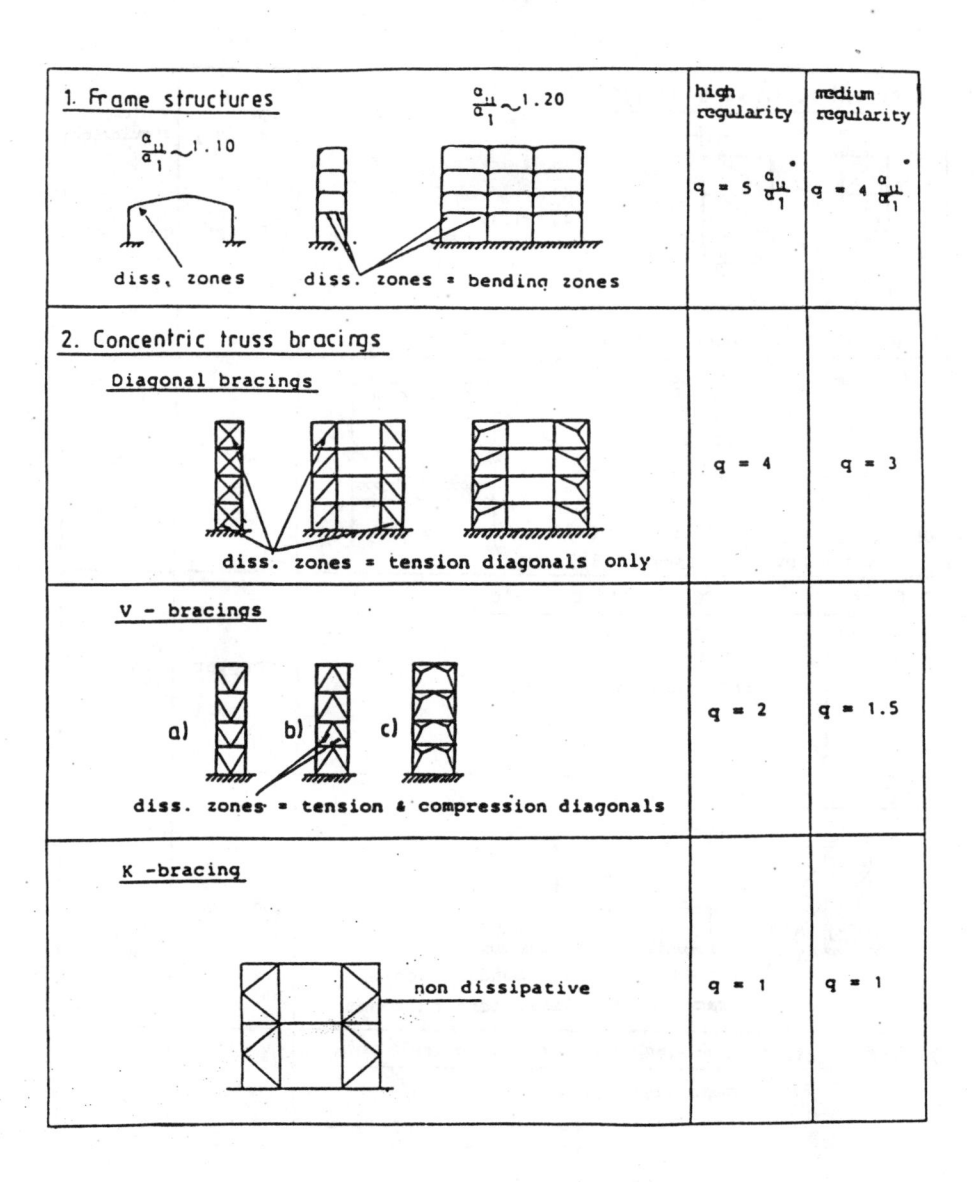

Structure type		high regularity	medium regularity
1. Frame structures $\frac{a_u}{a_1} \sim 1.10$ (diss. zones) ; $\frac{a_u}{a_1} \sim 1.20$ (diss. zones = bending zones)		$q = 5 \frac{a_u}{a_1}$	$q = 4 \frac{a_u}{a_1}$
2. Concentric truss bracings			
Diagonal bracings — diss. zones = tension diagonals only		$q = 4$	$q = 3$
V - bracings — diss. zones = tension & compression diagonals		$q = 2$	$q = 1.5$
K - bracing — non dissipative		$q = 1$	$q = 1$

Fig. 5. Eurocode 8 prescribed values for steel structures, according to different levels of regularity and different levels of ductility.

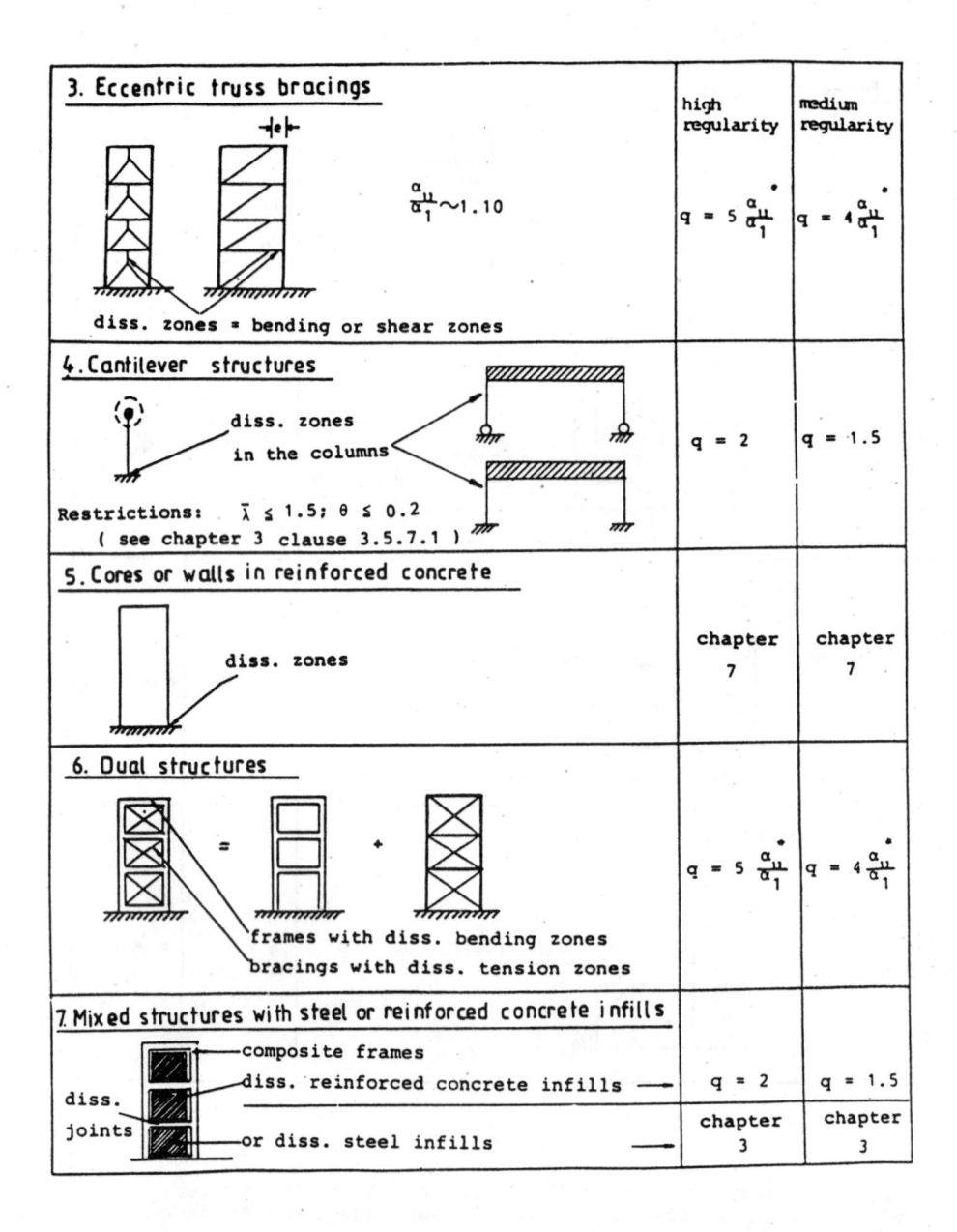

	high regularity	medium regularity
3. Eccentric truss bracings $\frac{\alpha_u}{\alpha_1} \sim 1.10$ diss. zones = bending or shear zones	$q = 5 \frac{\alpha_u}{\alpha_1}$	$q = 4 \frac{\alpha_u}{\alpha_1}$
4. Cantilever structures diss. zones in the columns Restrictions: $\bar{\lambda} \leq 1.5$; $\theta \leq 0.2$ (see chapter 3 clause 3.5.7.1)	$q = 2$	$q = 1.5$
5. Cores or walls in reinforced concrete diss. zones	chapter 7	chapter 7
6. Dual structures frames with diss. bending zones bracings with diss. tension zones	$q = 5 \frac{\alpha_u}{\alpha_1}$	$q = 4 \frac{\alpha_u}{\alpha_1}$
7. Mixed structures with steel or reinforced concrete infills composite frames diss. reinforced concrete infills diss. joints — or diss. steel infills	$q = 2$ chapter 3	$q = 1.5$ chapter 3

Fig. 6. Continuation of the previous figure.

EARTHQUAKE RESISTANT DESIGN OF LARGE PRESTRESSED BRIDGES

R.G. FLESCH [1]

1. INTRODUCTION

Most bridges all over the world (except special structures like suspension bridges and cable stayed bridges) have been designed for earthquakes using quasi-static approaches. In many cases simple design procedures are sufficient, but criteria and guidelines determining the limits of applicability are necessary. The work of BVFA/Structural Dynamics and Technical University Graz strongly concentrates on this point. Large prestressed bridges are typical for Alpine regions of Central Europe. In our opinion the main open problems are:

- which method of calculation must be used for a certain project. Which method gives the most realistic results, a high degree of safety with an acceptable amount of design work?

- in which situation travelling ware effects have to be considered?

From the practical point of view, response spectra calculations should be the standard design method for large bridges, but more research work is necessary to elaborate criteria for the limits of applicability. Results of the investigations carried out by BFVA and TU-Graz are given in [5-8].

Considering the serviceability limit state, the structure should not undergo severe nonlinear deformations. On the other hand, in regions of moderate to high seismicity, it is often uneconomic to design bridges for severe earthquakes without providing reliable means to dissipate significant amounts of seismic energy. This can be achieved e.g. by a ductile design of piers or by special energy dissipating devices.

The main aspects of displacement ductility of bridge piers are discussed in what follows:

* there is no clear evidence from real earthquakes about the behaviour of the potential plastic hinge zones of a bridge pier.

* work was done on curvature ducility of different pier sections, but a realistic estimation of the length of the plastic hinge is difficult. The length can be approximated by well known classical formulas (theoretical formulas or experimentally obtained empirical formulas for relatively short cantilever piers, e.g. [1.4]) but more research work is necessary especially for high piers with large hollow-rectangular cross sections.

---

1 - BVFA-Arsenal/Structural Dynamics, Austria

* in addition to the problems mentioned above, a considerable part of the total displacement at the top of the pier can be due to deformations of components which remain elastic after the formation of plastic hinges. Such elastic contributions result from rotations of the foundation and deformation of elastic bearings. Further, especially in the case of large bridges the deformation of the pier will be different to the deformation of a cantilever beam with a concentrated mass at the top, which is the classical assumption for the calculation of the plastic hinge length.

* in general, the appropriateness of ductility concepts is debatable especially for high piers. For bridges, ductility is considered to be efficient only if the plastic hinges can be repaired after the event. Hence, it is concluded that the behaviour should be nearly elastic.

For the example presented in chapter 2 and 3 a behaviour factor $q = 1.5$ was selected. Using the classical formulas, an attempt was made to estimate the available displacement ductility for pier 2 and 3.

2. FUNDAMENTAL CONCEPT FOR THE EARTHQUAKE RESISTANT DESIGN

2.1. Philosophy

The seismic design philosophy of bridges is based on the general requirement that communication should be maintained with appropriate reliability, after the design seismic event. The fundamental requirements are defined in EC 8/part II-bridges:

- non collapse requirement (ultimate limit state)

- serviceability limit state. A sufficient capacity for seismic- and other loads must be attainable by repair work after the design earthquake.

The repair of plastic hinges of large piers could become difficult, hence nearly elastic behaviour ($q = 1.5$) is required.

The concept of distribution of earthquake loads to the piers and abutments is shown in fig. 1. All piers are framed to the girders. At the abutments, the girders are assumed to be fixed in transverse direction, while spring-damper devices are acting in longitudinal direction. A spring constant equivalent to the bending resistance of pier 3 was selected for the devices.

2.2. Codes

The design is based on the following codes:

(1) EC no.8, especially part II - bridges (draft 1990)
(2) ÖNORM B 4040 (Austrian Code) general principles on reliability for structures
(3) ÖNORM B 4015, part 1 + 2 Austrian Seismic Code, draft 1990

2.3. Modelling

2.3.1. Model of exicitation

For the calculations, response spectra elaborated for Central Europe (by Grossmayer [9]) were used. The spectra (for rock, for vertical and horizontal excitation) are given in fig. 2. These spectra are in close agreement with a simplified response spectrum used in ÖNORM B 4015 and with the spectrum of EC 8.

According to EC 8/part II a viscous damping ratio ξ = 5 % was used for all modes.

The spectra were calibrated to a maximum acceleration of 0,4g which could be the acceleration of the maximum credible earthquake in Austria (the maximum design acceleration beeing 0,12g for seismic zone no. 4).

2.3.2. Model of structure

The original FE-Model of Lavant bridge [5, 6, 7] was simplified. The FE-mesh is shown in fig. 1. Using programm ARS/GENF (in PC-version) the bridge was modelled as a space frame with 46 nodal points, 23 beam elements for the superstructure and 28 beam elements for the piers. The cross section properties for the beams and piers were taken from the original model. Further, the moduli of elasticity obtained by global fitting of the original model to the test results were used. As it is assumed that the bridge should be constructed using the cantilevering method, piers 2-4 are designed as twin piers, with deep coupling beams (see fig. 1). Spring-damper devices acting in longitudinal direction are provided in points 1 and 46. Fixed end conditions are assumed in points 5, 12, 16, 23, 27, 34, 38 and 44.

Hence, foundation and soil influence were disregarded in the FE-model.

2.4. Material properties

concrete for piers: B 500, cube strength f_c = 50 MPa
$E = 4.96 \cdot 10^7$ kN/m² (from model-fitting to test results)
γ = 2,5 t/m³

concrete for girders: B 600, cube strength f_c = 60 MPa
$E = 5.53 \cdot 10^7$ kN/m² (from model-fitting to test results)
γ = 2,5 t/m³

steel: BSt 550 f_y = 550 MPa

Material properties are considered to be characteristic values. It is assumed, that an overstrength factor γ_{oH} = 1.15 and a material safety factor of γ_s = 1.15 cancel each other out, hence steel will really yield at 550 MPa.

2.5. Method of analysis

The bridge was modelled using program ARS/GENF in the PC-version. Then, with module DYNA, 45 eigenfrequencies and modeshapes were calculated. The eigenfrequencies are given in tab. 1. Using the response spectra described in chapter 2.3.1. the modal responses were obtained for modal damping ratios $\xi = 5\%$. The modal responses were combined using the SRSS combination rule.

3. ANALYSIS

3.1 Program of the investigations

Starting from the structural model of Lavant bridge the simplified FE model given in fig. 1 was developed.

Using response spectra, the maximum bending moments, shears and displacements were calculated, assuming excitation in transverse as well as longitudinal direction. More details of the calculations are given in chapter 2. The results are presented in tab. 2. Further, it was assumed that during a severe earthquake the joints between the coupling beams (elements 117 and 118) and twin pier 3 could faile. Hence a variant of the model without element no 117 and 118 was investigated. The results are also given in tab. 2.

3.2. Results for pier 2 and 3

3.2.1. Estimation of curvature ductility

The maximum bending moment for transverse excitation occured at pier 2 and 4. It was found to be 649 MNm for the elements 105 and 108.

154 bars $\emptyset 32$ were selected for the tension zone. Yielding of the longitudinal reinforcement will start under the maximum bending moment.

The available curvature ductility of T-shaped shear walls was investigated by Keintzel [1]. Parameters equivalent to [1] were defined in fig. 3. With $N_d = 7.1 \cdot 10^7 N$ (from permanent load) and $f_{ck} = 50/\gamma_c \sim 30$ MPa ($\gamma_c = 1.5$; material safety factor for concrete) the factor $n = 0.28$ was obtained. It is known from [1] that curvature ductility is mainly influenced by γ, n and $\varepsilon_{s,max}$ while the reinforcement ratio has only a minor influence. Hence, with $n = 0.28$ and $\gamma = 0.2$ from [1] the available curvature ductility was found to be $\mu_Q = 4$. In a second step, the cross section was improved by adding two stiffening walls (see fig. 4). The stiffening was considered necessary especially to improve the ductile behaviour in longitudinal direction. A curvature ductility of at least $\mu_Q = 5$ is estimated for the improved cross section.

The maximum bending moment for longitudinal excitation occured at pier 3 and was found to be 442 MN_m for the elements 113 and 116. 370 bars $\emptyset 32$ were selected for the tension zone. Yielding will start under the maximum bending moment.

With n = 0.24 and γ = 0.5 an available curvature ductility of μ_Q = 18 was found from [1].

3.2.2. Estimation of plastic hinge length

Several formulas for the calculation of the plastic hinge length are given in literature [1, 2, 3, 4] and in EC 8/part II. According to the comparison of experimental and theoretical plastic hinge lengths [2, 3, 4] for n < 0.3, l_p is given by:

$$l_p = (0.08l + 6d_b)(0.5 + 1.67n) \tag{1}$$

with l ... height of the cantilever beam
d_b ... diameter of longitudinal reinforcing bars
n ... normalized normal force

For the improved cross section (fig. 4) a plastic hinge length of l_p = 11m was obtained from equ. (1).

3.2.3. Displacement ductility

For a cantilever pier with the height l the displacement ductility μ_s can be estimated using equ. (2), (see also [1]):

$$\mu_s = 1 + (\mu_Q - 1) \frac{3l_p(l - 0.5l_p)}{l^2} \tag{2}$$

with μ_Q ... curvature ductility
l_p ... length of the plastic hinge

For transverse excitation, with μ_Q = 5 and l_p = 11m, from equ. (2) the displacement ductility μ_s = 1.82 is obtained (fig. 4). For longitudinal excitation, with μ_Q = 18 and l_p = 11m, the displacement ductility μ_s = 4,5 is obtained.

3.2.4. Discussion of the results

According to EC 8/part II, clause 4.4, a bridge designed after the EC 2 provision will behave as a low ductile structure. Without any additional design or special details a behaviour factor of 1.5 can be assumed.

In our example the structure was designed to start yielding under the given earthquake load (a_{max} = 0.4g).

The displacement ductility was obtained using the classical formulas, which is a highly simplified approach. There is no practical evidence of an adequate ductile behaviour of large piers with large hollow-rectangular cross sections.

The available displacement ductility evaluated in chapter 3.2.3. was within $1.82 \leq \mu_s \leq 4.5$. Hence, it is concluded, that a behaviour factor $q = 1.5$ is realistic for both directions, providing a maximum earthquake capacity (without severe damages) up to the maximum acceleration of $a_{max} = 0.6g$.

For both directions (without areas A_1^* in fig. 4), altogether 524 bars $\phi 32$ (geometrical ratio $\rho_L = 4.3$ %) were used for the longitudinal reinforcement.

Transversal reinforcement was designed after ÖNORM B 4200/part 8: links $\phi 14$ with two legs at 100 mm distance.

P-Δ effects can be disregarded because of the relatively small maximum displacement $d_{max} = 0.1m$.

4. CONCLUSIONS

A simplified more regular model was derived from the original model of Lavant bridge. Comparing the results, for the same maximum acceleration no substantial differences of the resulting maximum forces and bending moments were detected.

It was assumed that the bridge should behave at least "nearly-elastically" under the design earthquake, because ductility is considered efficient only if plastic hinges of piers can be repaired after the event, which is difficult in the case of large piers.

It was shown, that in principle the bridge could be designed to remain within the elastic range also under a maximum acceleration $a_{max} = 0.4g$ which is very conservative for the Alpine Region.

Further, although more detailed studies are necessary, a displacement ductility factor of $\mu_s = 1.5$ is considered to be available even in the case of large piers.

5. LITERATURE

1. MÜLLER,F.P., KEINTZEL,E., Erdbebensicherung von Hochbauten, 2. Auflage, Berlin, Ernst & Sohn, 1984.

2. ZAHN,F.A., PARK,R., PRIESTLEY, M.J.N., Design of Reinforced Concrete Bridge Columns for Strength and Ductility, Research Report No. 86-7, Dept.of Civ.Eng. Univ.of Canterbury, 1986.

3. PRIESTLEY,M.J.N., PARK,R., Strength and Ductility of Concrete Bridge Columns under Seismic Loading, ACI Structural Journal, V.84, No.1, 1987.

4. MANDER,J.B:, PRIESTLEY,M.J.N., PARK,R., Behaviour of Ductile Hollow Reinforced Concrete Columns, Bull.New.Zealand Nat.Soc.f.Earthqu.Eng., V.16, No.4, 1983.

5. FLESCH,R., et al, Earthquake investigations of Lavant-bridge. Proc. 9.WCEE, Vol. VI, Tokyo-Kyoto, 1988.

6. KLATZER,A., Vergleich der Antwortspektrenmethode und der Zeitverlaufsmethode am Beispiel der Erdbebenuntersuchung des Talüberganges Lavant, Diplomarbeit, TU-Graz, 1988.

7. FLESCH,R.G., KERNBICHLER,K., KLATZER,A., Earthquake resistant design of bridges, Proc. 14.European Reg.Sem., EAEE, Ossiach, 1988.

8. FLESCH,R.G., KERNBICHLER,K., KLATZER,A., The evaluation of the earthquake resistance of an arch bridge based on in-situ testing, 9.ECEE, Moscow, Sept. 1990.

9. GROSSMAYER,R.L., FRITZE,R., Design response spectra for Central Europe, Vol. 2, Proc. 8.WCEE, San Francisco, 1984.

Tab. 1.

Modal frequencies:

Nr	Hertz						
1	0.317	13	2.090	25	3.594	37	5.842
2	0.420	14	2.186	26	3.716	38	6.267
3	0.498	15	2.222	27	3.864	39	6.304
4	0.635	16	2.310	28	4.947	40	6.398
5	1.076	17	2.326	29	4.963	41	6.847
6	1.195	18	2.518	30	5.061	42	6.918
7	1.236	19	2.620	31	5.174	43	6.973
8	1.270	20	2.643	32	5.202	44	6.997
9	1.408	21	2.862	33	5.258	45	7.015
10	1.589	22	3.190	34	5.538		
11	1.878	23	3.394	35	5.603		
12	2.038	24	3.513	36	5.808		

Pier	Point	Transvers. excitation			Longitud. excitation			Longitud. excitation without el. 117 and 118		
		M_{max} [MNm]	V_{max} [MN]	d_{max} [mm]	M_{max} [MNm]	V_{max} [MN]	d_{max} [mm]	M_{max} [MNm]	V_{max} [MN]	d_{max} [mm]
1	3	26,5	8,8	41	193,2	8,2	65	192,3	8,2	64
	4	204,3	9,4	13	25,0	8,5	28	24,5	8,5	28
	5	433,6	9,4	0	226,5	8,5	0	225,4	8,5	0
2	13	20,6	9,4	71	169,8	7,7	65	164,2	7,2	65
	14	463,2	9,4	46	243,4	7,7	46	236,2	7,2	45
	15	269,2	16,8	23	211,3	12,0	22	202,7	11,4	21
	16	649,0	16,8	0	429,0	12,0	0	409,9	11,4	0
3	24	14,8	8,2	102	173,6	7,9	66	211,3	8,5	66
	25	406,0	8,2	59	250,0	7,9	46	263,9	8,5	49
	26	274,4	14,7	24	215,4	12,3	22	192,0	11,5	24
	27	624,0	14,7	0	438,8	12,3	0	442,0	11,5	0

Tab. 2

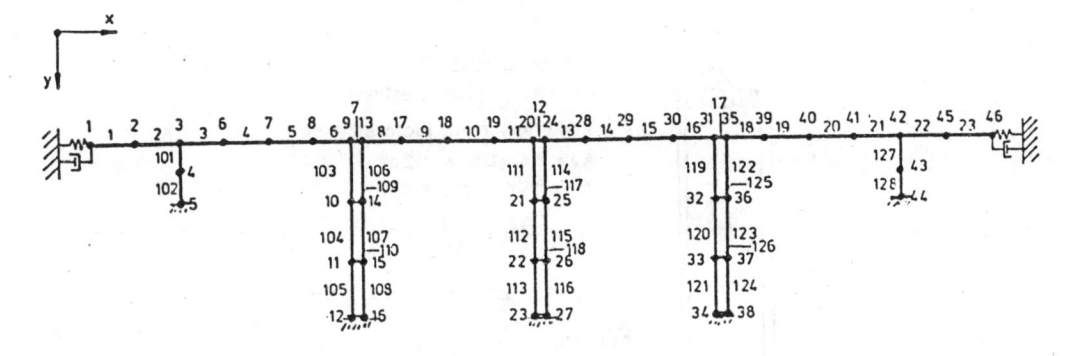

Fig. 1   Simplified model of Lavant-bridge

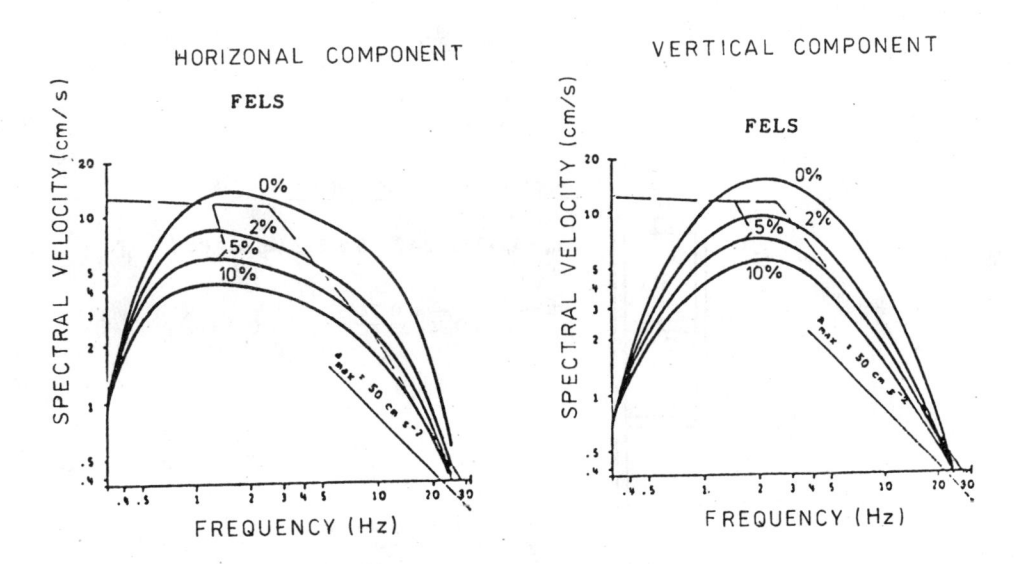

Fig. 2   Response spectra from [9]

201

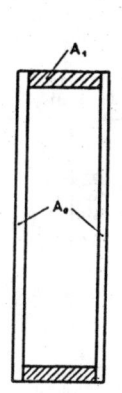

available ductility
(model: I-cross section)

most important factors: γ, n

$A_1 = 2.4 \times 0.5 = 1.2 \, m^2$

$A_o = 2 \times 0.3 \times 10 = 6.0 \, m^2$

$\gamma = \dfrac{A_1}{A_o} = \dfrac{1.2}{6.0} = 0.2$

$A_b = A_o + 2A_1 = 6.0 + 2.4 = \underline{8.4 \, m^2}$

Fig. 3

Estimation of ductility for y-direction

$f_{ck} = 30 \, MPa$
$N_d = 7.1 \cdot 10^7 N$

$n = \dfrac{N_d}{f_{ck} \cdot A_b} = \dfrac{7.1 \cdot 10^7}{3 \cdot 10^7 \cdot 8.4} = \underline{0.28}$

from Fig. 8: $n = 0.28$, $\gamma = 0.2 \longrightarrow \mu_\phi = 4$

$L_p = (0.08 \cdot 150 + 6 \cdot 0.032)(0.5 + 1.67 \cdot 0.28)$
$= 11.8 \, m$

$\mu_s = 1 + (\mu_\phi - 1) \dfrac{3 L_p (L - 0.5 L_p)}{L^2}$

$= 1 + 3 \dfrac{3 \cdot 11.8 \, (150 - 5.9)}{150^2} = \underline{1.68}$

improvement cross-section

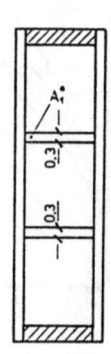

$A_b = A_o + 2A_1 + 2A_1^* = 8.4 + 2 \times 2.4 \times 0.3 = \underline{9.84 \, m^2}$

$n = \dfrac{7.1 \cdot 10^7}{3 \cdot 10^7 \times 9.84} = \underline{0.24}$

Fig. 4

Estimation of ductility for y-direction / improved cross section

from Fig. 8: $n = 0.24$, $\gamma = 0.2$ (or greater due to A_1^*)

$\longrightarrow \mu_\phi = 5$

$L_p = 12.19 \cdot (0.5 + 1.67 \cdot 0.24) = 11.0 \, m$

$\mu_s = 1 + 4 \dfrac{3 \cdot 11 (150 - 10.5)}{150^2} = \underline{1.82}$

ASEISMIC DESIGN METHODS FOR INDUSTRIAL PLANT

Mr JP Newell[1] & Mr GP Roberts[1]

INTRODUCTION

Contemporary safety standards and financial requirements for maximum availability of industrial plant processes call for the application of aseismic design methods which provide a realistic assessment of seismic performance. Contracts for new-build plant are increasingly being awarded on a turnkey basis wherein the design-and-build organisation is required to address the seismic risk to the plant. Designers, prime contractors and equipment suppliers are therefore being required to consider seismic loading as a major load case on these projects.

This paper draws upon experience of aseismic design and assessment programmes on a number of contemporary major construction projects in the power, defence and transportation industries. The various aseismic design and assessment methods available for plant and structures are explained and guidance is given on the choice of appropriate methods. The procedures are applicable to a range of industrial plant types including power, chemical, offshore and similar works.

ASEISMIC DESIGN PHILOSOPHY

The most common aseismic design philosophy is based on the deterministic approach to the definition of the seismic loading and response of the plant. The plant is designed to meet chosen performance criteria. The full probabilistic approach is not yet developed to the status of a design tool. The most expedient and effective approach is to define the seismic hazard on a probabilistic basis and to adopt a deterministic design method for the plant items and structures. A safe design is ensured by a conservative calculation of structural response and an assessment of stresses against code allowable values.

Plant risk assessment

The seismic hazard is often one of a number of possible natural and man-made hazards to which the plant may be exposed. The first stage in the plant risk assessment is therefore to assess the possible plant unavailability and safety implications arising from these various hazards. Such hazards may include earthquake, flooding, extreme wind and rain, aircraft impact and the like. The particular significance of earthquake attack is that the earthquake affects all parts of the plant together, whereas events such as flooding place a demand on limited plant sub-systems, such as site drainage and pumping.

The initial plant risk assessment enables an appropriate earthquake 'return period', or exceedance probability, to be chosen for aseismic design of the plant. Depending on the perceived level of acceptable risk to the plant occupants, off-site personnel and to the plant itself, typical earthquake exceedance probabilities may lie in the range 10^{-2} to 10^{-4} per annum. This design-level seismic event is often known as the 'Safe Shutdown Earthquake' (SSE) in the nuclear industry and the term has now found a wider acceptance in other types of plant. The plant may also be designed to withstand an 'Operating Basis Earthquake' (OBE), of lower magnitude than the

---

[1] WS Atkins Engineering Sciences, Bristol, UK

SSE, without significant interference with the plant operation.

In the case of particularly hazardous plant installations, the performance of the plant may be checked against events with an even lower exceedance probability than that of the SSE. In the UK, for example, new nuclear installations are designed to withstand an earthquake with an exceedance probability of 10^{-4} per annum and the reserve margins against collapse are identified as justification of capability to withstand more severe events.

Existing plant which was not seismically designed is generally assessed on a case-by-case basis having regard to the possible consequences of failure. The available design margins inherent in the plant are identified and a justification for the safe performance of the plant is formulated. Although the method has not yet found commonplace application, the concept of 'constant risk' or 'constant probability of failure' (Ref. 1) may be helpful.

Justification of plant performance

The plant performance under seismic loading must be justified and documented. A systematic approach is often adopted and the safe operation of the plant is documented in a 'safety case'. It is important that a detailed safety case is prepared at an early stage in the project, owing to the inherent vulnerability of structures and plant systems to seismic loading.

The responsibility for justification of the plant performance can fall within the brief of designers or contractors. Such organisations can therefore expect to be required to prepare detailed reports which describe their approach to aseismic design or assessment. These may be subject to independent assessment to ensure that an appropriate approach has been adopted.

Performance requirements

It is particularly important in seismic design to ensure that the performance requirements of the plant installation are clearly defined and these should ideally be completed early in the project. The required performance of the plant must firstly be defined in terms of the desired availability of the system when it is subjected to the seismic hazard.

Performance requirements for each main sub-system, eg power supplies, ventilation plant, pumping, communications, firefighting, civil works etc, are defined when the requirements for the system as a whole have been agreed.

Criteria are generally defined on a project-specific basis. In most cases, performance requirements fall into one of the following categories

  (a) no damage to plant systems and structures
  (b) limited damage or minor loss of function
  (c) no collapse or complete loss of function
  (d) no seismic requirement.

It is important to define the functional requirements of each of the primary plant and structural systems in terms of such categories. Fault tree analysis provides a rational basis for the choice of appropriate performance requirements. A common approach to new-build plant is to design to Category (a) or (b) and to provide reserve capability to Category (c).

Beyond-design basis events

The SSE forms the basis of normal plant design. Reserve capability may well be required beyond this level to ensure that the plant does not suffer failure 'just beyond the design basis'. Once designed to the SSE level, the plant is then checked to identify the seismic design margin. Depending on the type of plant, a margin of 40% on earthquake ground motion may be adequate.

DESIGN INTERFACES

The seismic analysis or assessment is normally divided into a number of successive stages. Some stages reflect natural interfaces between different engineering disciplines and others provide for analytical convenience. Design conservatisms are introduced at each interface which, when compounded, lead to over-design resulting in significantly increased capital cost.

The geotechnical engineer provides an assessment of the properties of the ground for input to the SSI analysis. His assessment of the shear modulus values must be realistic, rather than pessimistic, to ensure that the natural frequencies and response spectra of the soil-structure system are predicted with confidence. Any shift in the spectral peaks would be reflected in possible over or under-design of the plant.

An interface is often introduced, for design convenience, between the structure and the foundation. This enables the foundation to be designed in detail to resist the global seismic forces imposed by the structure, which are often derived from a separate fixed-base analysis of the structure.

Owing to the heavy plant and equipment content of most industrial plant installations, the interface between the plant and civil structure is important. The plant imposes a mass loading onto the structure, which affects its seismic response. Some plant, such as major cranage and process vessels, is both heavy and flexible. In these cases, it is necessary to model the dynamic flexibility of the plant in addition to its mass within the civil analysis. Secondary response spectra calculated at this interface are nearly always a source of conservatism.

DESIGN GROUND MOTION

The design ground motion is taken to apply at ground surface level in the free-field; ie distant from the plant installation. The ground motion is defined in terms of a response spectrum and peak ground acceleration (pga).

Seismic hazard curve

A seismic hazard study is carried out to an appropriate level of detail, having regard to the type and scale of the industrial plant under consideration. This is typically based on an assessment of historical seismicity in the area. A seismic hazard curve is then derived, which relates pga to probability of exceedance. Both horizontal and vertical ground motion must be considered.

Response spectra

Response spectra describe the frequency content of the design ground motion. This is derived from consideration of the type of earthquake which might be expected at the site, having regard to the geotechnical characteristics of the site; typically hard, soft or intermediate ground stiffness conditions. A typical ground response spectrum for a UK hard site is illustrated in Fig. 1.

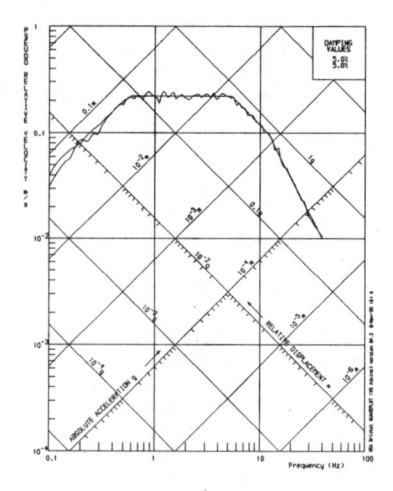

Fig.1. UK hard site response spectrum scaled to 0.25g

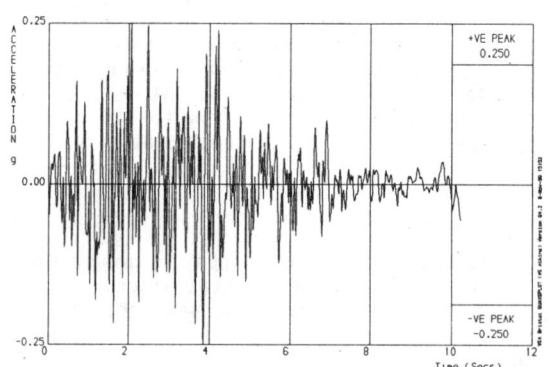

Fig. 2. Artificial UK hard site accelerogram

Accelerograms

Artificial accelerograms can be generated to provide ground motion data for input to time-history analyses of the plant and civil structures. Three independent accelerograms are required for the three coordinate directions. Suitable accelerograms are generated with computer programs, such as THGE (Ref. 2). A typical artificial accelerogram is shown in Fig. 2, which matches the spectrum shown in Fig. 1. These figures were plotted using the PC program QUAKEPLOT (Ref. 3).

It is common practice to use a single set of 3 artificial accelerograms which envelope the design response spectrum, to carry out a structural analysis (Ref. 4). For complex 3D structures, particularly if a non-linear analysis is performed, it is prudent to check the sensitivity of the seismic response by using additional accelerograms.

ASEISMIC DESIGN AND ASSESSMENT OF PLANT AND EQUIPMENT

Plant and equipment which is contained within buildings and structures is subjected to the seismic motions of such structures. This motion is typically narrow banded, reflecting amplification (or attenuation) by the dynamic response of the structures. Aseismic design or evaluation of plant and equipment is carried out by shake-table testing or by analysis, and it utilises secondary response spectra which represent the dynamic response of the buildings.

Prediction of secondary response spectra

Secondary response spectra are calculated at main floor levels in the buildings and, particularly, at the mounting positions of essential equipment.
Time-domain and frequency-domain methods are both used in practice to derive secondary spectra from the ground response spectra. The two techniques are equivalent, although the time-domain method is the more conventional method. The spectral peaks are normally broadened by 15% in frequency to allow for minor modelling sensitivities.
Sensitivity analyses are carried out to determine, for example, the effects of different site conditions and material properties on the secondary response spectra. The secondary spectra from these sensitivity analyses are normally enveloped together to provide conservative design spectra for plant and equipment. Arguably, the spectra corresponding to variations in assumed ground conditions might best be quoted individually to avoid over-design of major plant items. Indeed, care should be exercised in enveloping spectra together to reduce the effect of compounded conservatisms.
The three spatial components of the earthquake response are normally combined prior to issue of the secondary response spectra to plant designers. The stresses in the plant and equipment are then analysed and the stresses due to the three directional secondary response spectra combined again. This again is a conservative approach which can be improved by retaining the spatial components of the seismic responses in the plant, until the final stress analysis stage.

Aseismic design and assessment by analysis

The qualification of plant items and equipment by analysis is carried out using conventional structural and mechanical engineering procedures. Cranes, piping systems, vessels and cable racking systems are common examples of equipment qualified by analysis.
Holding-down bolts and anchors are designed or assessed by analysis. Experience of equipment performance during earthquakes shows that equipment exhibits an inherent ruggedness, provided that it is adequately restrained. The effects of long-term loss of preload must also be allowed for in designing fixings anchored in concrete. Contractors experience some considerable difficulty with post-drilled concrete anchors where provision has not been made at design stage for reinforcement-free pockets.

Aseismic design and assessment by shake table testing

Shake-table testing is normally required for electrical equipment to demonstrate that a required functional performance has been met.
Full-scale testing is performed by subjecting equipment to base motions which conservatively simulate the specified secondary response spectrum motion. This normally takes the form of shake-table testing to a specified

'required response spectrum' (RRS). A 'test response spectrum' (TRS) is developed from the measured shake-table motion and this is required to envelope the RRS.

The TRS normally envelopes the RRS by a substantial margin, as illustrated in Fig. 3.

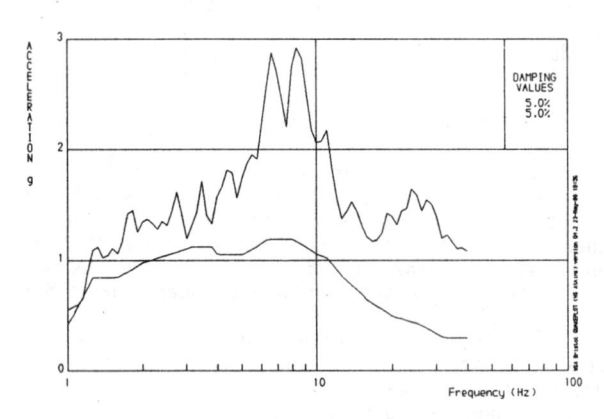

Fig. 3. Typical TRS and RRS from shake table testing

ASEISMIC DESIGN AND ASSESSMENT OF CIVIL STRUCTURES

The seismic qualification of civil structures is carried out by analysis. Generally, the simplest method should be used that will adequately predict the seismic response. Almost invariably, trade-offs are made between the level of modelling detail and the appropriate resources that are available within the overall project programme.

Modelling aspects

It is essential to define at the outset the scope and level of detailed design information that will be required from the seismic analysis. For example, seismic design forces in the seismic load resisting system may be required from the primary structure analysis, whereas secondary structures may require floor-level accelerations for localised design purposes. These considerations determine the type of mathematical model of the structure that will be required.

A prime objective is to represent the global mass and stiffness distributions of the structure, since the global dynamic response of the structure is governed by these factors. The dynamic characteristics of secondary structural elements tend not to affect the global seismic response and these may generally be omitted from the structural model. Beam finite element models are most generally used, but detailed plate finite element models of complete buildings have also found application.

In most cases, the best approach to structural modelling is to utilise a two-stage procedure (Refs. 5 & 6). In the first stage, the global seismic behaviour of the structure is represented and in the second stage the localised response of structural elements is modelled. This is sometimes known as the global/local modelling approach. An example of this approach is illustrated by the global (multi-stick) model of a complex concrete shear wall building shown in Fig. 4.

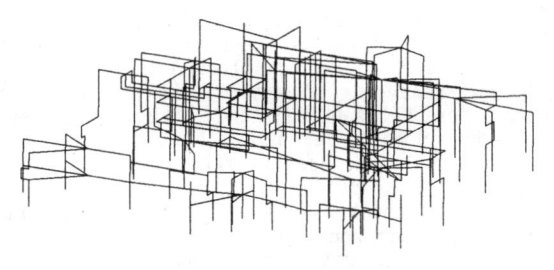

Fig. 4. Stick model of shear wall building

Analysis methods

Seismic analysis is carried out by one of the following methods:

(a) static analysis
(b) equivalent-static analysis
(c) dynamic analysis

Static analysis is the most conservative approach. The acceleration response of the structure is calculated from the peak of the ground response spectrum and is multiplied by a 'static coefficient', typically 1.5.

Equivalent-static analysis treats the structure as a single degree of freedom system. The natural frequency of the single degree of freedom system is calculated and the acceleration is determined from the ground response spectrum at this frequency. Seismic forces are calculated by multiplying the structural mass by this acceleration. A static coefficient may also be applied.

Dynamic analysis may be carried out by hand calculations, finite element modelling or problem-specific computer programs. Analysis may be carried out by response spectrum or time-history methods and may be linear or non-linear.

Most commonly, a linear finite element analysis using the response spectrum method is employed. This approach is generally the most efficient of the methods mentioned here in producing realistic seismic design forces in a form suitable for the designer.

Load combinations

Seismic analysis is often regarded as a specialist activity which cannot be covered by the conventional design office. The treatment of the seismic loadcase should however be an integral part of the design process and the seismic loading should take its place alongside the other types of loading that the structure is required to withstand. Appropriate load combinations must be considered and these are defined in the design codes of practice.

SOIL-STRUCTURE INTERACTION STUDIES

Geotechnical conditions at the site can induce a modification of the design ground motion in the vicinity of the building. In such conditions, a site response analysis and a soil-structure interaction (SSI) analysis may be required to define a modified input ground motion.

In the case of a hard rock site, neither type of analysis is generally required and the free-field ground motion may be input directly to the structural analysis. An SSI analysis is typically required for soft and medium sites, and this tends to be a specialist activity.

The compliance of the ground can cause rocking of the building on its foundations, giving rise to large displacements towards the top of the building. Such dynamic behaviour is normally the governing condition for the sizing of seismic gaps between adjacent structures, particularly when considering beyond-design-basis events. It is also important in the generation of secondary response spectra for the qualification of plant and equipment.

The simplest and most common SSI approach is the impedance method using frequency-independent springs and damper elements to represent the ground. The shear modulus information required to calculate the spring and damper constants is based on the strain-reduced values derived from a site response analysis.

Owing to the natural variability of ground conditions at a site, sensitivity studies are carried out for a range of geotechnical properties. The shear modulus, which represents the ground stiffness, is usually varied in the range 50% to 200% of the best estimate value (Ref. 5). If justified by the quality of the site investigation, this range may be narrowed to 67% to 150% of the best estimate value.

More complex analyses may also be carried out using 2D and 3D SSI models, which enable layered sites and complex foundation shapes to be represented.

DESIGN OF PRIMARY STRUCTURE

The primary structure is normally designed from a structural analysis model, rather than from an SSI model which only requires a coarse representation of the structure. SSI effects are usually embodied in a modified ground response spectrum for input to the structural analysis.

Foundation design

Basemats are normally designed using a plate or grillage finite element model on an elastic foundation. Elastic continuum and Winkler spring methods are both used to represent the compliance of the ground. Continuum methods offer a number of advantages, including the modelling of a layered site, but such methods require additional computational effort.

An equivalent-static analysis of the foundation is carried out with the seismic forces from the structure applied, together with the other imposed load conditions, such as wind, thermal, live and dead load. It is important to ensure that the seismic moments and shears in the basemat are calculated separately for the three spatial components of earthquake loading. The seismic actions in the basemat are then combined by the SRSS or 100:40:40 combination rule prior to combination with the basemat actions from the other load cases.

Redistribution of earth pressures due to localised basemat uplift must be incorporated in the design. Twist moments should be combined with the two orthogonal direct moments using, for example, the Wood-Armer method to derive design moments for top and bottom reinforcement.

Piled foundations under seismic loading require special attention to ensure that lateral loading at the pilehead can be accommodated in addition to the conventional axial loading. The relative deflections between the structure and the surroundings must be checked.

Conventional earthworks are normally designed using an equivalent static approach. Stability is commonly the governing criterion and liquefaction should also be checked if a build-up of pore water pressure is possible.

Design of walls

In-plane and out-of-plane moments and shears are required for the conventional design of walls. In-plane actions are most readily obtained from a finite element beam 'stick' model representation of the wall. Out-of-plane actions are normally assessed by hand on the basis of a two-way spanning panel under lateral loading. The lateral loading is obtained by multiplying the panel mass by the spectral acceleration corresponding to the natural frequency of the wall panel and also by a 'static coefficient'. Out-of-plane actions must also include for the effects of inter-storey drift and equipment attached to the wall.

This approach is conservative because the loading is based on the secondary response spectrum at the floor level corresponding to the top of the wall, rather than at the middle. The secondary spectrum itself embodies some conservatism. A further conservatism arises from the assumption of a uniform lateral loading distribution across the face of the wall.

Design of floors

Hand calculation methods are normally used, but detailed plate finite element models are also employed if secondary response spectra are required for the qualification of floor-mounted equipment. The seismic loads applied from any fixed equipment on the floor must also be included in the floor design.

Steel frameworks

The natural frequencies of steel-framed buildings tend to lie towards the low-frequency end of the ground response spectrum. Such structures are therefore vulnerable to amplification of the response spectrum in this frequency range by SSI effects.

Steel roof structures and in-structure plant support frames are commonly designed with seismic forces derived from secondary response spectra. Such structures generally appear to be over-designed, owing to the level of conservatism embodied in secondary response spectra. Whilst steel structures have an inherent capability for ductile energy absorption, elastic instabilities and end-connection failures must be guarded against by design.

ASSESSMENT OF EXISTING PLANT INSTALLATIONS

The seismic assessment of existing plant installations centres on the definition of acceptable performance requirements. These are based on the function of the structure in relation to its contribution to plant availability and safety and may also have regard to the desired future life. The starting point for a typical approach is described in Refs. 4 and 7.

Typically, plant items and structures not designed to SSE levels will not have sufficient inherent design margins to withstand normal loads in combination with the SSE, where this is assessed by the conventional design methods described above and where normal design criteria are applied.

In such cases, the conservatisms in the conventional analysis and design assessment process may be minimised by reducing the number of interfaces in the analysis, or by the use of more complex analysis. For example, non-linear time-domain analysis may be used to quantify the amplitudes of seismic movements and the nature of post-elastic behaviour. A safety case is then put forward on this basis.

The offshore structure shown in Fig. 5 was assessed for its resistance to seismic loading. The conventional response spectrum analysis method, using a linear analysis, indicated that the structure would fail by pull-out of the raking piles. Non-linear time-domain analysis indicated that the structure would, however, retain its lateral stability, because pull-out failure of the raking piles would be a transient phenomenon offset by additional shear mobilised through the deck into the vertical piles.

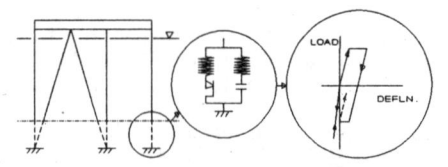

Fig. 5. Seismic assessment of an offshore structure

CONCLUSIONS

A typical approach to the aseismic design of industrial plant has been summarised. A method for the assessment of the seismic capability of existing installations has also been outlined.

It is important to define performance criteria for the plant and structures early in the project programme. The design should then be targetted on these performance criteria. It is helpful to document the seismic approach, ideally at an early stage.

Current seismic design practice is often conservative. Conservatisms arise in the specification of the ground motion, by division of the seismic analysis into distinct stages and at interfaces between different engineering disciplines.

These conservatisms offer the opportunity to assess the seismic capability of existing installations by the use of a rationalised approach utilising more realistic, but often more complex, analysis.

REFERENCES

1. American Society of Civil Engineers. Structural analysis and design of nuclear plant facilities. Manual 58, 1980.

2. THGE, Program for generation of artifial accelerograms, Belgonuclaire

3. QUAKEPLOT, Program for manipulation of response spectra and accelerograms, WS Atkins Engineering Sciences, Aztec West, Bristol, UK

4. Lawrence Livermore Laboratory. Recommended Revisions to NRC seismic design criteria. USNRC NUREG/CR-1161, May 1980.

5. Newell J.P. and Ray S.S. Design of concrete containments for seismic and thermal loads. Inst. of Nucl. Eng. Conf. on Nuclear containment, Cambridge, April 1987

6. Murray J.T., Ray S.S., Newell J.P. and Trott G.N. Study of variations in analysis and design methods for seismic loading combinations on the foundation of THORP Chemical Separation Plant at Sellafield, Cumbria. Inst. of Civil Eng. SECED Conf. on Civil engineering dynamics, March 1988.

7. Newmark N.M. and Hall W.J. Development of criteria for seismic review of selected nuclear power plants, USNRC NUREG/CR-0098, 1978.

SEISMIC FRAGILITIES FOR NUCLEAR POWER PLANT RISK STUDIES

L. Pečínka[1], J. Ždárek[1]

INTRODUCTION

Fragility analysis of structures and equipment is one of the important tasks in performing a seismic probabilistic risk assessment (PRA) of a nuclear power plant. Without reliable fragility data, the results and conclusions drawn from the fault tree and event tree analyses as to dominant contributors and accident sequences may be questionable. From the historical point of view, the fragility research was started as a part of Seismic Safety Margin Research Program founded by US NRC with a goal to develop the methodology for calculations of the probability of failure (fragility) of safety-related components in the reactor system which actively participate in the hypothesized accident scenarios.

Failure of components or structures is defined as either loss of functional operability or loss of pressure boundary integrity, as appropriate. Structures, i.e. reactor containment building, auxiliary/turbine building and crib house (intake structure) are considered to fail functionally when inelastic deformations under seismic load are sufficient to interfere with the operability of safety-related equipment attached to the structure. Failure (fragility) is characterized by a cumulative distribution function which describe the probability that failure has occured, given a value of loading. In the context of PRA studies, the loading may be spectral acceleration, zero period acceleration or internal forces resultant (such as moment or shear), depending on the component and failure mode under consideration.

---

[1] Nuclear Research Institute, Řež near Prague, CSFR

BASIC THEORY

According to previous chapter, Newmark /1/ defined the acceleration at failure by a relation

$$A_F = A_D F_S F_\mu , \qquad (1)$$

where

A_D ... design peak acceleration,
F_S ... factor accounting for limit load capacity,
F_μ ... factor accounting for the inelastic energy absorption.

The factor accounting for the limit load capacity is computed from

$$F_S = \frac{\sigma_{lim} - \sigma_{dead}}{\sigma_{seismic}} , \qquad (2)$$

where

σ_{lim} ... limit load or flow stress,

σ_{dead} ... static load due to weight, pressure, thermal, etc.,

$\sigma_{seismic}$ ... peak load induced by the seismic excitation.

Thus F_S scales the design acceleration to the failure acceleration, assuming all loads (or stresses) are calculated by a linear elastic analysis, since the peak load (or stress) is proportional to peak acceleration.

Before failure occurs, a significant amount of inelastic deformation (and hence energy absorption) takes place. In this inelastic response range, the stress increases much more slowly than the peak acceleration. Hence, the structural acceleration at failure is much higher than that predicted by the product $A_D F_S$ alone. This additional acceleration is accounted by the ductility factor F_μ introduced also by Newmark /2/ as a function of both the ductility of the component and the component damping. The ductility μ is usually estimated on the basis of engineering judgment and a knowledge of component construction details.

The statistical distribution of the acceleration at failure is obtained by assuming that the factor F_S and F_μ are lognormally distributed random variables, see /3/. If F_S and F_μ denotes of median values of F_S and F_μ and if β_S and β_μ denotes their logarithmic standard deviations, than the following relations are valid

$$\bar{A}_F = \bar{A}_D \bar{F}_S \bar{F}_\mu \quad . \tag{3}$$

APPLICATIONS TO NPP TYPICAL BUILDINGS

Because of no actual test to failure of typical NPP buildings exists, it is necessary to base the assessment of building fragilities on a comparison of analytically calculated loads with experimentally determined wall, slab and beam capacities. The acceleration at failure is then computed using the modified relation (1)

$$A_F = A\, F_S F_\mu F_R \quad , \tag{4}$$

where

A ... reference point acceleration for which stress resultants are known,

F_S, F_μ ... see relation (1),

F_R ... factor accounting for conservatisms in the method of analysis from which the acceleration and stress resultants were obtained.

For the assessment of strength factor F_S the following relation is valid

$$F_S = \frac{L_{lim} - L_{dead}}{L_{seismic}} \quad , \tag{5}$$

in which $L_{lim} = L_{ult}$ is limit state (failure mode) representing a state of undesirable structural behaviour. The term L_{dead} and L_{seism} are the calculated static and seismic loads respectively. In following two typical NPP buildings and their limit states are analysed.

A/ Shear-wall construction (turbine building etc.)

The typical shear-wall construction is given in Fig. 1.

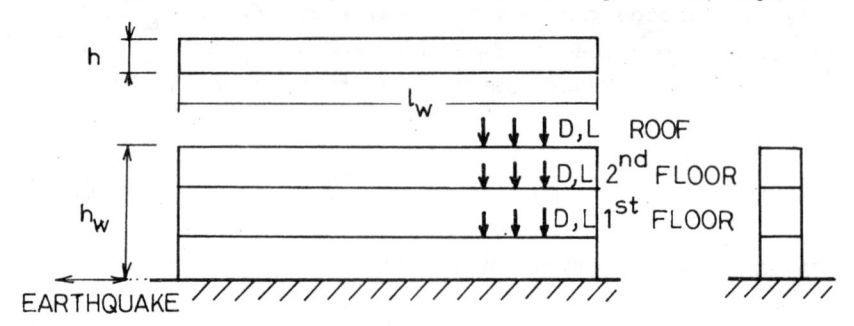

Fig. 1. Representative shear wall structure

Two main limit states i.e. bending and in-plane shear are defined below.

α/ <u>The flexure limit state</u> is reached if a maximum concrete compressive strain at the extreme fibre of the cross-section is equal to 0.003, while yielding of reinforcement steel is permitted. A typical flexure limit state surface is shown in Fig. 2. In this figure point "a" is determined from a stress state of uniform compression. Points "c" and "c'" are so called balanced points at which a concrete compression strain of 0.003 and a steel tensile strain f_y/E are reached simultaneously. Points "e" and "e'" correspond to

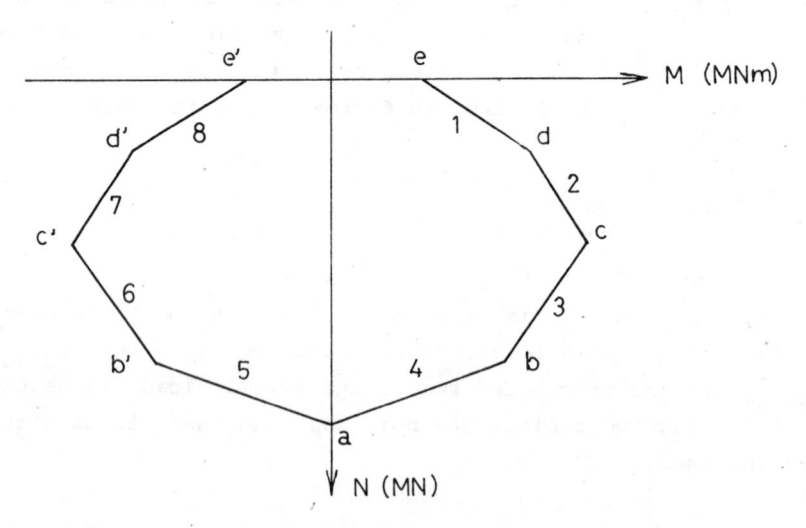

Fig. 2. Flexure limit state surface

zero axial force. Lines "abc" and "ab'c" represent compression failure and lines "cde" and "c'd'e'" represent tension failure.

β/ <u>The shear limit state</u> is reached when diagonal cracks form in two directions; following the formation of the diagonal craks either concrete crushes or reinforcement bars yield and fracture. The ultimate shear strength of a shear wall, V_n, expressed in force/area is /4, 5/

$$V_n = V_c + V_r , \qquad (6)$$

in which

V_c ... contribution of concrete to the unit ultimate shear strength,

V_r ... contribution of reinforcement to the unit ultimate shear strength.

For the V_c term following relation is recommended /4, 5/

$$V_c = 8.3 \sqrt{f'_c} - 3.4 \sqrt{f'_c} \left(\frac{h_w}{l_w} - \frac{1}{2} \right) + \frac{N}{4 l_w h} . \qquad (7)$$

Since the effectiveness of the horizontal and vertical reinforcement varies for different height to lenght ratios, the following equation for V_r is recommended

$$V_r = \left(a \rho_h + b \rho_v \right) f_y , \qquad (8)$$

where

ρ_h ... horizontal reinforcement ratio,
ρ_v ... vertical reinforcement ratio,
a, b ... constants, see /5/.

The total ultimate shear strength $L_{ult} = L_{lim}$ is computed as

$$L_{ult} = V_u h \, d, \qquad (9)$$

where

h ... see Fig. 1,
d = 0.8 l_w for rectangular walls.

The typical shear limit state surface is shown in Fig. 3. In this figure, lines 9 and 12 are governed by Eqs. (6) + (8), lines 10 and 11 are governed by equation

$$V_u = 0.25 \, f'_c , \qquad (10)$$

which defined the unit ultimate shear strength for the case of diagonal crushing failure.

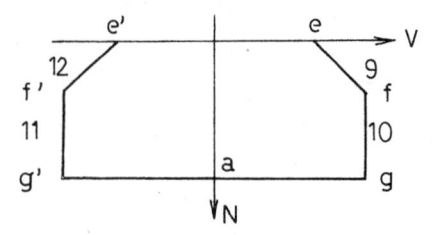

Fig. 3: Shear limit state surface

B/ Reinforced concrete containment.

Containment structure consists of a vertical cylinder with a hemispherical dome on the top, which are reinforced with hoop, meridional and diagonal rebars. Two failure modes, namely flexure failure and the tangential shear failure are typical.

/ The flexure failure mode had been analysed by Shino - zuka at all /6/. The FEM had been used for the construction of flexure limit state surface. The limit states are defined similar to shear wall case.

/ For the tangential shear mode, the ultimate shear strength V_u, expressed in force /area is/7/

$$V_u = V_c + V_{so} + V_{sd} , \qquad (11)$$

where

$V_c = 0$ ... unit tangential shear strength provided by concrete,

$V_{so} = \rho_o f_y (1 - f_m / f_y)$ ... unit tangential shear strength provided by orthogonal (hoop and meridional) reinforcement. f_m is the orthogonal membrane stress and ρ_o is the orthogonal reinforcement ratio,

$V_{sd} = \rho_d f_y$ ... unit tangential shear strength provided by diagonal reinforcement ρ_d is the 45° diagonal reinforcement ratio.

Oesterle /8/ suggests the following maximum unit shear strenght in the form

$V_u = 0.25 f'_c$.

The L_{ult} is then $L_{ult} = V_u bt$, the V_u is taken as (11) the smaller of that from Eqs. (10) or (11), t is the wall thickness and b is the unit lenght of cross-section.

C/ Fragility analysis is the modern, probabilistic oriented branch of classical mechanics. In Material Research Division of NRI Řež is in present used for seismic margin assessment of safety significant nuclear piping as a part of leak before break methodology application to CSFR NPP's 191. The aim of this paper was to explain the applications in structural engineering.

REFERENCES

1. NEWMARK, N.M. CORNELL C.A.: On the Seismic Reliability of NPP's. ANS. Topical Meeting on Probabilistic Reactor Safety, Newport Beach, CA, May 1977.
2. NEWMARK, N.M.: Inelastic Design of Nuclear Reactor Structures and its Implications of Design of Critical Equipment. SMiRT Paper K4/1, 1977 SMiRT Conf., San Francisco.
3. FREUDENTHAL, A.M. et al: The Analysis of Structural Safety. J. of the Structural Division, ASCE, STI, February 1966, pp. 267 + 325.
4. SHIGA, T. et al: Experimental Study on Dynamic Properties of Reinforced Concrete Shear Walls. 5th World Conf. on Eartquake Engng., Rome, Italy, 1973.
5. HWANG, H.H.M.: Probability Based Load Combination Criteria for Design of Shear Wall Structures. NUREG/CR 4328, US NRC, Washington, USA, 1986.
6. SHINOZUKA, M. et al: Reliability Assessment of Reinforced Concrete Containment Structures. Nucl. Engng. and Design, 80 (1984), pp. 247 + 267.
7. HWANG, H.H.M.: Fragility Assesment of Containment Tangential Shear Failure. Res Mechanica 30 (1990), pp. 291 + 301.
8. OESTERLE, R.G.: Tangential Shear Design in Reinforced Concrete Containments - Research Results and Applications. Nucl. Engng. and Design 79 (1984), pp. 161-168.
9. ŽĎÁREK, J. et al : Methodology of LBB project. Internal Report NRI, (will be published).

INFLUENCE OF EMBEDMENT ON THE FIRST
RESONANCE OF SOIL-STRUCTURE SYSTEM

A.G. Tyapin[1]

INTRODUCTION

After the Chernobyl accident NPP safety has been widely discussed in the USSR. Underground siting or deep embedment of view of mitigating possible consequences of accidents. Meanwhile a number of strong earthquakes in the last years has led to the reestimation of the seismic component of the NPP safety (1). Thus it is important to know, whether embedment could influence seismic loads on the NPP structure (first of all, the reactor building). Of course, a very little effect should be expected in the case of the outcropped rock as the NPP foundation (SSI) effects essential for the seismic response of heavy rigid structures. The idea is to use wave damping in the soil to mitigate response accelerations.

NUMERICAL EXAMPLES

Three-layered soil profile with the properties, put into Table 1, was considered first. Medium soil was underlaid by

Soil profile for the first sample   Table 1

Number of layer	Thickness m	Shear waves velocity, m/s	Poisson's ratio	Internal damping
1	18.0	240	0.45	0.03
2	24.0	600	0.49	0.05
3	30.0	720	0.49	0.05

bedrock. Calculations, performed using method from (2), proved the cutoff frequency of such a foundation to be about 2.15 Hz. Standart NPP reactor building (see Fig. 1), consisting of three upper structures (internal boxes with reactor equipment, containment shell and the external building) on the common foundation of 20 m height, was taken as an example of structure. The basement was embedded into the upper soil layer. Five variants of the embedment depth: 0, 4.5, 9.0, 13.5, 18 m - were considered. Plain-strain model was studied. FEM mesh for the second variant of the embedment depth is shown in Fig. 1.

The second sample with another building of the same kind and with a single layer, underlaid by bedrock, was studied by 3-D model. Thickness of the layer was 30 m, shear wave velocity 400 m/s, Poisson's ratio 0.494 (extremely high one to check effects discovered before), internal damping 0.05. Four variants of the embedment depth - 0, 10, 20, 30 m - were considered.

1- MEI, Moscow, USSR

The last variant in fact was to use the outcropped rock as a foundation for the NPP. In this case no SSI effects occur.

METHODS USED

Computer code NASSI, based on the ideas of (3), was used. Unlike (4), kinematic interaction is avoided even for the embedded structures. Scheme used (see Fig.2) is somewhat like well-known impedance method, but the excitation is not kinematic. Seismic forces, acting on the immovable basement, may be obtained either solving the additional problem (in this case every type of the seismic wave leads to a new problem), or using the reciprocy theorem (3) - through the integration of the free-field motion with the contact stresses in the set of impedance problems (they are solved while obtaining impedances of the basement). The later method makes it easy to consider different waves (body or Rayleigh /3/) without time-consuming calculations.

The response of the structure is obtained in three steps. On the first step SSI itself is investigated: impedances in the frequency domain are calculated together with the transfer functions from the free-field motion to seismic forces and moments shown on Fig.2. Structure is considered separately. It is described by a set of the eigenforms and eigenfrequencies of the upper structures on the fixed basement. Different upper structures may be investigated by different methods: thus, the containment shell was modelled using FEM, and two other structures - using the conventional frame model. On the second step algebraic system of the order 10 (in general; usually 3 for sway-rocking motion and 2 for the vertical motion) is solved for each frequency, thus obtaining transfer functions from the free-field motion to the response of interest. The third step in NASSI code may be done in several ways. If the excitation is given as a non-stationary random process (1), time-dependent response dispersions are obtained (5). If it is described by the accelerogram, special variant of the Fourier transform is used. It is different from the conventional FFT, because the so-called "quiet" time interval after (or before) the excitation in FFT for the particular cases of interest (low-damped low frequency resonances) proves to be about 15 s (and not 4, as commonly used) thus increasing the time of duration. To overcome this difficulty, accelerogram is presented as a result of multiplication of high-frequency function with a low-frequency amplitude function (standart), providing long "almost zero" time interval. The first function is then set into Fourier sum apart from the amplitude, thus leading to the irregular trigonometric sum for the initial acceleration. As the transfer functions can be used for arbitrary trigonometrical sums, the response is calculated simply enough.

NUMERICAL RESULTS

The simpliest seismic wave - vertically propagating shear one - was considered. Control motion was set at the surface. Fig. 3 shows the horizontal impedance for all the five depths in the first sample (curves 1-5 relate to the real part of the horizontal stiffness, curves 6-10 - to the imaginary part). Fig. 4 shows horizontal stiffnesses for the second sample. The vivid growth of the imaginary parts begins from the cutoff frequencies. Embedment leads to the considerable growth of the static stiffnesses. Dashed lines on both figures relate to the functions $\omega^2 M$, where M are the masses of the structures (different in two samples). Points of crossings Re C with $\omega^2 M$ mark resonant frequencies of the lowest sway-rocking resonances(approximately, of course, but rocking stiffnesses are high enough, so the first resonances are almost swaying).The deepest embedment leads to the first resonant frequency above the first natural frequency of the soil. The absolute values of the transfer functions are shown on Fig. 5 for the first sample. Increasing embedment leads to the mitigation of the first resonance and for the deepest embedment this resonance practically vanishes (due to the overcritical damping mostly). In the high frequency range however, there is another situation.

To estimate the overall influence of embedment, two artificial accelerograms were used for the first sample. Three floor response accelerations were calculated at the points x_1, x_2, x_3 respectively (see Fig. 1) together with horizontal force and bending moment at the floor x_1. Maximum values are presented in Table 2. Responses to two accelerograms are divided by slash.

Table 2
Maximum floor response accelerations and forces for the first sample

Emb. m	Accelerations, m/s^2			Force at x_1, 10^8 N	Moment at x_1, 10^{10} Nm
	x_1	x_2	x_3		
0	8.56/10.01	9.99/11.99	11.03/12.83	14.95/17.86	3.41/3.79
4.5	6.44/6.32	7.74/7.70	8.36/8.37	11.51/11.48	2.43/2.63
9.0	4.79/6.90	5.56/7.18	5.84/7.41	8.38/11.23	1.71/2.07
13.5	4.83/4.94	5.34/4.68	5.79/4.87	7.83/7.51	1.69/1.35
18.0	4.12/3.95	5.07/4.17	6.43/4.82	7.56/6.55	1.66/1.26

For the second sample a single accelerogram was used, (different from both used for the first one). Structural response this time was calculated in two directions (unlike the first building considered, the second one had different flexibilities in the two main planes; excitation in two directions was the same). Maximum floor response acceleration are shown in Table 3. The results in two planes are divided by slash.

In both cases the same effect was obtained: horizontal stiffness increased due to embedment together with the first frequency of the soil-structure system. The first resonance decreased, but high resonances increased. When the first frequency overcame the natural soil one, the first resonance vanished. Then higher resonances became essential for structural response. As they were caused by structural flexibility, they were badly damped by soil waves, so at the upper floors structural response in the case of "overembedment" was high.

Table 3

Maximum floor accelerations for the second sample

Floor height, m	Embedment depth, m			
	0	10	20	30
61.3	2.373/2.239	1.970/1.787	1.432/0.791	2.600/2.101
55.9	2.085/1.954	1.896/1.662	1.339/0.810	2.328/2.000
38.8	1.575/1.722	1.423/1.352	0.845/0.857	1.562/1.586
27.8	1.623/1.991	1.024/1.171	0.840/0.853	0.986/1.379
19.3	1.638/1.896	0.936/1.103	0.747/0.822	0.951/1.125
8.5	1.527/1.653	0.812/1.061	0.725/0.756	0.799/0.868

CONCLUSIONS

Influence of the embedment depth on seismic response of NPP is investigated by two SSI models. The case of layered soil underlaid by bedrock is studied, when the soil foundation has the cutoff frequency.

It is shown, that the embedment depth may be chosen so, as to drive the first eigenfrequency of the soil-structure system above that frequency (for the surface basements it is below). In this case the first resonance, essential for the response of rigid massive structures (like the NPP reactor buildings) will be damped not only due to the internal dissipation, but also due to the wave effects. These effects provide considerable (up to half) mitigation in structural response. However, if the embedment depth is about medium soil thickness and foundation stiffness is very high, the high-frequency resonances increase, so the overall response at the upper floors increase also. When the foundation is completely rigid, upper floor response is several times higher, than that in the case of the soft soil (even for the surface foundation). It means that there exists optimal embedment depth mitigating the response according to the given criterium (for example, maximum floor acceleration at the given floor, or some combination of floor responses; the best way is to mitigate the integral seismic risk /7/). In practice mitigation of the seismic response surely will not be the main point for the engineering decision. For the given embedment depth (deep enough to mitigate the first resonance) the optimization may be achieved using other ways, for example, changing flexibility of the structural elements.

ACKNOWLEDGEMENTS

The author is grateful to Prof. Yu. K. Ambriashvily and Ing. N. M. Yusipov (Atomenergoproject, Moscow, USSR) for scientific sponsorship.

REFERENCES

1. BOLOTIN, V.V.: New Problems in the Seismic Safety Design. In: Scientific Problems in Machinery (ed. K.V. Foloff). Moscow, Nauka, 1988. pp. 30-38 (in Russian).
2. TYAPIN, A.G.: Complex Roots of the Wave Equation for the Rayleig Waves in Multirayered Media Underlaid by Rigid Halfspace //USSR Academy of Sciences Proc. Structural Mechanics. 1989. No. 6, pp. 139-143 (in Russian).
3. TYAPIN, A.G.: Seismic Response Analysis for Structure on Rigid Basement Considering Soil-Structure Interaction// MEI Proceedings, 1984, Vol. 26, pp. 89-96 (in Russian).
4. KAUSEL, E. - WHITMAN, R.V. - ELSABEE, F. - MORRAY, I.P.: Dynamic Analysis of Embedded Structures // SMiRT-4. 1977. K 2/6.
5. TYAPIN, A.G.: Rigid Basement's Response Analysis for Wave Excitation in Soil // Structurel Mechanics and Calculations of Structures, 1983, No. 6, pp. 48-51 (in Russian).
6. LYSMER, J. - WAAS, G.: Shear Waves in Plane Infinite Structures // J. Eng. Mech. Div. ASCE 1972. Vol. 98. No. EM1. pp. 85-105.
7. BOLOTIN, V.V.: Statistical Simulation in Seismic Safety Design // Structural Mechanics and Calculations of Structures. 1981. No. 1. pp. 60-64. (in Russian).

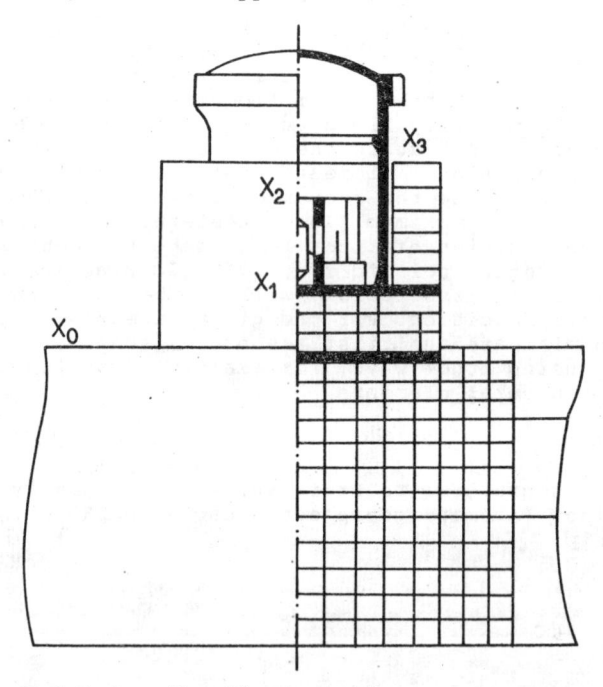

Fig. 1. Model for the first sample.

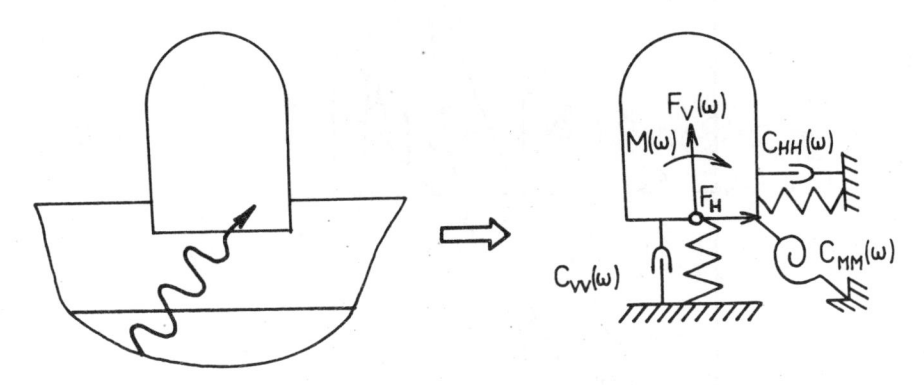

Fig. 2. Impedance SSI scheme for embedded structure and arbitrary waves.

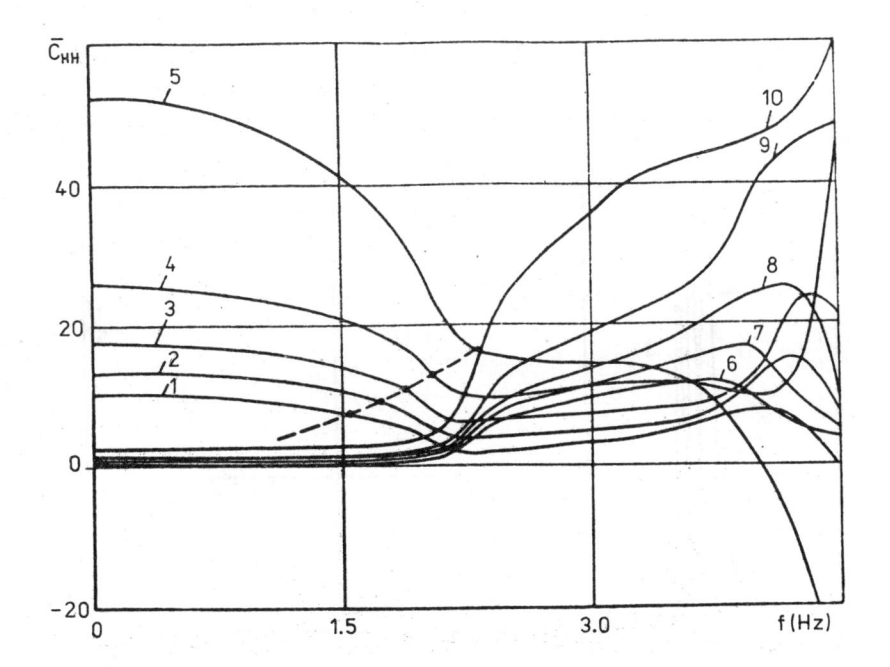

Fig. 3. Horizontal stiffnesses for the first sample. 1 ... 5 − Re C_{HH}, 6 ... 10 − Im C_{HH}.

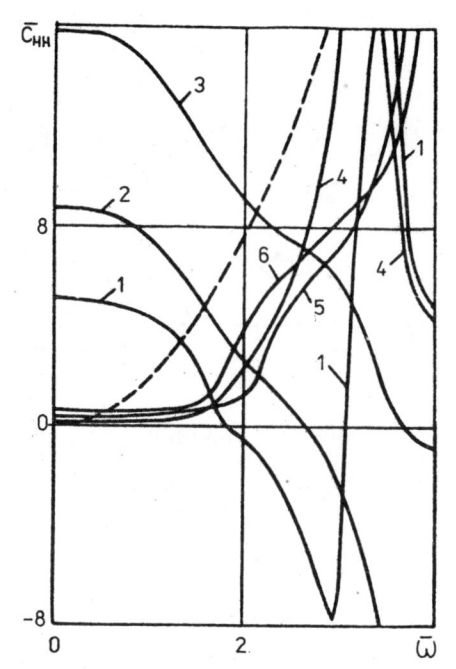

Fig. 4. Horizontal stiffnesses for the second sample.
1, 2, 3 - Re C_{HH}, 4, 5, 6 - Im C_{HH}.

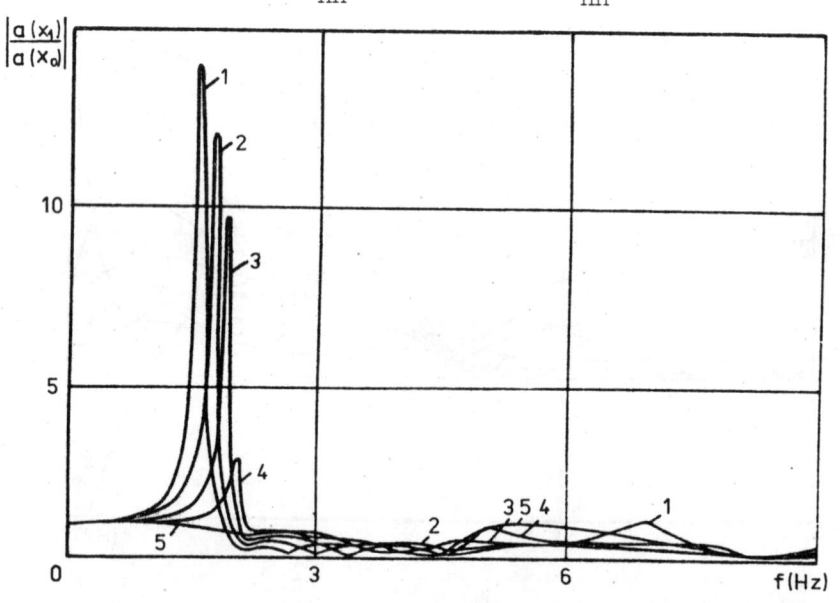

Fig. 5. Transfer functions from free-field accelerations to floor accelerations at point x (first sample).

T6

SPECIAL THEMES.

GEOTECHNICAL ASPECTS IN EARTHQUAKE ENGINEERING

Atilla M.Ansal, Hüseyin Yıldırım and M.Aysen Lav [1]

INTRODUCTION

It has been observed in the past earthquakes that one of the important factors controlling the structural damage is the local soil conditions. Soil layers as well as modifying properties of earthquake excitations, would also be affected by them and may cause important instabilities as in the case of liquefaction and slope failures or may experience large settlements due to dissipation of excess pore pressures in cohesive soils and due to densification in cohesionless soils.

The aim in this paper is to review very briefly some basic issues concerning the geotechnical earthquake engineering and summarize the research activities that has been going on at Istanbul Technical University during the last decade. In this context, an attempt will be made to review (a) behavior of cohesive soils under cyclic loading conditions; (b) liquefaction susceptibility of sandy soil layers; (c) effects of local soil conditions on earthquake characteristics on the ground surface.

BEHAVIOR OF COHESIVE SOILS UNDER CYCLIC LOADING

During earthquakes soil layers are subjected to cyclic shear stresses with different amplitudes and frequencies which will lead to cyclic deformations. These deformations are going to affect structures located on these layers and may cause damage. In addition, the change in the stress-strain and strength properties of soil layers during cyclic loading may have a significant influence on the stability of earthdams, embankments, retaining structures, and natural slopes.

In evaluating the behavior of soils under cyclic loads, one alternative is to consider stress-strain and shear strength properties separately. Dynamic shear modulus, damping ratio and their variation with shear strain may be regarded as the dynamic stress-strain properties of soils. Cyclic stress amplitudes and number of cycles leading to failure or excessive deformations may be defined as dynamic shear strength properties.

In this respect the behavior of clays subjected to cyclic loading has been studied by large number of researchers up to the present. Some of the early investigations reported in the literature were performed by Seed & Chan (1966), Thiers & Seed (1968, 1969). In these studies stress-strain and strength properties of clay samples were evaluated based on cyclic triaxial and cyclic simple shear tests. Another group of study reported in the literature were performed by Sangrey, Henkel & Esrig (1969), France & Sangrey (1977), Sangrey, et

---

1-Istanbul Technical University, Civil Engineering Faculty, Department of Geotechnical Engineering, Ayazaga, Istanbul, Turkey

al.(1978), and Sangrey & France (1980). In these investigations the cyclic stress-strain-pore pressure behavior of clays are studied based on cyclic triaxial tests performed at relatively slow rates. The cyclic behavior of clays are also studied by other researchers as Koutsoftas (1978), Matsui, Ohara & Ito (1980), and Ishihara (1985).

At Istanbul Technical University the behavior of normally consolidated saturated clays under cyclic shear stresses were first studied based on undrained cyclic simple shear tests conducted on samples prepared in the laboratory using kaolinite clay. The purpose was to investigate the effects of cyclic shear stress amplitude, frequency, number of cycles. An effort is made to develop an empirical procedure to evaluate the cyclic response of normally consolidated clay samples.

Following this initial phase, most of the cyclic testing was conducted on natural undisturbed samples. One of the major projects carried out involved a parametric study where different combinations of static and cyclic loads were used to evaluate shear strength properties of a natural marine clay. Four types of loading schemes were taken into consideration. In the first set of tests, a dynamic shear strength was evaluated for soil samples subjected to uniform cyclic shear stresses. In the second group of tests the samples were first subjected to uniform symmetric cyclic shear stresses under undrained conditions and later stress controlled static shear tests were performed without allowing the dissipation of accumulated pore pressures. The basic intention was to determine the reduction in the shear strength due to prior cyclic loading. In the third group of tests the samples were first subjected under drained conditions to static shear stress and later cyclic shear stresses were applied to study the effect of sustained shear stresses on the dynamic shear strength properties. The last set of tests are performed to determine the shear strength properties under simultaneously acting cyclic and quasi-static loads. All of the cyclic tests were conducted in a dynamic simple shear testing system.

The cyclic stress-strain-pore pressure behavior of clay samples were studied with respect to different shear stress amplitudes for sets of tests conducted at the same cyclic frequency. The response patterns observed from these tests indicate that it appears possible to consider a critical shear stress ratio which can be defined as "the critical level of repeated stress". This definition was first given by Larew and Leonards (1962) as the maximum level of repeated stress that will not lead to failure. Later Sangrey, Henkel, and Esrig (1969) have demonstrated the validity of this concept. If the clay sample is subjected to cyclic shear stresses with stress ratios larger critical level, pore pressure will accumulate continuously and the sample will undergo large cyclic shear deformations. However, if the applied stress ratio are smaller than the critical stress ratio, the accumulated pore pressure will be limited and the sample will experience relatively small shear deformations.

The other important parameter controlling the cyclic stress-strain characteristics of clays is the number of cycles. This parameter plays a crucial role especially in analyzing the behavior of soil layers under earthquake loads. Even if the applied cyclic stress are larger then the critical level, the accumulated pore water pressure and cyclic shear strain amplitudes may still be limited if the number of cycles are small.

One approach generally adopted in analyzing the response of soil layers under earthquake loads utilizes the concept of equivalent number of cycles determined to represent the effects of random cyclic stresses generated by an earthquake. It was observed that cyclic shear stress ratio versus shear strain amplitude relationships yield hyperbolic stress-strain curves. Under these conditions, it appears possible to define a yield point for each curve as the intersection of the tangents drawn to the initial and final portions of the curves. Since each stress-strain curve corresponds to a specific number of cycles, it appears realistic to consider the cyclic shear stress amplitude corresponding to this yield point as the cyclic yield strength for this number of cycles.

Effect of Cyclic Loading

One of the important issues encountered in the field is the reduction in the static shear strength due to cyclic loading. This aspect of the clay behavior has been investigated by various researchers up to the present. However, there are differences in the results reported in the literature in relation to the decrease of the static shear strength after cyclic loading. Thiers & Seed (1968, 1969) have presented data showing a significant decreases in the shear strength if the cyclic strain/static yield strain ratio is large. Castro & Christian (1976), Andersen, et al.(1980), and Koutsoftas(1978) have presented data indicating that the loss in shear strength is not so drastic even for large cyclic strain levels. However, Thiers & Seed (1969), Taylor & Bacchus (1969), and Lee & Focht (1974), and Sangrey & France (1980) observed that the static shear strength after cyclic loading may decrease as much as 50% as the cyclic stress amplitude approaches the critical level of repeated loading.

The results obtained from series of tests conducted under undrained conditions to evaluate the effect of cyclic loading on static shear strength have indicated that the strength reduction following cyclic loading without allowing the dissipation of excess pore pressures, was more dependent on number of cycles then cyclic stress ratio. And it appears possible to consider a critical number of cycles below which strength reduction would be limited and can be considered as independent of cyclic stress ratio. For larger number of cycles, it was possible to observe a more distinct pattern for reduction in the shear strength post cyclic loading as a function of cyclic stress ratio.

In order to evaluate the effect of prior cyclic loading on static shear strength properties a series of tests were carried out on normally consolidated samples subjected to different cyclic shear stress amplitudes for different number of cycles. The samples were first subjected to uniform cyclic shear stresses under undrained conditions. The accumulation of excess pore water pressure cyclic strain amplitudes were monitored continuously. Following cyclic loading the sample were kept under K_o stress conditions for an hour to assure stabilization and uniform distribution of pore pressure in the sample. Then the sample was sheared under stress controlled conditions. It was observed that the strength reduction following cyclic loading without allowing the dissipation of excess pore pressures, was more dependent on number of cycles then cyclic stress ratio. And it appears possible to consider a critical number of cycles below which strength reduction would be limited and can be considered as independent of cyclic stress ratio. For larger number of cycles, it was possible to observe a more distinct pattern for reduction in the shear strength post cyclic loading as a function of cyclic stress ratio.

Effect of Sustained Shear Stresses

During earthquakes a soil element located in an embankment or in a natural slope and already under the effect of sustained shear stress would be subjected to cyclic shear stresses. In order to study the response of a soil element under these conditions, a series of cyclic shear tests were performed on clay samples that were subjected to static shear stresses under drained conditions. The samples were allowed to reach equilibrium conditions under this sustained shear stresses for a day and later were subjected to increasing uniform cyclic shear stress amplitudes. Since it was reported by Ishihara (1985) that the shear stress - residual shear strain response in not affected very much form the magnitude of sustained shear stresses, all of the tests conducted at different axial consolidation pressures were first subjected to 25 % of consolidated undrained shear strength determined from static shear tests.

Cyclic and Monotonic Loading

One other alternative is the application of quasi-static and cyclic loading simultaneously. In this way it appears possible to estimate the reduction in the static shear strength during cyclic loading. The results obtained at this stage can also be used in evaluating the drivability of piles since the stress conditions are very similar. The controlling factor in this case is the cyclic stress ratio and the reduction in the static shear strength can be very significant even for relatively small cyclic stress ratios.

The cyclic behavior of a undisturbed, normally consolidated, marine clay has been studied by conducting several series of undrained cyclic simple shear tests with different shear stress amplitudes. The experimental results indicate that in the case of cyclic loading, the shear strength may be defined as a cyclic yield strength and a simple procedure is developed to estimate its magnitude as a function of number of cycles. The results also indicate that the effect of prior cyclic loading on static shear strength appears to become important over a critical number of cycles under which the stress reduction remains limited. If cyclic shear stresses are simultaneously applied along quasi-static shear stresses the static shear strength reduction may be significant and would be a function of cyclic stress ratio.

EVALUATION OF LIQUEFACTION SUSCEPTIBILITY

One of the major fields of research involves the evaluation of liquefaction potential of saturated sandy layers located in seismicly active regions. In one case the liquefaction susceptibility of shallow sandy soil layers encountered along the soil profile were evaluated by carrying out a parametric study using seven different semi-empirical procedures developed based on SPT blow counts and grain size distributions. In addition cyclic simple shear tests were performed on undisturbed soil samples obtained by special sampler. The liquefaction susceptibility of sandy silt and silty sand layers encountered at the investigated site are evaluated based on methods proposed in the literature by :

Method 1 - Seed, Tokimatsu, Harder, and Chung (1985)

Method 2 - Taiping, Chenchun, Lunian,and Hoishan (1984)

Method 3 - Iwasaki, Tatsuoka, Tokida, and Yasuda (1978)

Method 4 - Ishihara and Perlea (1984)

Method 5 - Yokota (1980)

Method 6 - Yuqing, Fang, Quingyu, and Guoxin (1980)

Method 7 - Atkinson, Liam Finn, and Charlwood (1984)

The safety factors against liquefaction susceptibility were evaluated using these 7 semi-empirical procedures. In the case of Method-3, three alternative formulations and in the case of Method-6, two alternative procedures were carried out. As a result, liquefaction susceptibility is evaluated based on 10 different semi-empirical methods for the total of 244 locations where sandy layers were encountered in boring profiles.

One purpose of this study is to obtain a comprehensive picture about the liquefaction susceptibility of sandy soil layers, while the other purpose to conduct such a parametric study concerning 10 different methods proposed by different researchers, is to demonstrate the subjective nature of the liquefaction evaluation procedures. One of the major reasons for discrepancies among different methods is due to variations in the SPT testing procedures or more precisely due to deviations in the impact energy between the different SPT testing systems and techniques since the above cited semi-empirical liquefaction evaluation methods were developed in different countries where SPT testing procedures may differ significantly. In addition dissimilarities in data bases (site conditions and soil types) used to develop the empirical correlations play an important role in the divergence of the results obtained.

The large scatter was observed in the safety factors within each method as well as differences between different methods. Even though the presence of safety factors smaller than one indicate a liquefaction susceptibility, it was apparent that it may not be very dominant and wide spread.

From an engineering point, this general statistical evaluation of all methods separately and together indicates relatively limited liquefaction susceptibility at the site during a strong earthquake. In addition presence of gravel, silt and clay particles and pockets in sandy soil layers as observed in samples obtained from the site will be another important factor decreasing the influence of liquefaction.

Cyclic Simple Shear Tests

A more accurate and realistic approach to determine the liquefaction potential of saturated sand deposits is to conduct a set of cyclic tests, preferably, on undisturbed samples. For this purpose, sets of consolidated undrained, cyclic simple shear tests were performed on undisturbed soil samples obtained from the site. The samples were taken by a special sampler and were frozen before they were transported to the laboratory.

The cyclic simple shear tests were carried out on samples with different grain size distributions and different percentage of fines. The grain size characteristics of sandy soil layers encountered at the site vary within a large range. At some locations the fines content may decrease below 5% where the sand samples can be classified as SW or SP and at some locations fines content may increase as high as 50% where the sand samples can be classified as SC or SM. An effort is made to conduct sets of tests on samples with different fines content in order to have a general picture to evaluate the liquefaction susceptibility of the sandy soil layers encountered at the site.

Initial liquefaction was observed in 4 sets of tests conducted on samples which had fines content varying between 11% to 29%. The results of these sets of tests are utilized to determine the liquefaction strength of the sandy soil samples. As expected the increase in the fines content increased the liquefaction resistance. In 2 sets of tests pore pressure accumulation were limited and no liquefaction was observed even after large number of cycles or at large shear strain amplitudes. The main reason for this type of response is the presence of high percentage of fines in these samples. Taking into consideration the effects of testing technique and sample disturbance and in order to be on the safe side, it appears justifiable to use the lower bound curve in the calculation of the safety factors for liquefaction.

Liquefaction Susceptibility

At the north part of the site, the base rock is relatively shallow overlain by medium dense silty gravelly sand layers of approximate thickness of 10 to 12 m. The SPT blow counts varies between 3 and 12 along the depth of the sand layers. The grain size analysis performed on the samples from the upper part of these sand layers indicate that the fines content are around 10% and the gravel content is around 20% while for the samples obtained from the lower part the fines content is around 15% and gravel content is around 10 %. The safety factors based on laboratory tests are larger than 1 for the whole depth of the sand layers while empirical methods show a large scatter.

Around the south part of the site, the depth of the base rock increases from approximately 16 m to 30 m. The sand layers encountered in the soil profile are medium dense and contains high percentage of fines and some gravel at various depths. The SPT blow counts were between 3 and 14 along the top 15 m. The safety factors based on laboratory tests are larger than 1 for the whole depth of the sand layers. However, the value of the calculated safety factor is around 1.05 indicating a medium liquefaction susceptibility between the depths of 6 to 8 m.

Similar results were observed for other boring locations and according to these analyses summarized above based on both semi-empirical procedures and laboratory test results, the liquefaction susceptibility of the sand layers at investigated site was considered to be marginal. Even though the laboratory determined strength values were reduced significantly (in some cases reduction is in the order of 100%), the calculated safety factors for all boring locations were all larger than 1.0 indicating a low or no liquefaction potential.

A detailed investigation was conducted to evaluate the seismicity of the region and to determine the effects of local soil conditions on the earthquake characteristics at the ground surface as well as the liquefaction potential for the site based on parametric and experimental studies. The liquefaction potential of the shallow sandy soil layers were studied in detail utilizing 7 (with their alternatives 10) different semi-empirical methods and based on laboratory cyclic simple shear tests conducted on undisturbed samples. The results obtained from semi-empirical procedures show a large scatter, however, the overall evaluation of these findings indicate only marginal liquefaction susceptibility. The more sophisticated evaluation based on cyclic simple shear tests and site response analysis supports this conclusion that the effect of liquefaction at the site would be negligible. This is mostly due to the relatively large percentage of fines and gravel present in these layers. In addition the presence of gravel pockets will lead to a faster dissipation of pore pressure preventing liquefaction.

EVALUATION OF THE EFFECT OF LOCAL SOIL CONDITIONS

The factors controlling structural response during earthquakes may be considered in three groups as; earthquake source characteristics, local soil conditions, and structural features. The earthquake source characteristics represents the effects of geology and tectonic formations of the region. However, even in the case of a widespread city like Istanbul, the influence of these factors are on more macro level and would not be sufficient to explain the variations in structural damage that may be observed within relatively short distances. On the contrary, the local soil conditions which can be very different due to changes in the thickness and properties of soil layers, depth of bedrock and water table would have a more dominant impact on damage distribution.

From an engineering perspective it appears possible to investigate the properties of local soil layers to implement necessary preventive measures and to design structures minimizing the vulnerability. However, at the present age, neither epicenter location, magnitude, and time of an earthquake can be controlled nor can be predicted. This aspect of the problem introduces an uncertainty into the engineering design. One logical way to approach this natural randomness is to adopt a statistical analysis in estimating the probabilities and for selecting risk levels with respect to the importance of the structures and the corresponding financial investment.

An ideal way to perform such a statistical evaluation of earthquake variability is to utilize strong motion records taken at same locations during different earthquakes. However, this type of data is very scarce if not completely unavailable. In studying this phenomena to estimate the effects and behavior of soil layers during a possible earthquake, one alternative is to use numerical models developed for site response analysis. In this situation the results obtained will be directly dependent on the characteristics of the input earthquake motion. Therefore one of the important stages in site response studies is the selection of an appropriate and realistic design earthquake. Experience and observations during the past earthquakes have shown that due to regional differences each earthquake would normally possess unique properties representing the local tectonic formations and earthquake source mechanisms. It was observed in some cases that even earthquakes occurring in the same fault zone with epicenters close to each other could have important differences. Therefore a statistical evaluation of this aspect of the problem could introduce a probabilistic interpretation enabling the design engineers to base their decision on better defined risk levels.

In some cities like Istanbul which has lived through many strong earthquakes and was demolished severely many times in its history, there may be no representative strong motion record since no major event has taken place during the last century. In these cases one the problems is the selection of a realistic strong motion record that would not yield overconservative and uneconomical results. Therefore the use of statistical and probabilistic analysis would allow to establish a relationship between risk levels and corresponding financial investment to decide on the level of allowable risk depending on the availability of the sources and the importance of the structures that are being analyzed.

The purpose was to evaluate the effects of randomness in strong motion records on site response analysis for different soil profiles. In this way it would be possible to understand the range of the influence of earthquake source characteristics. And depending upon the statistical distribution of these factors a better defined criteria can be adopted for selecting the design earthquake for soil amplification and microzonation investigations.

A total of 25 strong motion records (8 from Turkey with their two components, 1 from Yugoslavia, 7 from California, and 1 synthetic) were used to determine the effects of different earthquake characteristics. All of these strong motion records were scaled to peak acceleration levels of 0.15, 0.30, 0,45 g to study the influence of the earthquake magnitude. The differences are partly due to the differences in the source mechanisms and partly due to the site conditions where these records were obtained. Such broad range of variation in the characteristics of the selected earthquake records is believed to cover all the possibilities concerning a probable earthquake that may affect the city of Istanbul.

The site response analysis adopted in this study is based on the procedure developed by Schnabel, et al.(1972). A parametric study is conducted using 25 strong motion records scaled to different peak acceleration levels and different soil profiles representing the soil conditions at various parts of the old sector of Istanbul. An effort is made to choose soil profiles with different characteristics.

Effects of Earthquake Characteristics

The results of the site response analyses conducted indicate the importance of the earthquake characteristics on the response of soil layers. One alternative under this condition is to draw an outer envelope as a possible design spectrum such that sufficient safety can be achieved in the design and construction of structures against all probable variations in earthquake characteristics. However, it was observed that the properties of such a design spectrum is also very dependent on the magnitude of the earthquake or more specifically on the level of peak acceleration. Since the earthquake induced stresses and resulting strains are larger in the second case, the response of the soil layers given in terms of absolute acceleration response spectra show much larger variation in terms of spectral amplification and predominant soil periods.

One other way of demonstrating the effects of earthquake characteristics is to consider the variation of the calculated peak accelerations on the ground surface. The normalized peak acceleration values for different earthquakes and for the selected base rock peak acceleration levels have a significant scatter. On the other hand the degree of amplification in terms of peak accelerations is very dependent on the level of the base rock peak acceleration which in a way represents the magnitude of the earthquake input. As observed by various researchers with the increase in the earthquake magnitude the calculated soil amplifications decrease.

The influence of the characteristics of input earthquake and its magnitude is also reflected in the calculated soil predominant periods. In the case of stronger earthquake input the predominant soil periods increases. However, again this phenomena is very dependent on the properties of the soil layers.

The thickness of soil deposit in the selected soil profiles are approximately same but the geotechnical properties of the soil layers encountered at all locations are different. As a result the effects of earthquake input are significantly different. It is clearly evident that structures located on these layers would experience different earthquake forces. In order to make a more realistic evaluation of earthquake induced forces on structures, it is very essential to take into account the properties of local soil conditions.

Probability Analysis

An attempt is made to conduct a preliminary statistical analysis based on the calculated variations in peak accelerations at ground surface and predominant soil periods. In this way it is believed that the effect of the differences in the input earthquake characteristics can be taken into account with respect to the safety level required for structures located on these layers. It appears realistic to assume that the selected strong motion records represent the range of possible earthquakes that may take place in the near vicinity of Istanbul and the variation of peak ground acceleration and predominant periods can be modelled statistically by a normal distribution. Under these circumstances, it is relatively simple to calculate the probability of exceedence in terms of peak ground accelerations and predominant soil periods. At this stage it is justifiable to consider the variations in the probability separately with respect to the selected base rock peak acceleration levels since they represent approximately the seismicity of the region. The magnitude of a possible earthquake in a region should be estimated based on available seismological data with relation to adopted risk levels or return periods. After these analyses have been conducted then the above mentioned probabilities concerning the characteristics of possible earthquakes can be taken into account. However, in order to be consistent in the final evaluation of the earthquake characteristics at the investigated site, the probability level selected for the peak ground acceleration and predominant soil period should match the risk level adopted in the seismicity study.

An attempt is made by conducting an analytical study to evaluate the effects of input earthquake characteristics on response of soil layers. Although the dominant factor controlling the response of structures located on the ground surface is local soil conditions, it is evident from this study that earthquake characteristics also may play an important role.

From an engineering perspective, since it appears rather difficult, if not impossible, to make a deterministic evaluation concerning the earthquake characteristics, a statistical approach can be utilized to estimate the range of input earthquake effects and to calculate the corresponding probabilities.

REFERENCES

Andersen,K.H.,Pool,J.H.,Brown,S.F., and Rosenbrand,W.F.(1980) "Cyclic and Static Laboratory Tests on Drammen Clay. J. Geotech. Engng., ASCE, 106(GT5), 499-529.

Ansal,A.M. (1985) "An Endrochronic Model For Cyclic Behavior." Special Volume on Constitutive Models for Soils, Proc. of 11th International Conference on Soil Mechanics and Foundation Engineering, San Francisco, pp.123-127.

Ansal,A.M. (1985) "The Effects of Local Soil Conditions During Earthquakes." Proc. of 12th Regional Seminar on Earthquake Engineering, Halkidiki, Greece, 35 p.

Ansal,A.M. (1986) "Liquefaction and Reliquefaction." Proc. of 8th European Conference on Earthquake Engineering, Lisbon, Portugal, Vol. 2, pp.5.3/9-15.

Ansal,A.M. (1987) "Constitutive Relationships For Soil Dynamics." Strong Ground Motion Seismology, ed. E.Erdik and N.Toksöz, D.Reidel Publishing Com., Proc. of Nato Advanced Study Institute, Ankara, pp.535-544.

Ansal,A.M., Ansal,H. and Krizek,R.J. (1987) "Modelling Cyclic Elastic Behavior of Sands." Soil Dynamics and Earthquake Engineering, Vol.6, No.2, pp.90-99.

Ansal,A.M. and Yıldırım,H. (1988) "Shear Strength of a Marine Clay Subjected to Cyclic Loading." Proc. of 14th Regional Seminar on Earthquake Engineering, Austria pp.53-62.

Ansal,A.M. and Erken,A. (1989) "Undrained Behavior of a Normally Consolidated Clay Under Cyclic Shear Stresses." ASCE, Journal of Geotechnical Engineering Division, Vol.115, No.7, pp.968-983.

Ansal,A.M. and Tuncan,M. (1989) "Consolidation in Clays due to Cyclic Stresses." Proc. of 12th Int.Conf. Soil Mechanics and Foundation Engineering,Vol.1,pp.3-6, Rio do Jenerio, Brazil.

Ansal,A.M. and Yıldırım,H. (1989) "Dynamic Shear Strength Properties of Golden Horn Clay." Proc. of Session on Influence of Local Conditions on Seismic Response, 12th International Conference on Soil Mechanics and Foundation Engineering, pp.121-126, Rio de Jenerio, Brazil.

Ansal,A.M. and Günes,A.M.(1990) "The 1894 Earthquake of Istanbul." Proc. of Third International Symposium on Historical Earthquakes in Europe, Liblice by Prag, Czechoslovakia, pp.263-271.

Ansal,A.M. and Erken,A. (1990) "Liquefaction Potential of Silty Sand Deposits." Proc. 9 th European Conference on Earthquake Engineering, Moscow, USSR, Vol.4B, pp.71-80.

Ansal,A.M. and Lav,A.M.(991) "Effect of Variability of Input Motion Characteristics on Ground Response Spectra." Proc. 4th Int.Conf.on Seismic Microzonation, Stanford, USA, Vol.2, pp.131-138.

Ansal,A.M. (1991) "Evaluation of Liquefaction Susceptibility".(1991) Proc.5th Int.Conf. on Soil Dynamics and Earthquake Engineering, Karlruhe, Germany

Ansal,A.M. and Lav,A.M.(991) "Effect of Earthquake Characteristics on Response of Soil Layers". (1991). Proc.5th Int. Conf. on Soil Dynamics and Earthquake Engineering, Karlsruhe, Germany,

Atkinson,G.M., Finn,L.W.D. and Charlwood,R.G.(1984) "Simple Computation of Liquefaction Probability for Seismic Hazard Application." Earthquake Spectra, Vol.1(1) pp.107-123.

Castro,G. and Christian,J.T.(1976)"Shear Strength of Soils and Cyclic Loading. J.Geotech.Engng.,ASCE,102(GT9),887-894.

Idriss,I.M. and Seed,H.B.(1969) "An Analysis of Ground Motion During the 1957 San Francisco Earthquake." Bull.Seismological Soc. America, Vol.58(6), pp.2013-2032.

Ishihara,K. and Perlea,V. (1984) "Liquefaction-Associated Ground Damage During the Vrancea Earthquake of March 4, 1977", Soils and Foundations, Vol.24(1), pp.99-112.

Ishihara,K.(1985) "Stability of Natural Deposits During Earthquakes. Theme Lecture." Proc. of 11th Int. Conf. on Soil Mechanics and Foundation Engng., San Francisco, 2 (1985) 321-376.

Iwasaki,T.,Tatsuoka,F.,Tokida,K., and Yasuda,S.(1978) "A Practical Method for Assessing Soil Liquefaction Potential Based on Case Studies at Various Sites in Japan, pp.885-896, Proc. 2nd Int.Con.on Microzonation for Safer Construction-Research and Application, San Francisco.

Koutsoftas,D.C.(1978) "Effect of Cyclic Loads on Undrained Strength of Two Marine Clays." J.Geotech.Engng., ASCE, 104(GT5) 609-620.

Larew,H.G.and Leonards,G.A.(1962) "A Repeated Load Strength Criterion." Highway Research Board, 41,529.

Lee,K.L. and Focht,J.A.Jr.(1976) "Strength of Clay Subjected to Cyclic Loading." Marine Geotechnology 1(3), 165-185.

Matsui,T., Ohara,H.,and Ito,T.(1980) "Cyclic Stress-Strain History and Shear Characteristics of Clays." J.Geotech. Eng., ASCE, 106(GT10) 1101-1120.

Sangrey,D.A. & France,J.W.(1980) "Peak Strength of clay Soils After Repeated Loading History." Proc. of Int.Symp. Soils Under Cyclic and Transient Loading, Swansea, Balkema, Rotterdam, 1, 421-430.

Sangrey,D.A., Castro,G., Poulos,S.J., and France,J.W.(1978) "Cyclic Loading of Sands, Silts and Clays." Proc. of Specialty Conference on Earthquake Engineering and Soil Dynamics, ASCE, Pasadena,CA 836-851.

Sangrey,D.A., Henkel,D.J., and Esrig,M.I.(1969) "The Effective Stress Response of a Saturated Clay Soil to Repeated Loading." Can. Geotech. J.,6(3) 241-252.

Schnabel,P.B.,Lysmer,J.,and Seed,H.B.(1972) Shake - A Computer Program for Earthquake Analysis of Horizontally Layered Sites, EERC Report No.72-12, Uni.of California, Berkeley.

Seed,H.B. and Chan,C.K.(1966) "Clay Strength Under Earthquake Loading Conditions." J. Soil Mechanics and Foundations., ASCE, 92(SM2) 53-78.

Seed,H.B., Tokimatsu,K., Harder,L.F., and Chung,R.M. (1985) "Influence of SPT Procedures in Soil Liquefaction Resistance Evaluations, ASCE, J Geotech. Engng. Div., Vol.111 (GT12) pp.1425-1445.

Taiping,Q., Chenchun,W., Lunian,W., and Hoishan,L. (1984) "Liquefaction Risk Evaluation During Earthquakes." Vol.1, pp.445-454, Proc. Int.Con. on Recent Advances in Geotechnical Earthquake Engineering and Soil Dynamics, St.Louis.

Taylor,P.W.and Bacchus,D.R.(1969)" Dynamic Cyclic Strain Tests on a Clay. Proc. of 7th Int.Conf. of Soil Mechanics and Foundation Engng., Mexico City, 1, 401-409.

Thiers,G.R. and Seed,H.B.(1969) "Strength and Stress-Strain Characteristics of Clays Subjected to Seismic Loading Conditions." ASTM STP 450, Vibration Effects of Earthquakes on Soils and Foundations 3-56.

Thiers,G.R. and Seed,H.B.(1968) "Cyclic Stress-Strain Characteristics of Clays.",J. Soil Mechanics and Foundations, ASCE, 94(SM2) 555-569.

Valera,J.E. and Donovan,N.C. (973) "Incorporation of uncertainties in seismic response of soils." Proc.5th World Conf. Earthquake Engng., Rome, Vol.1, pp370-379.

Yokota,K.(1980) "Evaluation of Liquefaction Strength of Sandy Soils, Vol.3,pp.121-124, Proc. 7th WCEE, Istanbul.

Yuqinq,W.,Fang,L.,Quingyu,H., and Guoxin,L.(1980) "Formulae for Predicting Liquefaction Potential Clayey Silt as Derived from Statistical Method." Vol.1, pp.227-234, Proc. 7th WCEE, Istanbul.

STATIC AND DYNAMIC ENGINEERING SOIL PARAMETERS EVALUATED BY MEANS OF BOX TESTS

J. Benčat[1]

INTRODUCTION

Since a complete knowledge of the state of stress in the subbalast and subgrade and the knowledge of their elastic parameters is necessary for predicting the contribution of these layers to track setlements, the box tests can provide important information on track structure system field performance. This research was performed as a part of a larger project (Federal Transport Ministry, ČSFR) dealing with dynamic behavior of the track system subjected to vehicle loading.
In the frame of this research, the box of the static and dynamic soil parameters were performed in the laboratory of the Department of Structural Mechanics, University of Transport and Communications Žilina.

THEORETICAL APPROACH

Theory of an elastic half-space (Timoshenko 1951, Johnson 1985) provides formulation for the evaluation of elastic moduli E_r, E_d. Pushing the rigid circular plate into the elastic half-space with uniform pressure (Hertz pressure) enables to develop the relationship for the evaluation of the resilient modulus $E_r = E_r (r, w, \nu)$ as follows:

$$E_r = (1 - \nu) \frac{\pi}{2} \cdot \frac{r \cdot p}{w} \qquad (1)$$

where: r = Radius of rigid circular plate, mm
p = Static or dynamic pressure at the contact area of the rigid circular plate and the half-sprace, MPa
w = Uniform normal displacement of the contact area, mm
E = resilient modulus, MPa
ν = Poisson´s ratio

Eq. (1) is utilized in both the static and the dynamic experimental tests for the evaluation of static and dynamic resilient moduli or soils in situ of by means of box test. Based on the examination of the available models of the truck substructure and consideration of its desired features, theoretical model based on an elastic approach provides satisfactory results for practical aims.

MEASURING PROCEDURE

There exist several standards and prescription for the static and dynamic loading tests (in situ only) in the branch

---

1 - Department of Mechanics, Faculty of Civil Engineering, VŠDS Žilina, ČSFR

of the civil engineering activity in ČSFR. E.g. Č.S. Standard ČSN 73 F-deral Transport Ministry prescription ČSD-S4-Railway Subgrade (comprehend prescription for static loading tests (SLT) of the subbalast and subgrade), etc. To introduce dynamic loading test (DLT) for in situ measurement of the resilient moduli it was necessary to find the relationship between static (E_r) and dynamic (E_d) moduli of the soils and substitute materials (e. g. granulated slag) which are used as a substructure material. Up to now, principles of the method of sleeper subgrade construction arrangement has been used at ČSD (Czecho-Slovak Railways) using modulus E_r for each layer. To avoid the expensive experiments in situ both the static and the dynamic tests of soil parameters, the box tests were carried out in the laboratory. Several tests have been validated by comparison with field measurements from a test track, too. A special test facility was constructed for this purpose. It incorporated a wooden box (Fig. 1) with dismensiones 320x290x120 (cm), to examine the correlation between static and dynamic resilient moduli of the subbalast and subgrade under simulated field conditions. Figure 1 also shows numbered layers of the subgrade and subbalast and provides the description of the construction of the box and steel frame of the hydromechanical equipment for the static loading test. At the contact areas both the bottom and side walls of the box were covered by soft composite material with thickness 5,0 cm and with resilient modulus E_r = 10,5 MPa. Before starting the static test, the optimal regime of the electrodynamic plate vibrator to achive optimal degree of the soil compacting according to Proctor Standard test (P.S.T.). For each layer after compacting specimens for standard laboratory test of the soil geotechnical properties e. g. grain characteristics, specific gravity G_s, unit weight (ρ), moisture (w), void ratio (e), compaction level, degree of saturation (S_r), Poisson ratio (ν), etc.

For cohesionless soils the optimal degree of compacting was expressed by volume mass (ρ) with the application of TROXLER device (radiometric probe - JGP 104). Figure 2 shows the location of the points ($G_1 \ldots G_{13}$) for taking specimens of soil at each layer of the substructure. The subgrade was successively created by three layers of 20.0 + 15.0 + 15.0 = = 50.0 cm of the cohesive soils and the subballast was created by the combination of the cohesionless soils imposed on the subgrade either with geofabrics of without the geofabrics. During the test the properties of the tested soils were approximately constant. Six combinations of soils and substitute materials which created the layered substructure were carried out. For each combination of soils the subgrade was made as a sandy loam (MS) approximately with parameters ρ = 1735.0 kg/m³ W_{OPT} = 13.9 %. The layers of the subballast were successively created by gravel sand, sand and slag. The layers of the subballast in each test have thickness of 20.0 + 15.0 = 35.0 cm.

Static Loading Test (SLT)

The static resilient modulus E_r (MPa) of the tested soils was evaluated by measuring the rigid circular plate vertical displacement w (mm) due to hydromechanical equipment. The

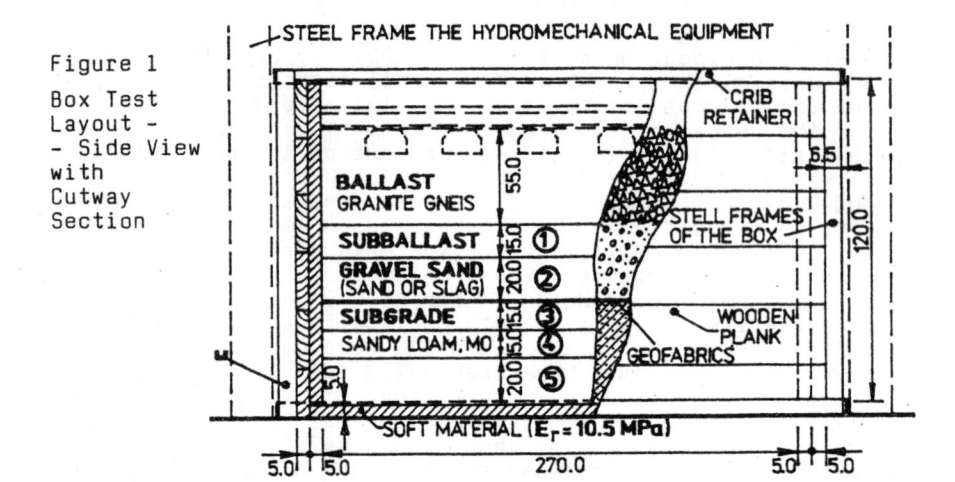

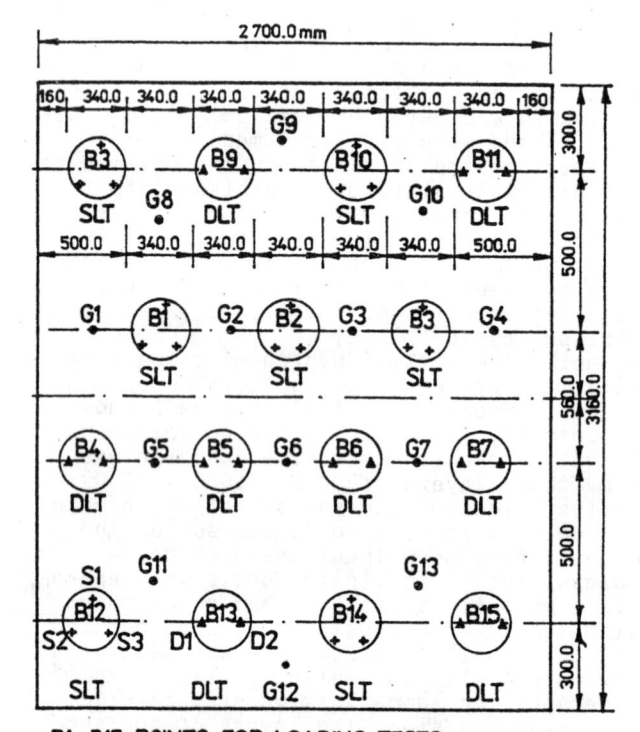

Figure 1
Box Test
Layout -
- Side View
with
Cutway
Section

Figure 2
Box Test Layout -
- Top View with
Position of
Measured Points

B1...B15 POINTS FOR LOADING TESTS
G1...G3 POINTS FOR TAKING SPECIMENTS (•)
SLT = STATIC LOADING TEST, S1,S2,S3...MEASURED POINTS (+)
DLT = DYNAMIC LOADING TEST, D1,D2...MEASURED POINTS (▲)

contact area of the plate for each test was A = 1000.0 cm^2.
Figure 2 shows the location of points B1 ... B15 for the SLT.
The resulting vertical displacement w of the circular plate
for each test was obtained as an average value of the displacement values measured in the points S1, S2 and S3 which were
situated at the top of the plate. The vertical displacement
were measured by inductive displacement transducers which were
conjugated with the signal amplifier and via computer recorded
and printed. In Fig. 3 is the interpretation of the displacement time history of the typical static loading test (T7-1/4).

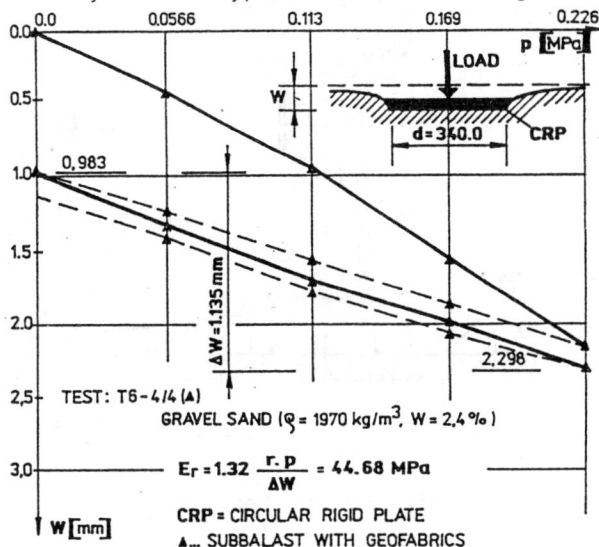

Figure 3 Displacement Time History of the Static Loading Test

The resilient moduli E_r were calculated according to eq. (1)
where w = Δw is the measured value of the vertical displacement
for the second cycle of the SLT due to pressure p = 0.22 MPa.
This approach is sufficient because after a number of load
repetitions, the soils behave nearly elastically as comfirmed
by the preceding tests. Then we can define the resilient modulus E_r as a repeated deviator stress (Seed 1965) divided by
the recoverable strain and it does not usually change significantly after a large number of cycles.

Dynamic Loading Test (DLT)

Dynamic resilient modulus E_d(MPa) was evaluated in the same way as a static modulus E_r, but the dynamic load was performed by impact loading test device, Fig. 4. This device consists
of the rigid circular plate (1) with the contact area
A = 1000.0 cm^2, dropping weight (2) with mass Q = 12.5 kg, in-
dention for setting the heigth of the weight (3), springs (4),
plunger (5), guide rod (6), casing (7) and safety pin (8).

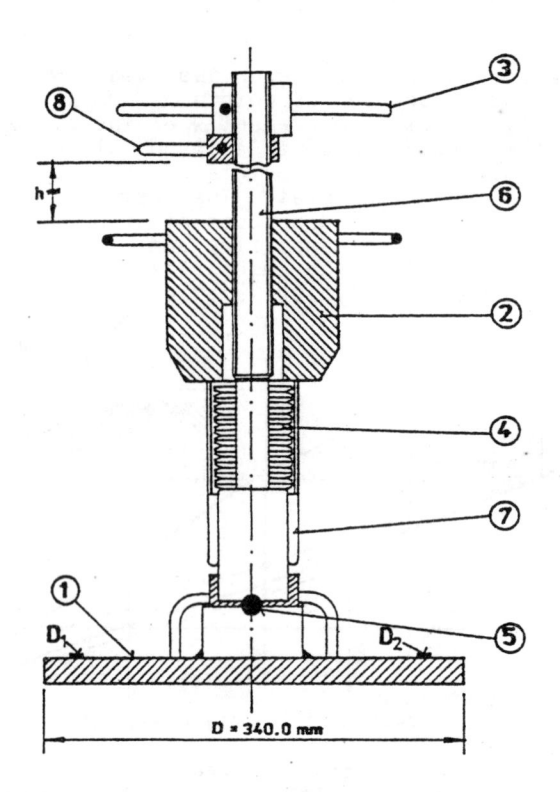

Figure 4 Dynamic Loading Test Device

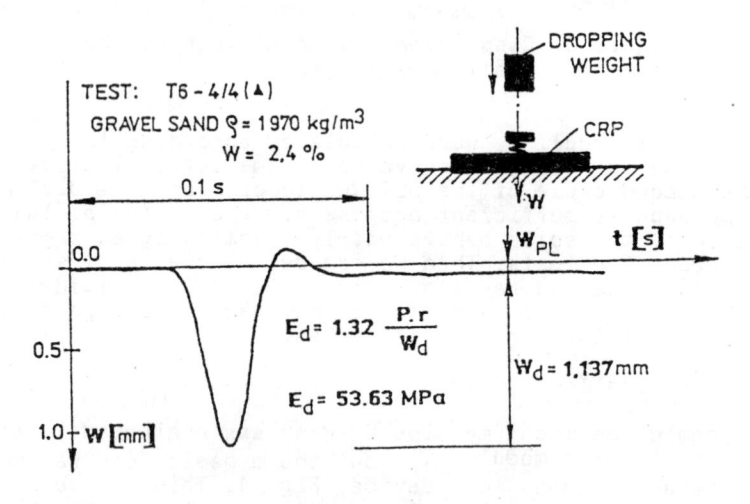

Figure 5 Displacement Time History of the Dynamic Loading Test

D1 and D2 are points where the dynamic vertical displacements were measured by inductive displacement transducers. Measuring of the dynamic deflections were performed by the same set of the apparatus as in the static tests. Figure 2 shows location of the points where DLTs were performed. In each dynamic test there were carried out 6 impacts in the measured spot caused by dropping weight from constant height h. The height h was set experimentally to achieve the constant area impact stress p = 0.22 MPa. In this case the dynamic moduli were calculated according to eq. (1), where w is the measured value, averaged from values of the displacement measurements at points D1 and D2 on the top of the circular plate and averaged from the 5 last values of the 6 performed impacts. Figure 5 shows typical time history of the dynamic deflection during the impact caused by dropping weight Q.

EXPERIMENTS RESULTS

Seven static loading tests and eight dynamic loading tests were performed in each layer of the soils creating a corresponding combination of the subballast and subgrade in the box. Finally, we have obtained 35 values of the E_r and 40 values of the E_d for the soil combinations in each box tests. There were carried out 6 combinations of soils with and without geofabrics at the top of the subgrade. Figure 6 shows the interpretation

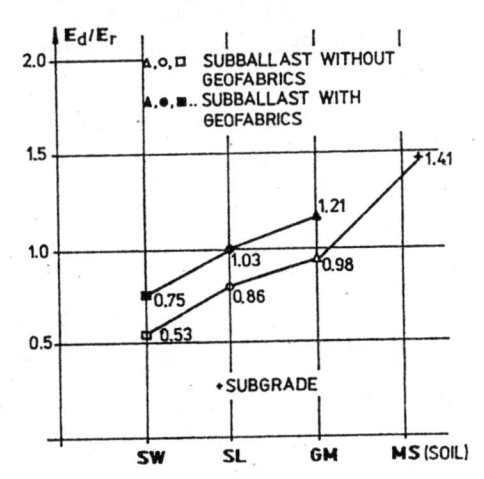

Figure 6 Ratio E_d/E_r for Individual Soils in the Box Test Combination

of the ratio E_d/E_r, which was calculated as a mean value of the statistical value collection of all corresponding tests.

The correlation coefficient k varied from 0.85 to 0.92.

CONCLUSION

The box test results confirmed the results obtained by in-situ tests by both the same ways and the same experimental set of the apparatus. The advantages of the box tests consist in the constant conditions during the performance of the box tests, possibility to change combinations of soils creating railway substructure and in the lower expenses in performing the study of the track structure material properties. The results will be utilized in performing DLT in the evaluation bearing capacity of structural layers of the substructure present and newly - built Czecho-Slovak railway networks.

REFERENCES

Benčat, J. (1989), "Research in Railway Structure Dynamics" Research Report III-4-6/03.03. Volume II. University of Transport and Communications, Žilina (in Slovak)

Johnson, K. L. (1985) "Contact Mechanics." Cambridge University Press, Cambridge, New York, Sydney.

Seed, H. B., et al., (1965), "Prediction of Pavement Deflections from Laboratory Repeated Load Test", Report No. TE-65-6, Soil Mechanics and Bituminous Materials Research Laboratory, University of California, Berkley, Calif.

Timoshenko, S. P., and Goodier, J. N. (1951), "Theory of Elasticity", 3rd Edn. New York, London et al.: McGraww-Hill.

LIQUEFACTION AND CHANGE OF MICROSTRUCTURE OF THE SOILS UNDER DYNAMIC LOADING

B.Svoboda [1]

INTRODUCTION

The variation of the geotechnical properties at dynamic loading well known in earthquake regions are observed also in regions with heavy vehicle traffic or in the vicinity of large excavated pits. Interest is centered first of all on fills, including sands, artificial fills from ash and overconsolidated clay sediments. A special laboratory apparatus was proposed and tested for observation of the variations in mechanical properties of soils dynamic stress, non cohesive soils and also the conditions for their liquefaction.

On the basic of many tests executed at the special dynamic laboratory, it has been eastablished that dynamic properties of cohesive and non-cohesive soils are essentially different depending on their inner structure and excess pore water pressure.

Non-cohesive soils, in general, consist of particules (grains). The remaining space is called porosity or a void ratio. It may be occupied by air, or by air and water. The water in the pores of such soils affect the type of response to dynamic loading. Usually, the grains are arranged in such manner that the soil may assume a more compact structure (consolidation). Release of high energy (e.g. earthquake or blast) causes liquefaction in loos, fine grained soils.

The following information is useful in estimating the behaviour of soil, as well as in evaluating its physical properties:

TABLE 1- Results of the static and dynamic tests

STATIC TESTS	DYNAMIC TESTS
porosity and void ratio water content degree of saturation unit weight of the soil specific gravity of soil particles microstructure grain size distribution relative density consistency of the clays bulk modulus shear modulus	limit porosity limit vibration velocity number cycles for disturbance dynamic shear modulus dynamic bulk modulus

Liquefaction potencial related to particle size (Shannon, 1972, Svoboda.1989) is on fig.1.

---

1 - STAVEBNÍ GEOLOGIE, Prague, ČSFR

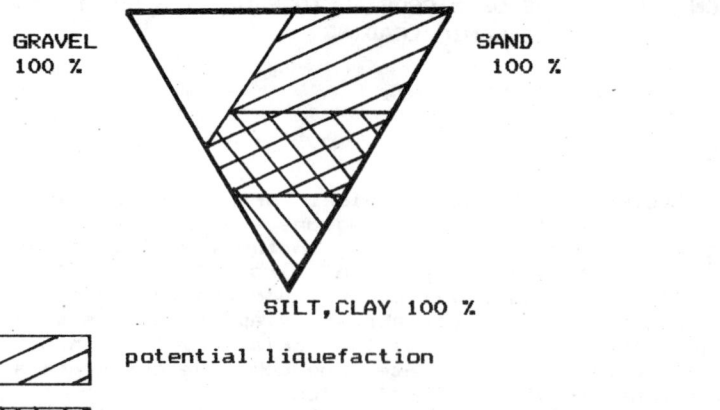

Fig.1 Liquefaction potential.

- potential liquefaction
- intermediate zone
- only overconsolidated clay with funicular (flokulation) structure

LIQUEFACTION

For liquefaction in saturated, non-cohesive soils the following conditions must be fulfilled:
a) The porosity or initial void ratio must be higher than the limit.
b) The vibration energy (expressed in vibration velocity) must be higher than the limit.

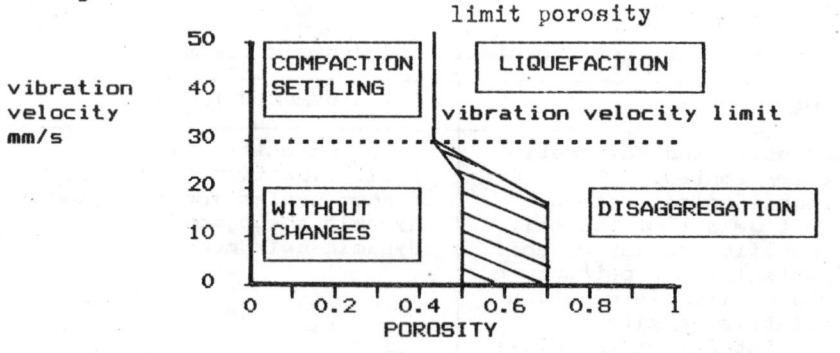

fig.2 Condition of the liquefaction, compaction and dissagregation.

On fig.2 these condition (Svoboda 1982, 1987. 1988, 1989) are expressed from various landing dumps. They determined by a simple apparatus composed from a cylinder fixed on a dynamic table.

248

The soils are disaggregated, when the vibration velocity is lower than the limit and the porosity is higher than the limit.

The soils are compacted and/or differential settling follows, when the vibration velocity is higher then the limit and porosity lower than the limit.

Given that soil has tendency to liquety for a shear test, it is possible to use a special dynamic triaxial apparatus according to the CIU - method with dynamic surcharge in vertical direction and with sensitive measurement of strain, pore water pressure and deviator stress. It is also important to fix the number of cycles for sampling disturbance.

Let us suppose a parent element of soil according to fig.3 and shear stress $\tau_{1.3} = \tau_{3.1} = 0$.

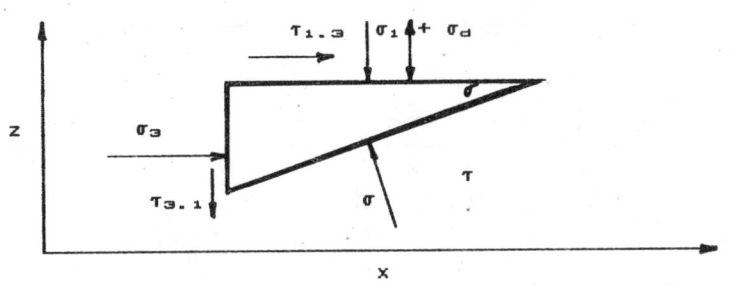

Fig.3 - Element of soil.

For equilibrium of the element the following equation is utilized:

$$\sigma_1 = \frac{\sigma_1 + \sigma_d + \sigma_3}{2} + \frac{\sigma_1 + \sigma_d}{2} \cos 2\delta \quad (1)$$

$$\tau = \frac{\sigma_1 + \sigma_d - \sigma_3}{2} \sin 2\delta \quad (2)$$

On the test is made so that the deviator stress according to the equation (3) equals zero:

$$\sigma_1 - \sigma_3 = 0 \quad (3)$$

Then we can express equation (1) and (2) in the following form

$$\sigma = \sigma_3 + \frac{\sigma_d}{2} + \frac{\sigma_d \cos 2\delta}{2} \quad (1,3)$$

$$\tau = \frac{\sigma_d}{2} \sin 2\delta \qquad (2,3)$$

On their results we can draw Mohr's circle according to the fig.4 or stress trajectory (Castro, 1987) fig.5.

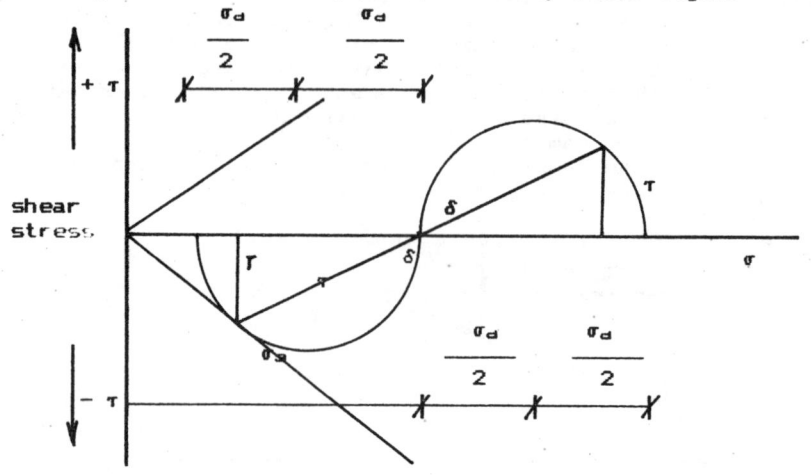

Fig.4 - Mohr's circle

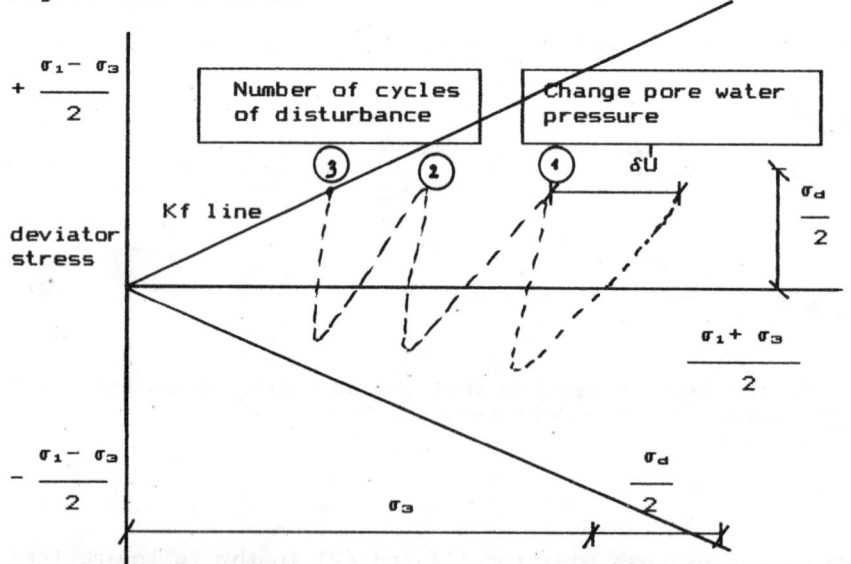

Fig.5 - Stress trajectory.

The magnitude of the initial porosity on lateral consolidation cell pressure is seen on fig.6, and the results of fig.6 correspond with fig.2.

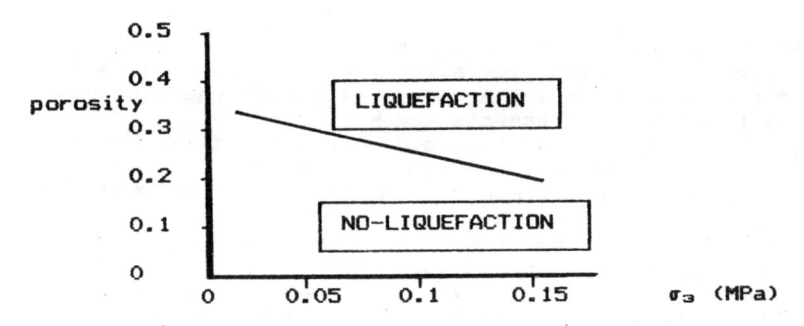

Fig.6 - Initial porosity of the liquefaction ash.

Furthemore a very important property is establishing the dependence of shear strain amplitude in percentage on excess pore water pressure (δu) for given number of loading cycles (N) and lateral overburden pressure (σ_3).

On figure 7, this relationship is given for ashes and as well as sands (according to Gazetas 1987 and Castro 1987). From fig.7 it is clear that the greater changes of pore water pressure occur in ashes by rather than in sands.

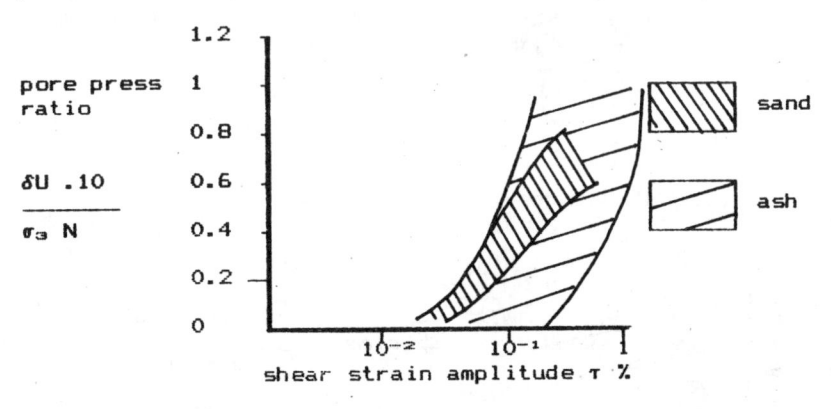

Fig.7 - Dependence of shear strain amplitude on the pore pressure ratio.

DYNAMIC CONSOLIDATION

It is well known to engineers that loose granular soil (sand) can be compacted by artificial vibration. Lateral vibration, acting on a large area is more effective than vertical vibrations. Similar compaction is caused by earthquakes, resulting in excessive settling.

If the porosity is lower than the limit and vibration velocity greater than the limit, the soil becomes a vite compact (fig.2).

If the porosity is higher than the limit and vibration

velocity is greater than the limit, at first the soil is compacted and then it is liquified after some cycles of dynamic loading and excess pore water pressure (fig.4,5) when the dynamic stress trajectory intersects the Kf line.

For building foundations and road structures it is important to know the variations of the bulk module at dynamic stress. These conditions are modeled in the dynamic oedometer. In non-cohesive soils, the module does not decrease for further loading cycles at the same vibration energy, but this does occur in cohesive soils. This is explained by the fact that non-cohesive soils create, by vibrations at a given higher than the limit of a relatively stable sructure, which is turn is deformed by higher energy again (fig.9).

In cohesive soils, which are sensitive to dynamic stress (fig.1) funicular (flokulation) structures collapse in successive steps (Svobodová 1987) - fig.8.

Fig.8 - Funicular structure.

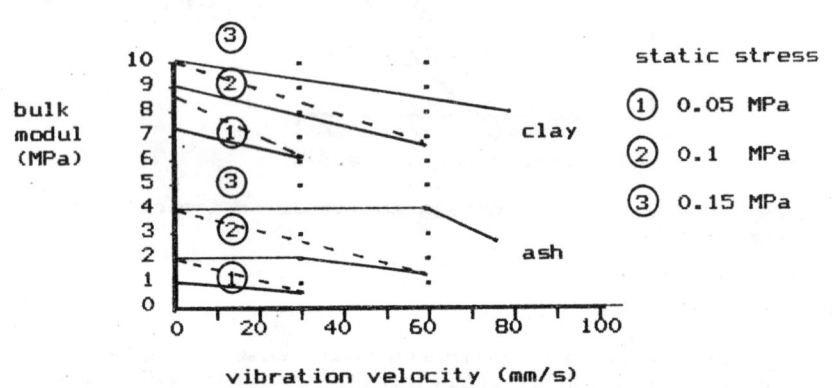

Fig.9 - Dependence vibration velocity on the bulk modul

SOIL AS FOUNDATION MATERIAL

If the building is too rigid and the soil is soft, the problem can be viewed as combination of sway and rocking - fig.10.

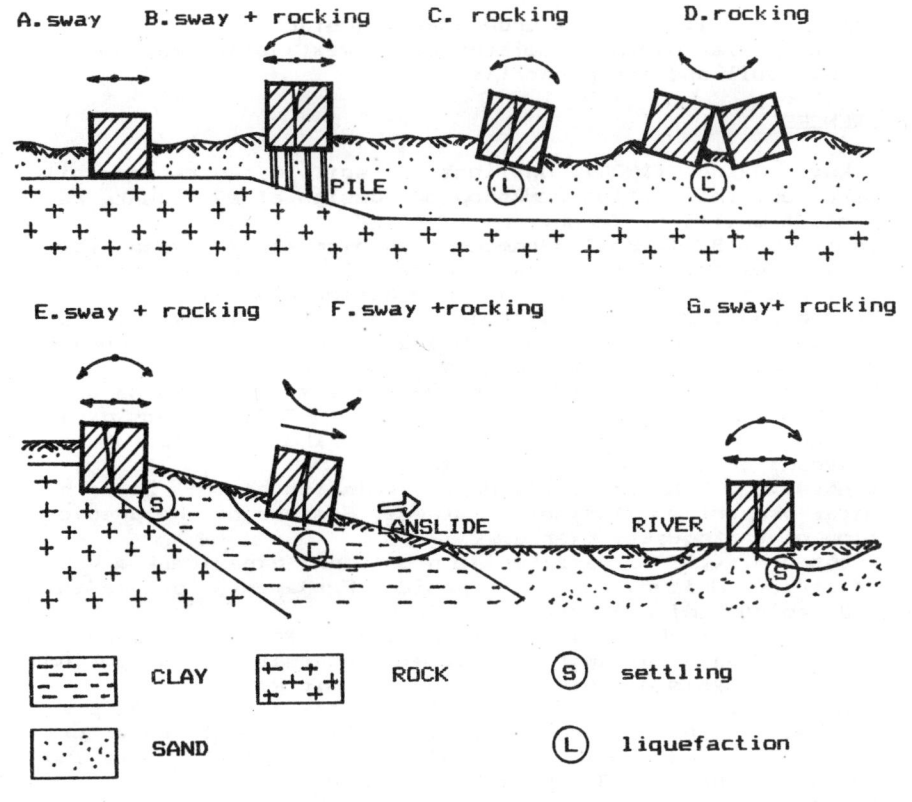

Fig.10 — The geological conditions sway and rocking of the buildings.

Individual foundations are highly vulnerable to damage, due to differential settling (fig.10, E,G) or liquefaction (fig.10, C,D,F), caused by earthquakes. Groundwater conditions (pore water pressure) is an important factor for differential settling or liquefaction.

Liquefaction potencial is related to particle size (fig.1), porosity, vibration velocity (fig.2) and excess pore water pressure (fig.4,5). In the case of liquefaction, landslides and damage of the building occur the slopes (fig.10, F).

Consolidation takes place in sand or overconsolidated clay with funicular structure (fig.8,9,10,E,G) in the absence of earthquakes.

CONCLUSIONS

Building cracks (caused by sway, rocking, or a combination of sway and rocking) may be observed in the earthquake regions if they are constructed on differential by settled rocks or soils, or soil that has undergone liquefaction.

The special dynamic geotechnical investigation may determine this soil and its properties.

REFERENCES

1. Alkut, Aytun (1986), Outlines of soil dynamics nad soil structure interaction workshop on design of earthquake resistant building, Jemen.
2. Castro,G (1987), Soil dynamics : An overview in Dynamic behaviour of soil during earthquake - linquefaction, in Cakmak, A.S.: Soil dynamic and liquefaction-Elservier.
3. Gazetas,G.(1987), Soil dynamics: An oveview in dynamic behaviour of foundations and buried structures.- in Banerjee D,K, and Butterfield R. Editors, Elservier.
4. Svoboda,B.(1987,1988,1989) Etapové zprávy o pracích na úkole VU III-3-4, provedených v letech 1987-89. Přenos dynamického rozruchu prostředím různých fyzikálních vlastností, Stavební geologie Praha (in Czech).
5. Svoboda,B., Svobodová,M.(1988), Variation of soil geotechnical properties at dynamic stress. Seminar on the prediction of earthquake, Lisboa.
6. Svoboda,B. (1989), Razžiženije grunta pri dinamičeskom vozdějstvii, Meždunarodnyj symposium Fundamenty pod mašiny s dynamičeskymi nagruzkami, Leningrad.
7. Svoboda,B. (1989), Liquefaction of the soil and damage of the building at dynamic stress. 4th International Symposium on the Analysis of Seismicity and Seismic Risk, Bechyně.
8. Svoboda, B. (1991) Variation of the soil geotechnical properties at dynamic loading. Second International Conference on recent Advances in Geotechnical Earthquake Engineering and Soil Dynamics. St. Luis Missouri Rolla.

DYNAMIC METHODS FOR EARTHQUAKE-RESISTANCE STUDY OF UNDERGROUND STRUCTURES

Mubarakov Ja.N.[1] Sagdiev H.[2]

INTRODUCTION

Design and construction of spatial systems of underground structures in regions with high seismicity are carried out on the basis of calculation results of structure simulation in the form of the most simple scheme of rod, beam or frame structure types. It leads to unfounded increase in building material expenditure and to complication of construction technology. Accounting of constructive peculiarities, deformation characteristics of structure elements and ground conditions of the site makes it possible to economize building materials and to ensure earthquake-resistance of underground structure elements.

In connection with above-mentioned this paper gives the principal approaches to dynamic theory of stress-strain state study of underground structures and the methods of seismic design. The evaluation of seismic stress state of underground cylinder shell by two methods based on different approaches was carried out. The field of application and practical use in solving the problem underground structure seismic safety was defined.

PRINCIPAL STATEMENTS

In general statement the problem of stress-strain state study of underground shell-type structures is reduced to combined integration of shell theory equation

$$L_m(u, v, w) = X_m \quad (m = 1, 2, 3) \tag{1}$$

and three-dimensional theory of soil elasticity

$$\bar{L}_n(u_1, u_2, u_3) = P_n \quad (n = 1, 2, 3) \tag{2}$$

with fulfilment of certain conditions on soil-structure interaction. In formulae (1) and (2) u, v, w and u_1, u_2, u_3 are the components of shell and soil displacement vectors; $X_1 \div X_3$ and $P_1 \div P_3$ are the components of surface and volume forces; L_m and $\bar{L}_n$ - differential operators.

When solving concrete problems some other necessary relations of shell theory of elasticity and wave propagation in soil theory and conditions of their interaction with underground objects are used.

In above-mentioned approach to investigation methods of

1, 2 - Institute of Mechanics and Earthquake-Resistance of Structures of Uzbek Academy of Sciences, Tashkent, Uzbekistan, USSR shell and soil stress-strain state depending on the statement and the aims of the problem different contact conditions are accepted. For example, at rigid contact kinematic and dynamic conditions have the form:

$$u = u_1, \quad v = u_2, \quad w = u_3 \tag{3}$$

$$X_1 = -X - \sigma_{zx}, \; X_2 = -Y - \sigma_{zy}, \; X_3 = -Z - \sigma_{zz}, \; (r=r_0)$$

here r_0 - outer radius of the shell; X, Y, Z - components of outer loads, applied to outer surface of the shell; σ_{zx}, σ_{zy}, σ_{zz} - components of stress vector $\bar{\sigma}$ in soil media. In the case of free sliding of ground media on the contact surface without take-off, it has the following form:

$$w = u_3, \quad \sigma_{zx} = 0, \quad \sigma_{zy} = 0, \tag{4}$$

$$X_1 = -X, \; X_2 = -Y, \; X_3 = -Z - \sigma_{zz}, \quad (r = r_0).$$

In the surface places where the shell is taken off from the soil the conditions may be as follows:

$$X_1 = -X, \quad X_2 = -Y, \quad X_3 = -Z, \tag{5}$$

$$\sigma_{zx} = 0, \quad \sigma_{zy} = 0, \quad \sigma_{zz} = 0, \quad (r = r_0)$$

Depending on the intensity of loading process by waves of different types a complex process of shell-soil mutual influence with a necessity of combined use of conditions (3÷5) may take place.

To get practical results in above-mentioned approach is a very complex problem and it requires the application of complicated mathematical apparatus allowing to obtain formal solutions.

That is why in underground structure stress-strain state study another approach based on simplified suppositions is used. Here stress vector $\bar{\sigma}_z$ on shell-soil interface is substituted by force vector $\vec{Q}$ of soil-structure interaction:

$$\vec{Q} = \vec{q}\, U, \tag{6}$$

here, $\vec{q}$ - vector operator with components q_x, q_y, q_z, which should be determined. If the operator is taken as a constant for normal constituent Q_z it follows from (6) the Winkler model [1] or Pasternak model [2] when $q_z = q + q_0 \nabla^2$, where ∇^2 - is two-dimensional Laplace operator on the contact surface. At proper conditions superposed on radial displacements the relation (6) includes integral models too ([3,4] and others).

Nowadays the soil-structure interaction model [5] based on the idea of the possibility of structure displacement relative to soil at dynamic disturbances (earthquakes, underground explosions etc.) is widely used. According to [5, 6] the strength of soil-underground structure interaction has the form

$$\vec{P} = K\vec{U}, \tag{7}$$

here K - complex operator which determines elastic, viscoelastic and elasto-plastic characteristics of soil-structure interaction.

In simplified approaches the coefficients have a complex composition and depend not only on physical parameters of the media surrounding the structure, but also on physical, geometric parameters and deformation form of considered system. The absence of obvious connection in formulae (6) and (7) with geometric and elastic parameters of soil and structure makes the choice of numerical values more difficult. Coefficients values are obtained on the basis of experiments and empirical dependences. In [5,6] they are built with account of separate essential geometric and physical parameters of soil and structures.

In connection with this a considerable practical interest is presented in papers [7, 8], where coefficients are defined on the basis of combined solution of shell and soil equations, proceeding from the rigorous mathematical statement of media interaction problem.

STUDY RESULTS

The main points of mentioned approaches in working out the design methods are the definition of their application fields. To reveal the possibility of these approaches application to the study of underground structure stress-strain state on the influence of dynamic (seismic, explosive and so on) loads we shall consider the problem of plane wave interaction with cylinder shell in elastic statement [8].

It is supposed that the wave is propagating perpendicularly to longitudinal axis of cylinder structure. In this case the plane problem of dynamic theory of elasticity may be considered with account of incident, reflected and radiated waves.

Presenting longitudinal wave in the form of incident potential

$$\varphi_{in} = \varphi_o e^{-i\omega t - iKr\cos\theta}, \tag{8}$$

the problem is reduced to the solution of two Helmholtz system equations relatively to longitudinal (φ) and transverse (ψ) potentials

$$\nabla \varphi_j + \alpha_j^2 \varphi_j = 0,$$

$$\nabla \psi_j + \beta_j^2 \psi_j = 0; \quad j = 1, 2, \tag{9}$$

here $\alpha_j^2 = \omega^2/c^2_{\alpha j}$, $\beta^2 = \omega^2/c^2_{\beta j}$; ω - is the frequency of incident disturbance, $c_{\beta j}$ and $c_{\alpha j}$ - longitudinal and transverse waves velocities; $j = 1$ for soil, $j = 2$ for structure; ∇ - Laplace's operator. Wave equation solutions (9) for the shell, the layer and the media surrounding them were obtained at stationary and non-stationary actions [8, 9]. For shell (layer)-media contact the boundary conditions (3) or (4) were accepted; and the inner surface of the shell (layer) is free from stresses.

In the second approach the considered problem on the basis of seismodynamic theory [5, 6] according to (7) is reduced to the solution of the equation system

$$\frac{\partial^2 u_\theta}{\partial \theta} - a^2 \frac{\partial^3 u_\theta}{\partial \theta^3} - a^2 \frac{\partial^4 u_r}{\partial \theta^4} - (1 + \frac{R^2 K_r}{D}) u_r =$$

$$= \frac{R^2 \mu}{D} \frac{\partial^2 u_r}{\partial t^2} + L(u_{r1}, u_{\theta 1}) + \frac{R^2 \mu}{D} \frac{\partial^2 u_{r1}}{\partial t^2}, \qquad (10)$$

$$\frac{\partial^2 u}{\partial \theta^2} + a^2 \frac{\partial^2 u_\theta}{\partial \theta^2} - \frac{\partial u_r}{\partial \theta} + a^2 \frac{\partial^3 u_r}{\partial \theta^3} - \frac{R^2 K_\theta}{D} u_\theta =$$

$$= \frac{R^2 \mu}{D} \frac{\partial^2 u_\theta}{\partial t^2} + L_2(u_{r1}, u_{\theta 1}) + \frac{R^2 \mu}{D} \frac{\partial^2 u_{\theta 1}}{\partial t^2},$$

here

$$L_1(u_{r1}, u_{\theta 1}) = a^2 \frac{\partial^3 u_{\theta 1}}{\partial \theta^3} - \frac{\partial u_{\theta 1}}{\partial \theta} + a^2 \frac{\partial^4 u_{r1}}{\partial \theta^4} + u_{r1};$$

$$L_2(u_{r1}, u_{\theta 1}) = \frac{\partial u_{r1}}{\partial \theta} - a^2 \frac{\partial^3 u_{r1}}{\partial \theta^3} + (1 + a^2) \frac{\partial^2 u_{\theta 1}}{\partial \theta^2}$$

are uninertial loads arising as the result of soil-structure interaction; $a^2 = h^2/12 R^2$; $D = Eh/(1 - \nu^2)$; E, h, R, μ, ν - elasticity modulus, thickness, radius, mass and Poisson's ratio of shell material respectively; K_r and K_θ - normal and tangential coefficients of shell-soil interaction.

Comparing the statements of the problem by two approaches based on the account of difraction process and simplifying supposition an essential difference becomes obvious from the point of view of mathematical simulation and their solutions. In the second case the problem solution at proper conditions is much more simple and it allowes to study in detail the underground structure vibrations depending on constructive peculiarities and the character of seismic motion.

An algorithm was worked out and the calculations of displacements and stress fields were made for two methods; they were obtained at stationary (and non-stationary) action of waves of wide spectrum frequencies. Fig. 1 shows the dependence of ring stress of the shell on wavelength-structure diameter ratio (λ/d). Results of stress numerical values analysis show the necessity of difraction wave account (stationary and non-stationary) at $d/\lambda \geq 1$. Considering an engineering problems of underground structure design on seismic waves action ($d/\lambda < 1$) and using the second approach the concrete numerical results with a sufficient may be obtained.

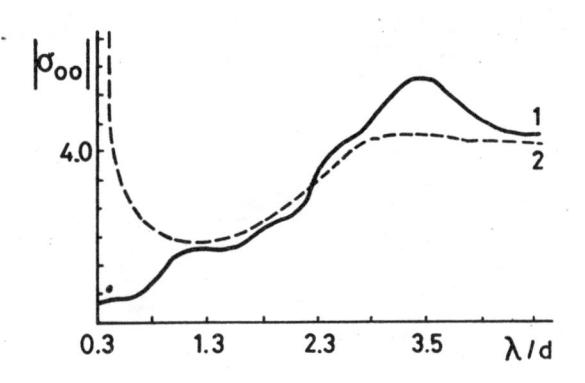

Fig. 1. Comparison of solution for different ratio λ/d (1 - wave solution, 2 - seismodynamic study solution).

CONCLUSION

Investigation results show the necessity of further development of the methods of stress-strain state study of underground structures under the action of seismic waves. The choice of solution methods for practical problems should be done on the basis of the aims and statement of studied problems.

REFERENCES

1. KORENEV, V.G.: Structures laying on elastic basement. In: Engineering Mechanics in the U.S.S.R (1917-1967), Moscow, Stroyizdat, 1969.
2. PASTERNAK, P.L.: The basis of a new design method of the basement on elastic foundation with the aid of two coefficients of bed. Moscow, Stroyizdat, 1954.
3. VLASOV, V.V., LEONTIEV, N.N.: Beams, plates and shells on elastic basement. Moscow, Fizmatgiz, 1960.
4. FILONENKO-BORODICH, M.M.: Several approximate theories of elastic basement//Sci. Reports of Moscow State University, 1940, V. 46.

5. RASHIDOV, T.R.: Dynamic theory of earthquake-resistance of complex systems of underground structures, Tashkent, Fan, 1973.
6. MUBARAKOV, Ja.N.: Seismodynamics of underground shell type structures. Tashkent, Fan, 1978.
7. ILGAMOV, M.A., IVANOV, V.A., GULON, B.S.: The design of shells with elastic aggregate. Moscow, Nauka, 1987.
8. MUBARAKOV, Ja.N., SAGDIEV, H., SAFAROV, I.: Reports of Uzbek Academy of Sci., Series of technical sci., 1986, N 6.
9. RASHIDOV, T.R., MUBARAKOV, Ja.N., SAGDIEV, H., SAFAROV, I.: Reports of Uzbek Academy of Sci., 1989, N. 6.

SEISMIC FORCES INDUCED BY MINING SHOCKS IN 5 STOREY CONCRETE WALL BUILDINGS

Roman Ciesielski[1] Wiesław Kowalski[2]
Edward Maciąg[1] Tadeusz Tatara[3]

INTRODUCTION

Underground and glory-hole mining exploitation of mineral resources causes a number of negative phenomena in the surrounding environment /e.g. mining shocks and quarry explosives and their effects on the ground surface and objects/. As for underground mining - there are two seismic regions in Poland showing high activity: The Upper Silesian Coal Mining District /GZW/ and The Legnica - Głogów Copper Mining District /LGOM/. Vibrations induced by underground mining shock are recorded first of all down in the mines. The aim of the mining seismograph networks is to create conditions for controlling the state of bumb threat to people working down in the mines. It is only in recent few years that more attention has been paid to surface vibrations induced by underground shocks and to their influence on objects. The collapse of a new 11 storey wall panel building in one of the LGOM towns /Polkowice/stimulied this interest. The analysis of the causes of the collapse showed that among others, a mining shock could have had some influence on it. Both in the GZW and in the LGOM mining exploitation is carried out under areas with dense development. Buildings in those regions, however, have not been designed for dynamic loads induced by mining shocks. For the evaluation of mining shock effects on surface objects continuous seismographic observations are necessary similarly as it is in seismic regions. Up till now only few vibration traces have been registred on the ground surface and in buildings.In the south of Poland there are many quarries that are in the neighbouring towns and villages. Some of them are in the towns.The quarry explosives induce the vibrations that act on the buildings. Methods of calculations of buildings subject to paraseismic vibrations called also technical /seismic vibrations caused by man - directly or indirectly/ can be similar to these used in the object design in the seismic regions. The paper deals with the evaluation of dynamic responses of selected prefabricated buildings with bearing concrete walls,typical for Polish housing. The method of modal analysis and response spectrum were used to the evaluation of building behaviour. For the GZW

---

[1] professor; [2] postgraduate student; [3] doctor
Department of Structural Mechanics, Cracow University of Technology, Cracow, Poland.

and LGOM areas the standard relative acceleration response spectrum was proposed which can be useful to the evaluation of seismic forces acting on the buildings.

ANALYSED BUILDINGS

Two 5 storey prefabricated apartment buildings with 1 storey basement typical of Polish precast concrete housing were subjected to dynamic analysis.

Precast concrete panel building No 1

This is an external 2 - staircase section of a 26 - staircase building. The floor on the basement is of multitube slabs 24 cm thick. The other floors were made of reinforced concrete slabs 14 cm thick. The stiffening walls are made of 14 cm thick slabs /panels/. The outside longitudinal curtain walls are multi - layered. Each storey is 2.7 m high. The building is founded on reinforced concrete strip footings. Typical floor plan of building 1 is shown in Fig 1.

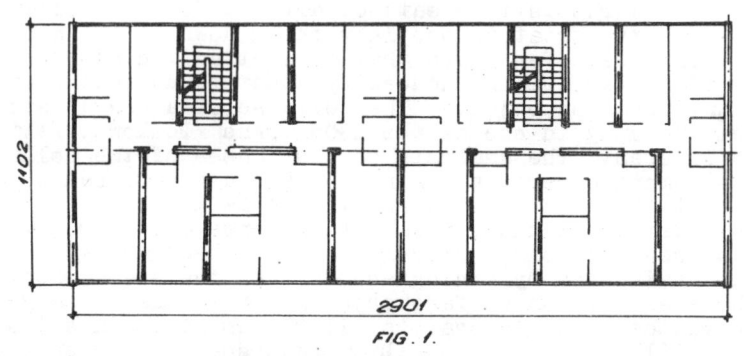

FIG. 1.

Precast concrete block building No 2

This is a one staircase section constituting a part of a 14 staircase building. Each floor is made of multitube slabs 24 cm thick. The stiffening wall system is a mixed transverse - longitudinal one; the bearing walls are 24 cm thick made of multitube slabs of "Żerań" type. Each storey is 2.8 m high. The longitudinal curtain walls are as in building 1. Typical storey plan of building 2 is shown in Fig.2.

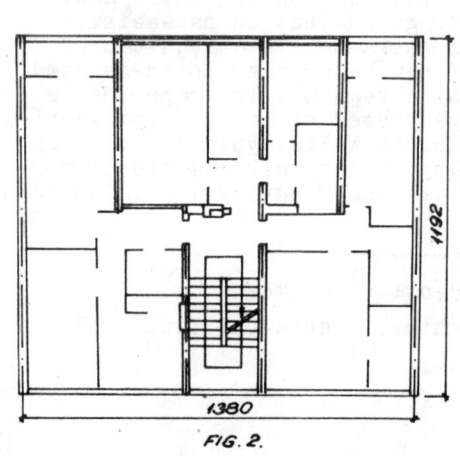

FIG. 2.

DYNAMIC PROPERTIES OF THE ANALYSED BUILDINGS

Dynamic properties of the discussed buildings were determined on the basis of measurement results of their natural vibrations and analytically by choice of dynamic models of the buildings. For theoretical determination of frequency and mode shape of natural vibrations of the investigated building a discrete model was adopted in form of a system of vertical cantilevers fixed at the height of the basement floor and hinged - joint with one another by floor diaphragms [4,10].The following factors were assumed to influence the building flexibility:

- bending and shearing of cantilevers,
- rocking of foundations and basement as a rigid body with regard to ground flexibility.

The stiffness matrix of the assumed model is determined by the formula

$$[K] = \sum_{i=1}^{n} [K_i] = \sum_{i=1}^{n} [D_i]^{-1} \qquad (1)$$

where: $[D_i], [K_i]$ - flexibility and stiffness matrices of i-th cantilever respectively.

In matrix $[K]$ in direction y longitudinal bearing walls curtain walls and ground flexibility were considered /unweakened state/ as well as only bearing elements /weakened state/. Adopting the model of building No 2 /precast block building/ the flexibility of vertical connections according to (1) was considered. The first two calculated natural vibration frequencies and the results from analysis of natural vibrations measurements of the discussed buildings were listed in Table 1. An essential influence on longitudinal stiffness of the buildings is exerted by non-structural elements. On the basis of not presented here data significant influence on values of natural vibration frequencies is exerted by ground flexibility especially in transverse direction.

TABLE 1

BUILDING	DIRECTION		FREQUENCIES OF NATURAL VIBRATIONS [Hz]		
			FROM CALCULATIONS		FROM MEASUREMENT
			f_1	f_2	f_3
1	2		3	4	5
1	x		3,3	25,3	3,3 - 4,1
	y	a*)	3,8	14,0	3,2 - 3,8
		b+)	2,1	9,1	
2	x		3,4	18,7	3,2 - 3,8
	y	a	3,2	13,0	3,1 - 3,6
		b	1,7	7,2	

*) THE UNWEAKENED STATE
+) THE WEAKENED STATE

From investigations of damping of the discussed buildings small values for critical damping fraction ξ are obtained /1.5 to 2% /.

CHARACTERISTICS OF APPLIED KINEMATIC FORCINGS

Fig 3 shows a representative trace of E-W ground surface vibrations recorded in the GZW excited by one of the stronger mining shocks of energy $E \cong 5 \times 10^7 J$. The trace of acceleration from Fig.3 will be treated as kinematic forcing for the model of building No 1 and No 2. Apart from the trace

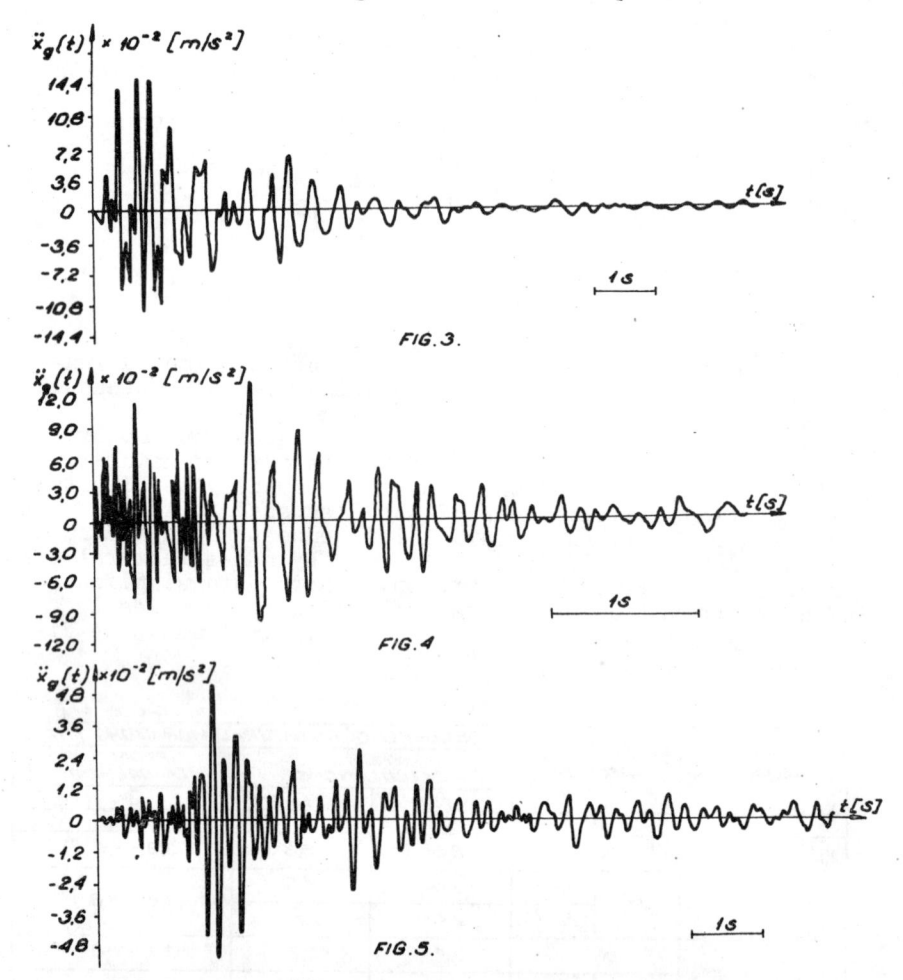

FIG.3.

FIG.4

FIG.5.

in Fig.3 two representative traces of ground vibration acceleration recorded in the LGOM will be used for calculations of seismic forces. Fig.4 and 5 show two vibration traces. As a forcing of vibrations of the model of the building No 1 representative trace of vibration acceleration caused by quarry blast was used too /Fig.6/. The calculated trace of seismic forces at the 5th storey of the building is shown in Reference [4]. The character of traces of seismic forces in case N=2 is very similar to these for N=1. In Fig.7a,b the example traces of seismic forces forced by vibrations in Fig.6 for cases N=1 i N=2 are presented respectively. The most disadvantageous distributions of seismic forces along the height of the building and bending moments at the bottom of the building (M_z) from forcing in Fig.3 are shown in Fig.8a,b for cases N = 1 and N = 2. Comparing values (M_z) for the case N = 1 and N = 2, a small influence of the second mode of natural vibrations on seismic forces is seen. Maximum accelerations from forcing in Fig 4 are several times smaller than from forsing in Fig.3 though a_{gmax} of ground vibrations of these two forcings are almost identical. Next calculated valuas (M_z) from forcing in Fig.6 are considerably smaller than from forcing in Fig.5, though a_{gmax} of ground vibrations of these two forcings are almost identical. The value of maximum forcing acceleration is not a decisive factor conditioning the building response; equally important is the "content" of vibration traces - the bands of dominant frequencies and

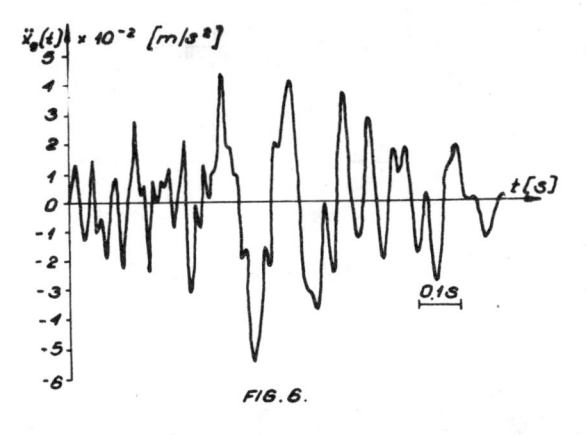

FIG.6.

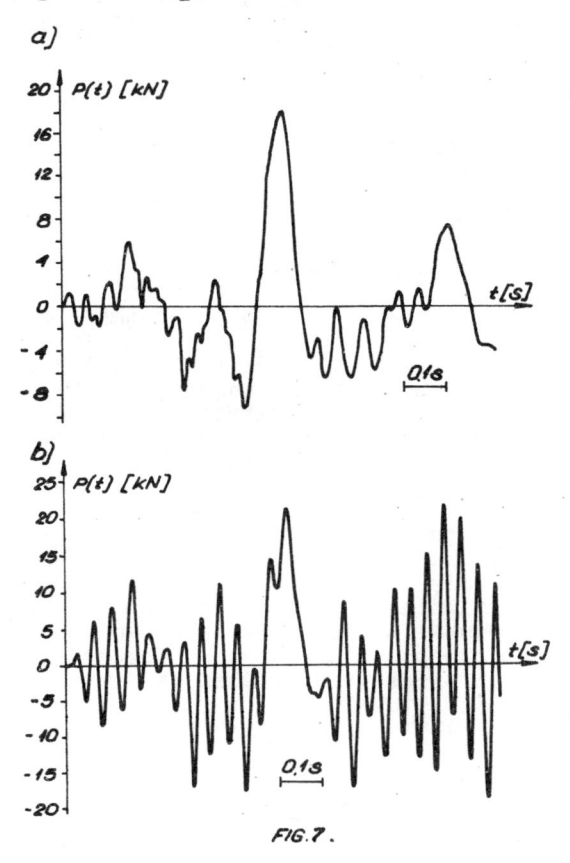

FIG.7.

FIG. 8

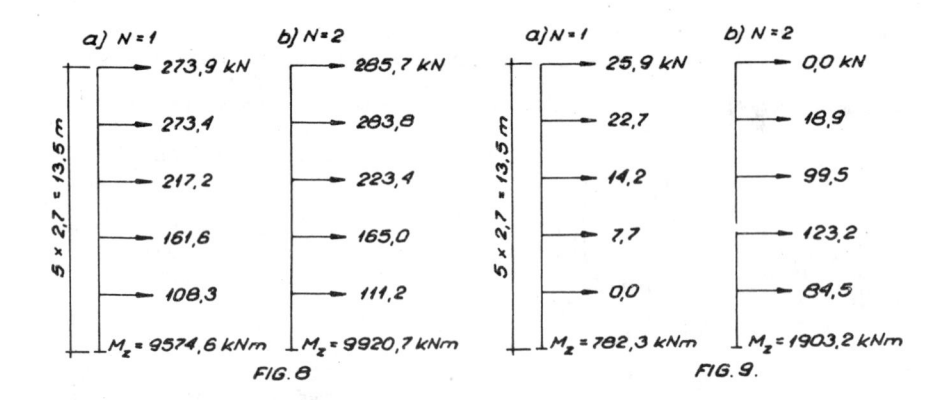

a) N=1: 273,9 kN; 273,4; 217,2; 161,6; 108,3; M_z = 9574,6 kNm
b) N=2: 285,7 kN; 283,8; 223,4; 165,0; 111,2; M_z = 9920,7 kNm
(5 × 2,7 = 13,5 m)

FIG. 9.

a) N=1: 25,9 kN; 22,7; 14,2; 7,7; 0,0; M_z = 782,3 kNm
b) N=2: 0,0 kN; 18,9; 99,5; 123,2; 84,5; M_z = 1903,2 kNm
(5 × 2,7 = 13,5 m)

their relation with respect /mainly/ to the fundamental frequency of natural vibrations of the object. Similar relations were obtained for seismic forces and (M_z) in the longitudinal direction of building No 1 as for the transverse one. The case of a weakened state of the building was also analysed in the longitudinal direction. Then it is assumed that non-structural elements /curtain walls/ do not influence the stiffness of the building. M_z from forcing in Fig.3 is then almost twice smaller than for the unweakened state of the building. The influence of second mode of natural vibrations on (M_z) is relatively small and causes its increase by about 6% with respect to case N=1. Seismic forces and M_z from forcing in Fig.4 prove to be several times smaller than from forcing in Fig.3. Calculations of the most disadvantageous distributions of seismic forces and moments M_z are shown in Figs 9ab for cases N=1 and N=2. From a comparison of Figs. 9a and 9b it is clearly seen that the second mode of natural vibrations has significant influence on seismic forces and (M_z). From forcing in Fig.3 an almost 7 times greater (M_z) was obtained than from forcing in Fig.4. For seismic forces and (M_z) similar relations were obtained in the longitudinal direction of building No2 as it was in the transverse direction. From an analysis of the weakened building No 2 in the longitudinal direction y similar relations were obtained as for the same state and direction in building No 1. Traces of ground vibrations were assumed as kinematic forcings exciting building motion. If traces of basement wall vibrations /at the level of basement floor/ were assumed as kinematic forcings then the calculated building responses were greater than the measured ones. In the traces of the basement wall vibrations /clearly visible on the response spectra/ frequencies of natural building vibrations proved to dominate. It follows from the comparison of calculated values of seismic forces from stronger shocks with code wind forces that the former ones can be several times greater than the latter ones.

APPLICATION OF RESPONSE SPECTRUM

One of the most convenient ways of partraying the maximum responses of dynamic system involves the response spectrum [8,9,11]. From the same forcings which previously were assumed seismic forces and (M_z) are determined by use of the acceleration response spectrum S_a according to [7].
The seismic forces are determined from the formula:

$$P_{ik} = m_k \eta_{ik} S_a \qquad (2)$$

where: P_{ik} - component of the force of inertia at point k corresponding to the i-th frequency of natural vibrations, k = 1,2,3,4,5; i = 1,2

$$\eta_{ik} = \phi_{ik} \frac{\sum_{j=1}^{n} m_j \phi_{ij}}{\sum_{j=1}^{n} m_j (\phi_{ij})^2} \qquad (3)$$

ϕ_{ij} - ordinate of the i-th mode of vibrations at point j of the building.
For: N=1 - $M_z = M_{1z}$; N=2 - $M_z = \sqrt{M_{1z}^2 + M_{2z}^2}$ (4)

Forces and moments calculated according to formulae (2) to (4) are compared with calculation results given previously and treated as "accurate". Fig.10,11,12 show the acceleration response spectrum (S_a) corresponding to the vibration trace in Fig. 3,4 and 6. Performing calculations by use of (S_a) close aproximation on (M_z) for building No 1 and No 2 is obtained. For comparison the values (M_z) in the building No 1 calculated by the "exact" way /dynamic analysis/ and response spectrum (S_a) are listed in Table 2 /from two representative forcings: from Fig.3 and 4 /.
It is evident that (M_z) calculated by use of (S_a) determines good approximation of the value calculated "exactly". For the model of building No 1/it is similar for building No 2/ in the longitudinal direction in a weakened state and in the case of forcing from Fig.4 the moments (M_z) calculated "exact" way and by use of (S_a) considerably differ for N=2. In case

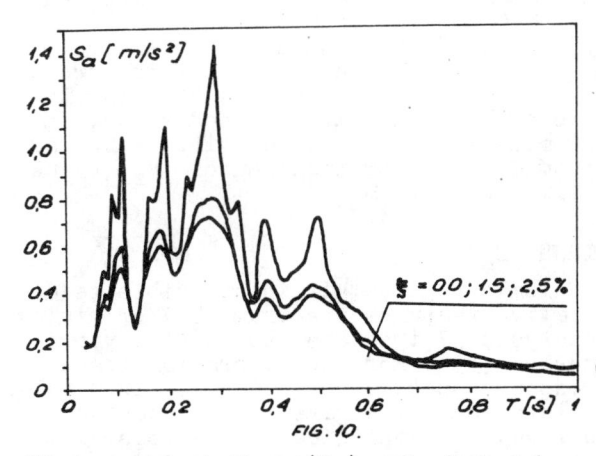

FIG.10.

FIG. 11.

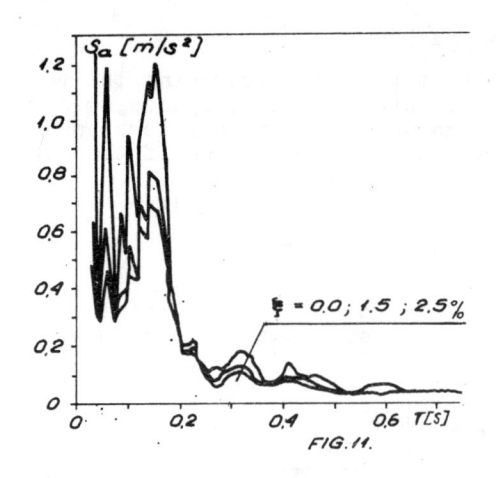

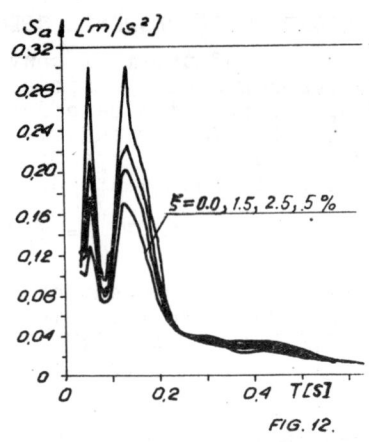

FIG. 12.

TABLE 2.

FORCING FROM FIG.	DIRECTION	VALUES M_z [kNm] CALCULATED BY:			
		MODAL ANALYSIS		SPECTRUM S_a	
		N=1	N=2	N=1	N=2
3	x	9573	9921	9542	9542
	y *	9850	10210	9914	9914
	y +	5009	5326	5088	5106
4	x	1782	1137	1633	1644
	y *	1069	1119	1142	1144
	y +	782	1903	765	867

* THE UNWEAKENED STATE
+ THE WEAKENED STATE

of significant influence of the second mode shape of natural vibrations on seismic forces and (M_z) determination of P_{ik} and (M_z) according to formulae (2) and (4) by use of (S_a) leads to decisively lowered values (M_z). Similar relations may occur in high buildings since their first two frequencies of natural vibrations are in the same bands as the first two frequencies of natural vibrations of the 5 storey buildings in weakened state. In case when only the first mode shape of natural vibrations has a decisive influence on (M_z), the approximation of (M_z) by use of (S_a) assures good estimation for (M_z).

STANDARD RESPONSE SPECTRUM (S_a)

Similarly, as in seismic regions, it is advisable to determine the standard relative response spectrum (S_a) for mining regions. For spectral analysis 27 traces of horizontal vibration components were adopted [3]. For each vibration trace absolute acceleration response spectra devided by maximum value of ground vibration acceleration /of a given trace/ were used, $S_a(T)/a_{gmax} = \beta(T)$; in this way the amplification factors also called relative acceleration response spectra useful for determination of seismic forces /similarly as in [5]/ were obtained. The smoothed curve "a" and "b" given in Fig.13 represent the mean values $\beta(T)$ for the region LGOM and GZW respectively and may be treated as amplification factor for these regions. It follows from the comparison of curves "a,b" in Fig. 13 that the band of dominating periods of ground vibrations induced by mining shock in the GZW is shifted to the right. It originates

FIG. 13.

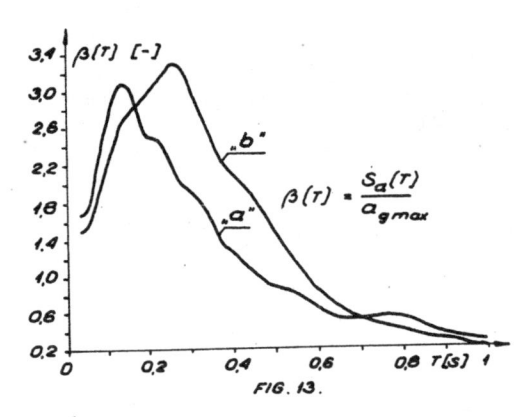

$$\beta(T) = \frac{S_a(T)}{a_{gmax}}$$

from different properties of ground layers lying on copper ore deposite in the LGOM and on coal deposite in the GZW. Comparing the curves in Fig.13 with the spectral curves adopted for earthquakes it can be seen that vibration traces from mining shocks include higher dominating frequencies /especially in the LGOM/ than earthquake records. Typical 5 storey wall panel buildings are characterized by fundamental natural vibration periods both in the direction of the transverse and longitudinal axes/ within the band 0.22-0.33 s, hence, remaining in the band of peak values $\beta(T)$ of the curve "b" in Fig.13. After slight modification /in some section the curves can be replaced by straight lines/ the curves in Fig.13 will be applicable for the evaluation of seismic forces in designing of new objects and verifying the existing buildings situated in the mining shock areas.

Building responses evaluated by use of these spectra for ξ =2.5% approximate well the measured building response. Value ξ = 2,5% was adopted on the basis of our own investigation of wall panel buildings (2).

CONCLUSIONS

- 5 storey buildings with load bearing concrete walls react mainly with the fundamental frequency of natural vibrations to kinematic forcings resulting from mining shocks.
- In buildings where the two first natural vibration frequencies are within the bands 1.5 - 2.5 Hz and 7 - 9 Hz the seismic forces from shocks in the LGOM should be calculated considering the two first modes of natural vibrations by modal analysis.
- The discussed buildings were not designed for horizontal seismic loads but only for code wind loads. Seismic forces from stronger mining shocks can be many times more intensive than code wind loads.
- A comparison of seismic forces determined on the way of dynamic analysis and by use of the response spectrum (S_a) according to Polish Code shows that results for seismic forces obtained in a simplified way /by use of (S_a) / are satisfactorily accurate if the building response is conditioned only by the fundamental mode shape of natural vibrations.
- Significant differences in acceleration response spectra (S_a) corresponding to traces of ground surface vibrations are noticeable in two regions of Poland: in the GZW and the LGOM. The same buildings react differently to ground vibrations of the same maximal acceleration a_{gmax} from these regions. Differences in values of seismic forces may be even several times greater.

- Investigations of typical 5 storey buildings with load bearing concrete walls carried out in full scale analysis of these buildings show that significant influence on their stiffness, hence also on frequencies of natural vibrations of these buildings in longitudinal direction, is exerted by non-structural elements /curtain walls/.
- Neglecting of non-structural elements in the dynamic analysis of the discussed buildings subjected to mining shocks reduces the expected seismic forces on the average by 50% and more.

REFERENCES

[1] Cholewicki A., Calculation of stiffening walls, /in Polish/, Research and Design Center of Building Engineering Warsaw,1980.
[2] Ciesielski R., Kuźniar K., Maciąg E., Tatara T., Experimental evaluation of vibration damping in different types of prefabricated buildings, /in Polish/, Proc. 5th Symp.: Seismic and paraseismic influences on structures, The Polish Academy of Sciences, Cracow, 1988, pp.195-207.
[3] Ciesielski R., Kowalski W., Maciąg E., Tatara T., Response spectra of mining shocks and their application for estimation of buildings reactions, 2nd Inter. Conf. on Traffic Effects on Structures and Environment, The Low Tatras, ČSRF, The Low Tatras, April 1991, pp.265-271/A.
[4] Ciesielski R., Maciąg E., Tatara T., Investigations of buildings with bearing concrete walls subjected to mining shocks, Inter. Conf. Buildings with Load Bearing Concrete Walls in Seismic Zones, Paris, 1991/06/13-14 pp.153-164.
[5] Civil Engineering in Seismic Regions /in Russian/, Soviet Union Code, 1982.
[6] Clough R.W., Penzien J., Dynamics of Structures, Mc Graw-Hill, 1975.
[7] Evaluation of harmfulness of building vibrations due to ground motion, PN-85/B-02170 - Polish Code, 1985.
[8] Hudson D.E., Response spectrum techniques in engineering seismology, Proc. 1st. WCEE, Berkeley, 1956, pp. 4.1-4.12.
[9] Hudson D.E., Some problems in the application of spectrum techniques to strong - motion earthquake analysis, Bull. Seism. Soc. Amer., vol.52, No 2, 1962, pp.417-430.
[10] Langer J., Klasztorny M., The influence of ground motion on prefabricated buildings, /in Polish/ Conf. Proc.: Experimental methods of investigation of mechanical properties of actual structures,Cracow - Janowice,1977,pp.43-52.
[11] Newmark N.M.,Horn M., Blume J.A., Kapur K.K., Seismic design spectra for nuclear power plants, J.Power Div., PO2 1973, pp.287-202.

SEISMIC EFFECT OF UNDERGROUND EXPLOSIONS

H. Sagdiev[1] E. Juhásová[2]

INTRODUCTION

During the strong earthquakes or large industrial explosions there is necessity to follow not only behaviour of structures on soil surface but also the behaviour of underground structures. For that purpose there exist different methods for the seismic response solution of underground structures using wave theory and different discrete calculation methods, e.g. [1], [3], [4]. The part of such works uses the approach concerning the influence of seismic motions from explosions on underground cylindrical structures. They are based on the difraction wave model according to the wave dynamics theory with consideration of soil and underground structure interaction. The cases of application of mentioned theories are usually limited and they should be controlled by experiments. In this paper we are dealing with the problem of analysis of explosion effects and with some approaches how to solve the response of buried structure.

EXPERIMENTAL INVESTIGATION AND DISCUSSION OF RESULTS

From experimental measurements of the effects of large explosions we will present two cases. In the first case there have been done measurements of the vibration of massive rock surface from the large explosions in Central Slovakia near Mochovce. The position of measured points can be seen in Fig. 1. [5], [6].

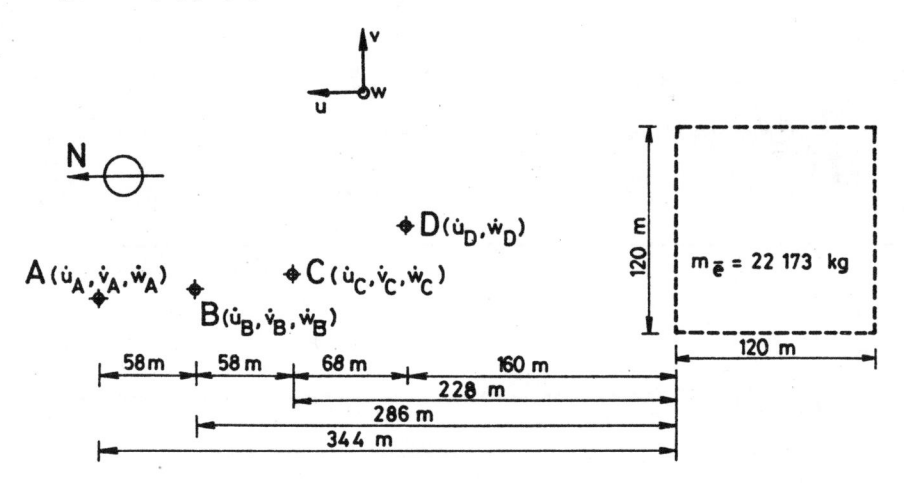

FIG. 1. DISPOSITION OF MEASURED POINTS DURING EXPLOSIONS IN MOCHOVCE

1 - IMISS UzAN, Tashkent, Uzbekistan, USSR
2 - ÚSTARCH SAV, Bratislava, ČSFR

Comparatively large amount of the charge was divided into 18 particular charges with time phase shifts à 0.024 s. The examples of obtained records are in Fig. 2. We can see there the time-history of velocities in points A, D both in horizontal and vertical directions. The phase shift of the front wave answers the wave velocity 3600 m/s. Prevailing frequency about 8.0 Hz is remarkable both in points A, D. Spectral distributions can be seen in Figs. 3, 4.

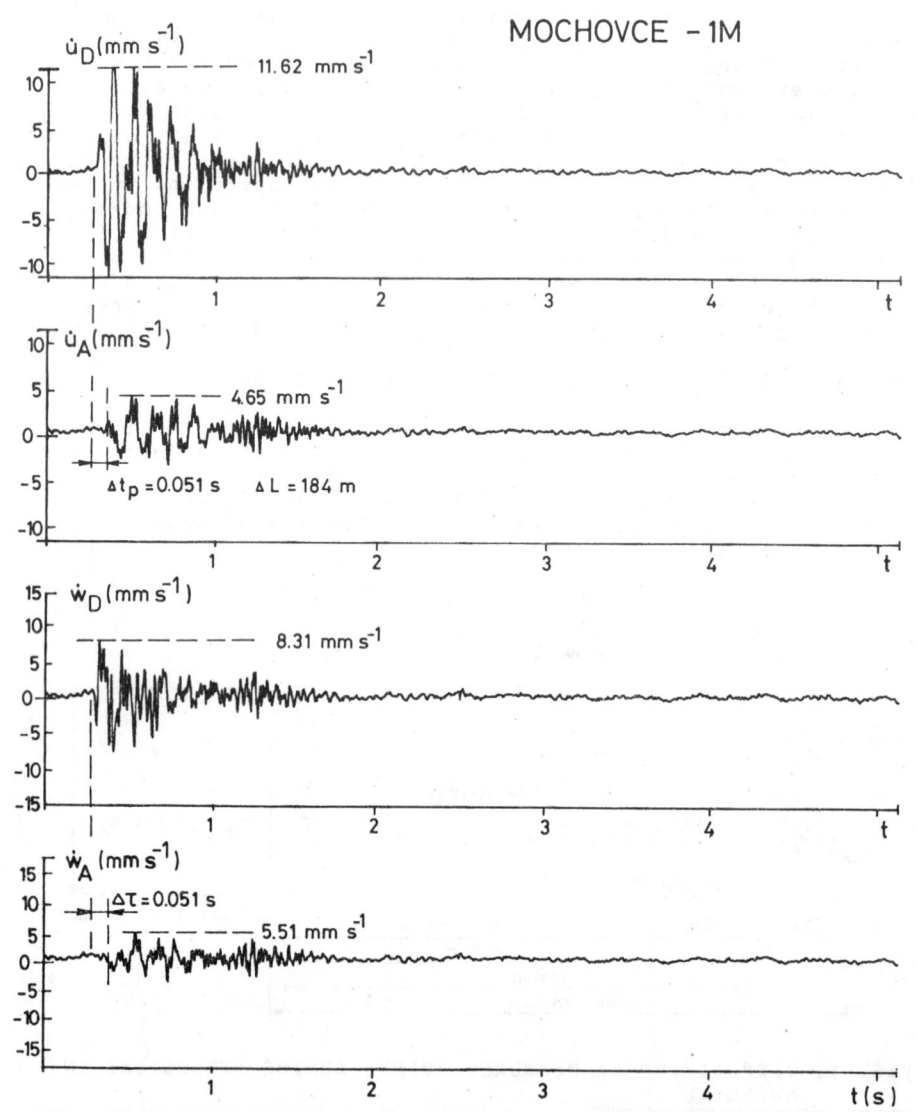

FIG. 2. TIME-HISTORY OF SEISMIC EFFECTS FROM EXPLOSIONS IN CENTRAL SLOVAKIA - REGION MOCHOVCE

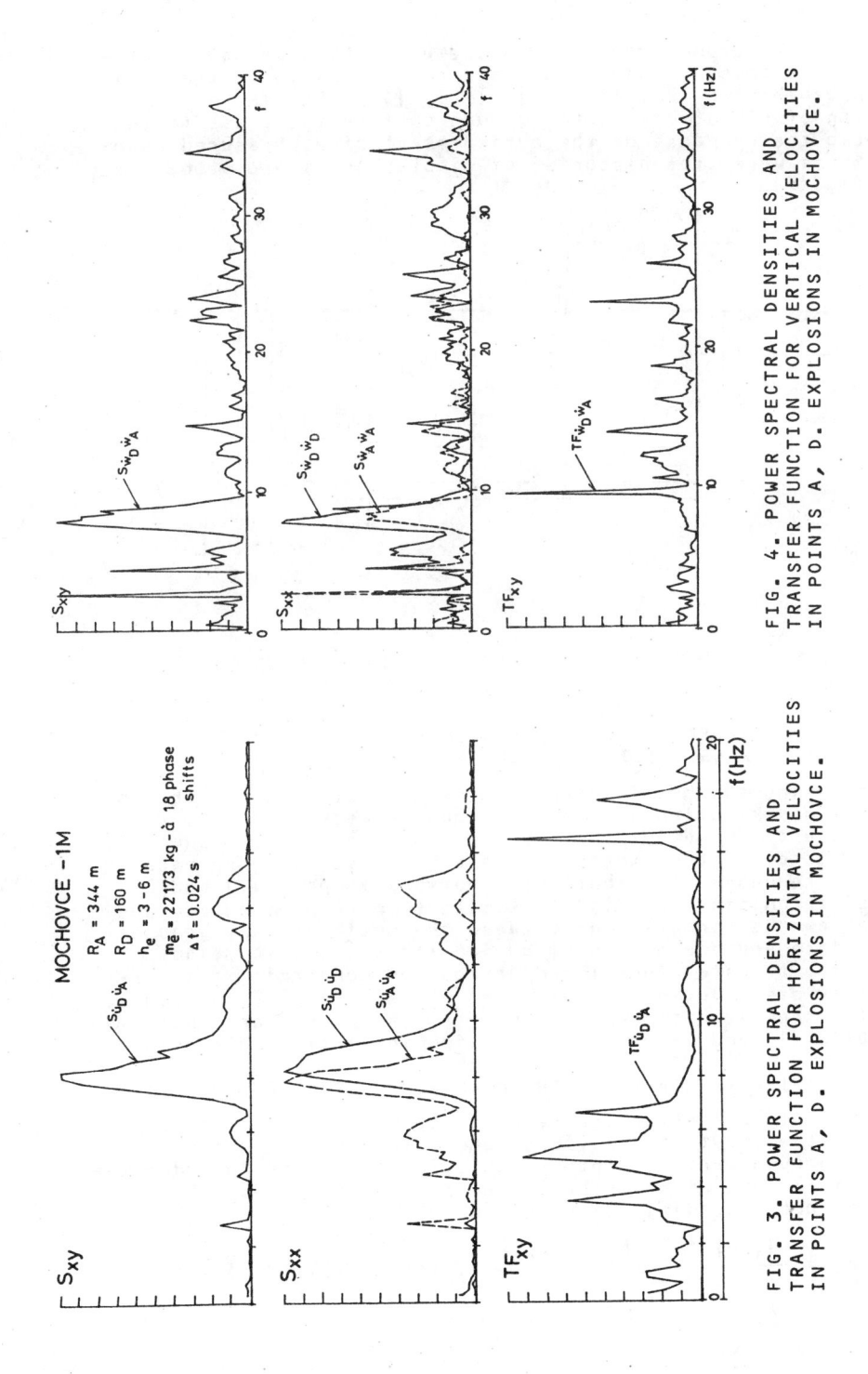

FIG. 3. POWER SPECTRAL DENSITIES AND TRANSFER FUNCTION FOR HORIZONTAL VELOCITIES IN POINTS A, D. EXPLOSIONS IN MOCHOVCE.

FIG. 4. POWER SPECTRAL DENSITIES AND TRANSFER FUNCTION FOR VERTICAL VELOCITIES IN POINTS A, D. EXPLOSIONS IN MOCHOVCE.

The second group of measurements which we have analysed are those obtained in field conditions in Central Asia in Uzbekistan [7], [8], [9], [10], [11]. Besides of measuring of vibrations on the ground surface there was follow also the response of the buried steel pipe. Measured quantities were time-histories of displacements and velocities (Fig. 5).

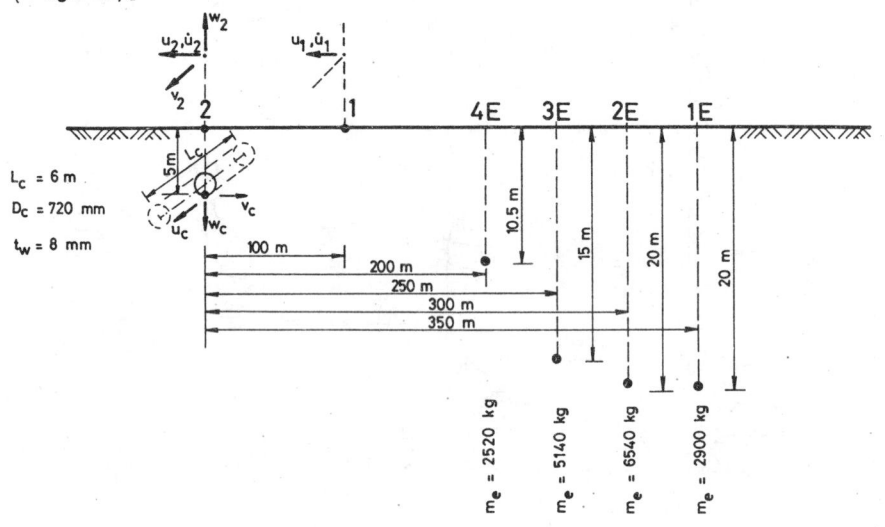

FIG. 5. DISPOSITION OF MEASURED POINTS DURING UNDERGROUND EXPLOSIONS IN UZBEKISTAN

Analysing the obtained records we can follow again the spectral distribution and correlations between points 1, 2 in frequency range 0-20 Hz. Due to near distance and another character of the explosion there is larger inconsonance in the frequency distribution of motions in points 1 and 2, as we can see from Fig. 6. Similarly we can calculate and follow the correlations between the motion on the ground surface and the motion of buried steel pipe. The example of the transfer function distribution concerning the vertical distribution of the pipe w_c is in Fig. 7. We can depict the natural frequencies of the pipe near 1 Hz, 10 Hz, 14 and 19 Hz for the motion in vertical direction.

SIMPLIFIED APPROACH FOR THE PIPE RESPONSE SOLUTION

Let us analyse the response of buried pipe considering that the input loading function is the seismic motion on the ground surface. Into calculation there are included the coefficients of soil-structure interaction k_x, k_y, k_z, determined according [9] using formulaes

$$k_i = \frac{s_i^o \bar{\sigma}_i - Q(k_{ci} - \bar{k}_{ci})}{\bar{s}_i A_i} \quad , \text{ for } i = 1, 2, 3 . \qquad (1)$$

274

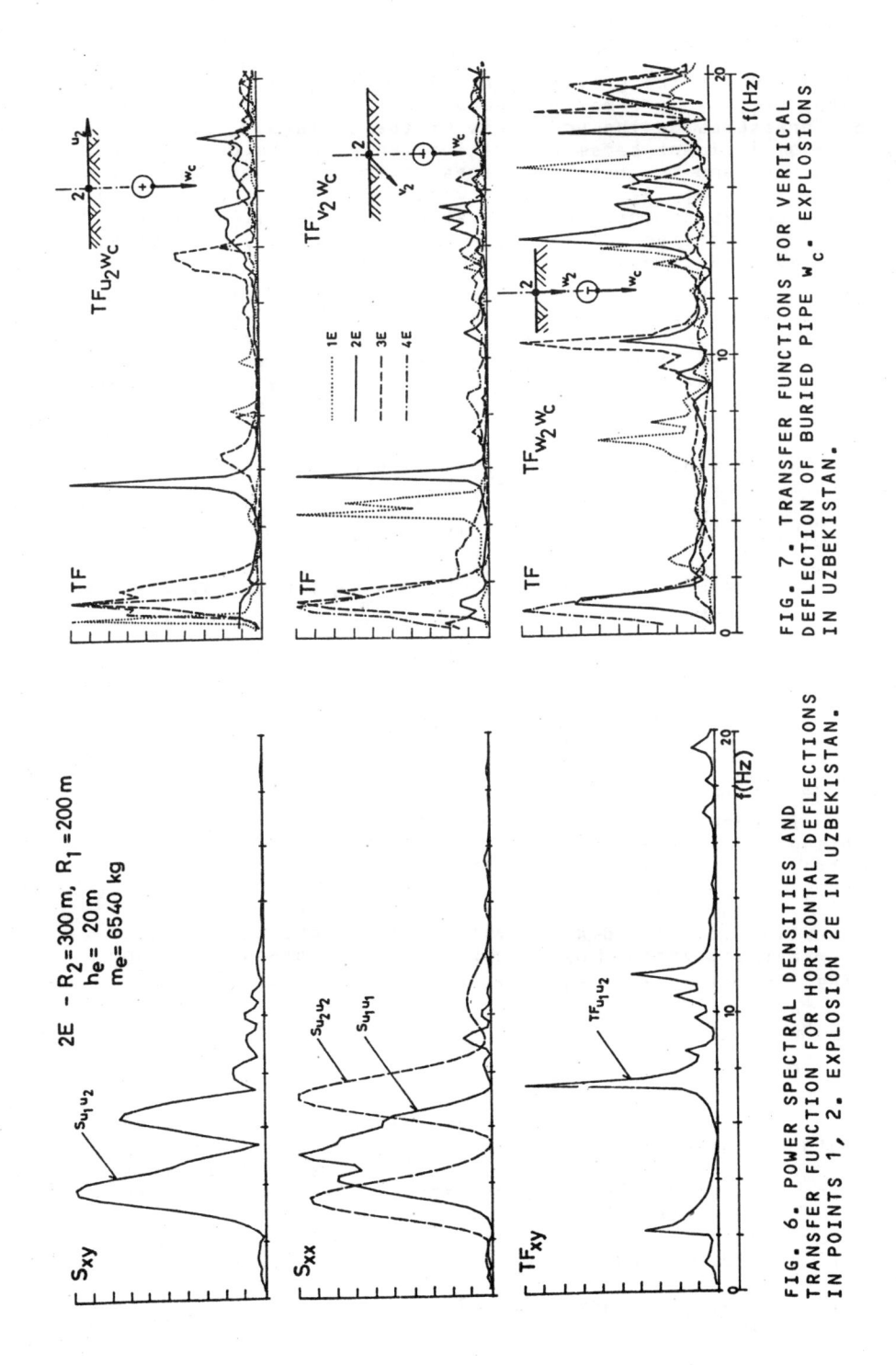

FIG. 6. POWER SPECTRAL DENSITIES AND TRANSFER FUNCTION FOR HORIZONTAL DEFLECTIONS IN POINTS 1, 2. EXPLOSION 2E IN UZBEKISTAN.

FIG. 7. TRANSFER FUNCTIONS FOR VERTICAL DEFLECTION OF BURIED PIPE w_c - EXPLOSIONS IN UZBEKISTAN.

In (1) Q is the gravity force of cylindrical shell structure, A_i, δ_i are maximum values of relative displacements and pressures on the unit area of the surface, s_i^0, $\bar{s}_i$ are area of incident seismic wave influence on the underground structure and the resisting area of the surrounding soil. Parameters δ_i, A_i, k_{ci}, $\bar{k}_{ci}$ were determined individually for each experiment. For the calculation we will use the medium coefficients k_i: $k_x = 20.96$ N cm^{-3}, $k_y = 50.96$ Ncm^{-3}, $k_z = 43.65$ N cm^{-3}.

When taking into account that the contact of the shell surface and the surrounding soil is preserved there is acting the process of soil-structure interaction through the whole surface of the shell. It results in relative displacements of the shell and in additional forces which are conditioned by relative displacements of the shell in relation to the surrounding soil.

Using mentioned assumptions we can write for the thin wall cylinder underground structure

$$\sum_{i,j} L_{ij} U_j = P_i + F_i + L_i(u_o, v_o, w_o), \quad i,j=1,2,3 \quad , \quad (2)$$

where $U_1 = u_o - u$, $U_2 = v_o - v$, $U_3 = w_o - w$, and U_j, u, v, w, u_o, v_o, w_o are the components of vectors of relative and absolute displacements of the shell and the soil. P_i are components of the soil-structure interaction forces, F_i and $L_i(u_o, v_o, w_o)$ are components of vectors of seismic inertial and non-inertial loading, L_{ij}, L_i are differential operators.

In general case we can describe the forces which are acting on the shell in the form (3), using the transfer matrix $\bar{C}$, unit vector $\vec{e}$:

$$\vec{F}(t) = f(t)\,\bar{C}\,\vec{e} \,. \quad (3)$$

In cylindrical coordinate system the components of the soil displacement vector (u_o, v_o, w_o) can be expressed through experimental values (u_1, u_2, u_3)

$$\begin{aligned} u_o &= u_2, \\ v_o &= u_1 \sin\varphi + u_3 \cos\varphi, \\ w_o &= u_1 \cos\varphi - u_3 \sin\varphi, \end{aligned} \quad (4)$$

where u_1, u_2, u_3 represent seismograms obtained during field explosions.

According to shell structure theory and to the closed shape of cylindrical structure the solution of the system (2) can be written in the form

$$u = \sum_{m,n} U_{mn}(t) \cos \lambda \xi \cos n\varphi, \quad (5a)$$

$$v = \sum_{m,n} V_{mn}(t) \sin \lambda \xi \sin n\varphi, \quad (5b)$$

$$w = \sum_{m,n} W_{mn}(t) \sin \lambda \xi \cos n\varphi . \qquad (5c)$$

By passing to the absolute displacements (u,v,w) and inserting (5) into (2) we obtain

$$\ddot{U}_{mn} + a_{11} U_{mn} + a_{12} V_{mn} + a_{13} W_{mn} = f^1_{mn} ,$$
$$\ddot{V}_{mn} + a_{21} U_{mn} + a_{22} V_{mn} + a_{23} W_{mn} = f^2_{mn} , \qquad (6)$$
$$\ddot{W}_{mn} + a_{31} U_{mn} + a_{32} V_{mn} + a_{33} W_{mn} = f^3_{mn} .$$

The description of coefficients in (6) and the next solution using integral Laplace transformation is described in details in [7].

Here we will present only some results which were obtained considering m=7, n=6. The numerical trapezoidal method of calculation was used with the time step $\Delta t = 0.01$ s. Comparison of theoretical and experimental results for explosions 2E, 4E can be seen in Fig. 8. Naturally we can see in this figure some differences between theoretical and experimental time-histories, what is influenced namely that we have used only plane model and simplified medium coefficients of soil-structure interaction. In spite of that we can state the used method as acceptable one for the rough forecasting of the seismic response of underground structure.

CONCLUSION

For the proper construction and maintenance of underground pipelines we must know more about the behaviour of the underground structures during strong seismic motions. The simplified methods for the solution of their seismic response can be used if there are available three component seismic motion records on the soil surface and some essential informations about soil properties.

REFERENCES

1. BOUWKAMP, J.G.-KANN, J.V. - CONSTANTINESCU, D.R.: On the hysteretic behaviour of a pipe elbow. In: Proc. 9ECEE, Moscow 1990. Vol.8, p.40.

2. HENRYCH, J.: The Dynamics of Explosion and Its Use. Academia, Prague, 1973. (In Czech)

3. ILIUSHIN, A.A. - RASHIDOV, T.R. - ISRAILOV,,M.S. - MARDONOV, B.M.: The effect of seismic wave on underground pipelines. In: Proc. The Friction, Wear and Lubrication Materials. Tashkent,1985. Vol. 3.2. p.128. (In Russian)

4. IWAMOTO,T. - WAKAI, N. - YAMAJI, T.: Observation of dynamic behaviour of buried pipelines during earthquakes. In: Proc. 8WCEE, San Francisco 1984. Vol. 7, p. 231.

5. JUHÁSOVÁ, E.: Seismic effects of explosion during construction of nuclear power plant Mochovce. ICA SAŞ, Bratislava, 1983. (In Slovak)

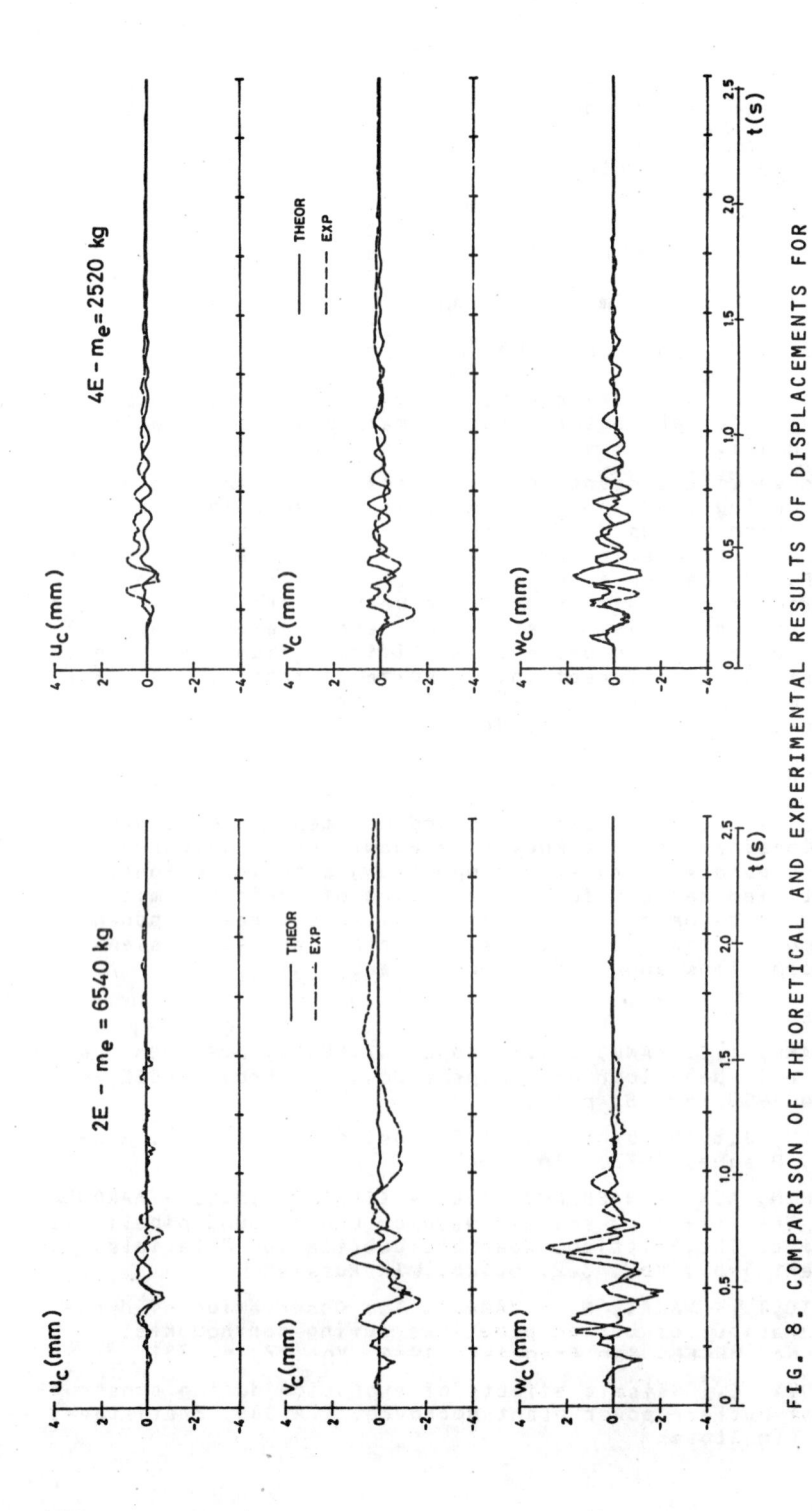

FIG. 8. COMPARISON OF THEORETICAL AND EXPERIMENTAL RESULTS OF DISPLACEMENTS FOR THE EXPLOSIONS 2E AND 4E. MEASUREMENTS IN UZBEKISTAN.

6. JUHÁSOVÁ, E. - POKORNÝ, M.: Characteristics of seismic motions caused by explosions. In: Dynamics of Engineering Structures, Smolenice, 1985. (In Slovak)

7. JUHÁSOVÁ, E. - SAGDIEV, H.: Seismic response of buried pipes using explosion loading. Stav. Čas. 1992. /In print/

8. MUBARAKOV, J.N.: Seismodynamics of Underground Shell Structures. FAN, Tashkent, 1987. (In Russian)

9. MUBARAKOV, J.N. - SAGDIEV, H.S. - RACHMONOV, B.S.: The investigation of interaction of underground shell structures and soil medium using explosion method. Bjuletin Inženernoj Sejsmologii, N. 13, 1989. (In Russian)

10. MUBARAKOV, J.N. - SAGDIEV, H.S. - SAFAROV, I.I.: The estimation of seismic stresses of underground shell structures at the explosion loading. Izvestia AN UzSSR, ser. Tech. nauk, N. 1, 1988. Tashkent 1988. (In Russian)

11. RASHIDOV, T.R. - CHOZMETOV, G.C.: The Seismic Resistance of Underground Pipelines. FAN, Tashkent, 1985. (In Russian)

THE EFFECT OF STRONG MOTION DURATION ON FIRST PASSAGE FAILURE OF STRUCTURES
Zbigniew ZEMBATY[1]

INTRODUCTION

Two general criteria of safety can be considered in structural mechanics:
a) the structure is said to be safe if its response due to all possible actions remains within a prescribed, safe domain,
b) the structure is said to be safe if the number of stress cycles has not reached a certain limit value.

For both criteria the reliability of structure can be defined as the probability of the safe state of structure during its lifetime. In seismic engineering, when the structure is subjected to non-stationary random excitations the first criterion leads to the applications of the solutions of the first passage problem whereas the second criterion leads to the fatigue analyses (mostly low cycle fatigue for inelastic vibrations).

In this paper the first criterion will be considered, and the effect of strong motion duration on the first passage probability will be analyzed in details.

Three main parameters of the seismic ground motion influence decisively the response of structures. These are intensity at the site, the shape of spectrum of excitations and the duration of seismic action. The intensity of shaking influence directly the level of structural response and its effect is obvious. The spectral content of excitations may significantly increase or decrease the response depending on the dynamic properties of structure. Strong motion duration, on the other hand, increases the probability of level crossings and consequently the probability of failure of the structure.

DEFINITIONS OF STRONG MOTION DURATION

Many definitions of strong motion duration can be found in the literature, [1-11]. In what follows a short review will be presented.

Consider the Arias [12] intensity of an accelerogram with length t_0

$$I_A = \int_0^{t_0} \ddot{x}^2(t)dt. \qquad (1)$$

This intensity can be treated as a measure of energy received by the structure and for not very long accelerograms can easily be correlated with other intensity scales (e.g. [4]).

---

[1]*Technical University, ul. Katowicka 48, 45-951 Opole, Poland*

Husid [1] analyzed the changes of cumulative Arias intensity

$$f_H(t) = \frac{\int_0^t \ddot{x}^2(\tau)d\tau}{\int_0^{t_0} \ddot{x}^2(\tau)d\tau} \qquad (2)$$

where t_0 is the length of the accelerogram.

Housner [2] applied the linear part of the Husid plot as the time of strong motion duration. Donovan [3] assumed the time required to build up 90% of intensity measured by the Husid plot. Trifunac and Brady [4] modified this proposal and analyzed the time between 5% and 95% of the value of Husid plot. The same assumption has been studied by Dobry, Idriss and Ng [5]. Bolt, in turn, analyzed the time between the first and last crossing of level 0.05g, [6]. McCann and Shah [7] applied more complicated criteria based on the analysis of the sign of cumulated energy released. Vanmarcke and Lai [8] analyzed the equivalent stationary part of accelerograms based on the analysis of crossings of a prescribed level by the excitation process. They found the strong motion duration to be proportional to the ratio of Arias intensity to the square of peak acceleration. McGuire and Barnhardt [10] compared four definitions of strong motion duration using least square regression analysis. They found the following equation to best fit the data of 50 USA accelerograms:

$$\ln t_d = c_1 + c_2 M + c_3 S + c_4 V + c_5 \ln R \qquad (3)$$

where t_d is the duration, $c_1, \ldots, c_5$ are constants, M – magnitude, S=0 or 1 for rock and alluvium sites respectively, V=0 for horizontal and V=1 for vertical component, R – distance to rupture surface of epicenter. In addition Trifunac and Westermo [11] indicated the importance of the depth of sedimentary deposits beneath the site on the duration of shaking.

General results of these research analyses reveal that duration increases as the Modified Mercalli Intensity decreases and is shorter for rock sites than for soils. However, all the authors indicate great scatter of the results and the difficulties in formulation of credible quantitative measures of the strong motion duration and its effect on the severity of shaking.

Another field of research is devoted to the effect of strong motion duration on the response of structures [13-17]. It should however be pointed out here that these effects are still not well reflected in the building codes.

Assume now that the excitation can be modeled as a uniformly modulated random process

$$\ddot{x}(t) = A(t)\ddot{\tilde{x}}(t), \qquad (4)$$

where $A(t)$ is a deterministic modulating function and $\tilde{x}(t)$ is a stationary random process. The spectrum of the stationary process can be described by the Kanai-Tajimi formula [18,19]

$$S_{\tilde{x}}(\omega) = \frac{1 + 4\zeta_g^2\left(\frac{\omega}{\omega_g}\right)^2}{\left[1 - \left(\frac{\omega}{\omega_g}\right)^2\right]^2 + 4\zeta_g^2\left(\frac{\omega}{\omega_g}\right)^2} S_0 \;, \qquad (5)$$

where ω_g and ζ_g are local site parameters.

Various modulating functions have been proposed to describe the non-stationary character of seismic ground motion. It can be a simple rectangular envelope:

$$A(t) = \begin{cases} 0 & \text{for} \quad t<0 \\ 1 & \text{for} \quad 0<t<t_0 \\ 0 & \text{for} \quad t>t_0 \end{cases} \qquad (6a)$$

or a "three-part" envelope introduced by Amin and Ang [20]:

$$A(t) = \begin{cases} 0 & \text{for} \quad t<0 \\ (t/t_1)^2 & \text{for} \quad 0<t<t_1 \\ 1 & \text{for} \quad t_1<t<t_2 \\ \exp[-\beta(t-t_2)] & \text{for} \quad t>t_0 \end{cases} \qquad (6b)$$

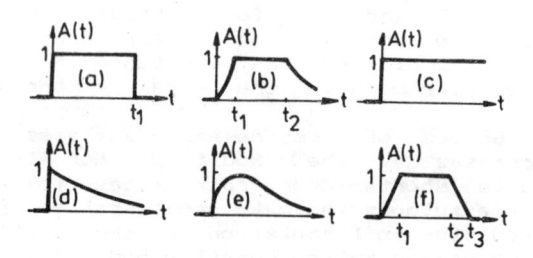

Fig. 1

These envelopes are shown in Fig. 1a,b together with some other proposals. Substituting eq. 4 into 1 and taking the mathematical expectation on both sides of this equation one obtains the mean Arias intensity

$$\langle I_A \rangle = \int_0^{t_0} \langle \dot{x}^2(t) \rangle dt = \int_0^{t_0} A^2(t) \int_{-\infty}^{\infty} S_{\tilde{x}}(\omega) d\omega dt. \qquad (7)$$

Substituting eq.5 as well as 6a and 6b into 7 one obtains following two formulae:

$$\langle I_A \rangle = t_0 \left[1+4\xi_g^2\right]\pi\omega_g S_0 /(2\xi_g) \qquad (8a)$$

$$\langle I_A \rangle = \left[t_2 - 0.8t_1 + 1/(2\beta)\right]\left[1+4\xi_g^2\right]\pi\omega_g S_0 /(2\xi_g) \qquad (8b)$$

Assuming constant mean Arias intensity one can now examine the dependence between duration t_0 or t_1 and intensity parameter S_0, and examine their influence on the results of crossing analysis for various types of structures. A similar analysis can be done for constant root mean square excitation (S_0=constant) and variable values of t_0 and $\langle I_A \rangle$.

THE FIRST PASSAGE PROBLEM

When estimating reliability of structures under stochastic dynamic loads one is faced with the calculations of the probability that the response of a structure will cross a certain threshold at least once during a prescribed time interval. This probability is known as the first passage probability. Consider the response of a certain point of a structure. Assume that it is a critical point of the structure with respect to asses the reliability (i.e. giving the least reliability). When the structure is excited by a random process then the response y(t) is also a random process. For stationary problem the probability U(t,λ) of at least one crossing of the level λ grows up from zero to one as time passes by. For nonstationary processes starting from zero for t=0 and after a short time decaying to zero, this probability equals

$$U = U(\lambda) = \lim_{t \to \infty} U(t,\lambda). \qquad (9)$$

On the other hand one can search for the level $\lambda = \lambda_{0.5}$ corresponding to probability of excursion equal to 0.5. This is median peak value and is a measure of peak response. The first passage problem has been the subject of considerable research in recent 20-30 years (e.g. [21-31]), but only approximate solutions have been found.

Generally there are three main approaches to the first passage problem:

       a) methods based on modeling the response as Markov vector, with the solutions of respective Fokker – Planck – Kolmogorov equation
       b) methods based on computations of response statistics and assumption that the response is Gaussian
       c) simulations.

In this paper the approach "b" described in details in Ref.[32,33,34,17], will be applied.

NUMERICAL EXAMPLES

Single degree of freedom system

Consider the equation of motion of a linear oscillator

$$\ddot{y} + 2\zeta\omega_0\dot{y} + \omega_0^2 y = -\ddot{x}(t) \qquad (10)$$

where ω_0 and ζ are natural circular frequency and ratio of actual to critical damping respectively. We shall consider the non-stationary response of this oscillator subjected to a rectangular pulse of a stationary random process (eqs. 4, 6a). Two assumptions will be analyzed.

First assume constant mean square excitation level. In this case the probability of failure increases with increasing time i.e. with increasing number of maxima and zero crossings. The same concerns peak values of the response. The sensitivity of this response to changes of the duration depends on the parameters of excitation and vibrating system.

Second, assume constant mean Arias intensity. In this case as duration increases the acceleration level of excitation decreases. On the other hand however, the increase of duration increases the first excursion probability or peak values of the response. Thus the final effect can not be predicted without performing a numerical analysis of the first excursion problem.

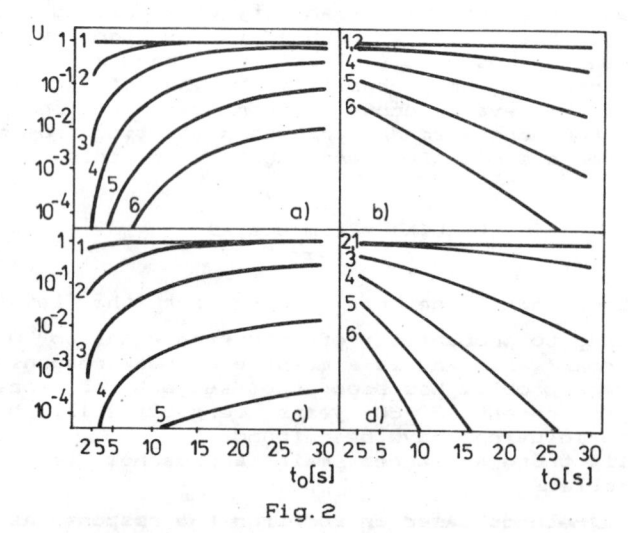

Fig. 2

In Fig. 2 the effect of strong motion duration on first excursion probability for six barrier levels: $\lambda=0.05\div0.30$m step 0.05m (plots 1-6) are presented. (a) for constant excitation level (damping ratio $\zeta=2\%$), (b) for constant Arias intensity ($\zeta=2\%$), (c) for constant excitation level ($\zeta=5\%$), (d) for constant Arias intensity ($\zeta=5\%$). Oscillator with natural period T=2s. It can be seen from this figure that, as could be expected the assumption of constant excitation level results in the increase of failure probability as duration increases (Fig. 2a & 2c). On the other hand however it can be seen from Fig. 2b & 2d that for constant Arias intensity the first passage probability decreases with increasing duration.

Vibrations of an industrial chimney

Consider the equation of motion of a tall, slender structure modeled as a continuous cantilever beam subjected to horizontal motion of the foundation (Fig. 3):

$$\frac{\partial^2}{\partial z^2}EJ(z)\left[\frac{\partial^2 y(z,t)}{\partial z^2}+\varkappa\frac{\partial^3 y(z,t)}{\partial z^2 \partial t}\right]+\mu m(z)\frac{\partial y(z,t)}{\partial t}+m(z)\frac{\partial^2 y(z,t)}{\partial t^2}=-m(z)\ddot{x}(t) \quad (11)$$

where

$y(z,t)$ is the relative horizontal displacement,

$\ddot{x}(t)$ is the acceleration of transverse motion of the base,

$EJ(z)$ is the flexural stiffness,

$m(z)$ is the mass per unit length

$\varkappa, \mu$ damping parameters.

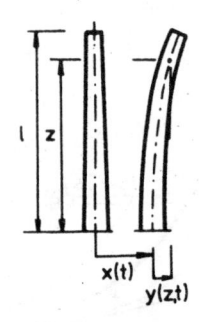

Fig. 3

Equation (11) can be solved by applying the modal superposition method:

$$y(z,t) = \sum_{j=1}^{\infty} V_j F_j(z) \int_0^t h_j(\tau) \ddot{x}(t-\tau) d\tau$$

where

$h_j(t)$ is the impulse response function corresponding to the jth vibration mode

$V_j, F_j(z)$ are modal participation coefficients and eigenfunctions of the structure respectively.

Fig. 4

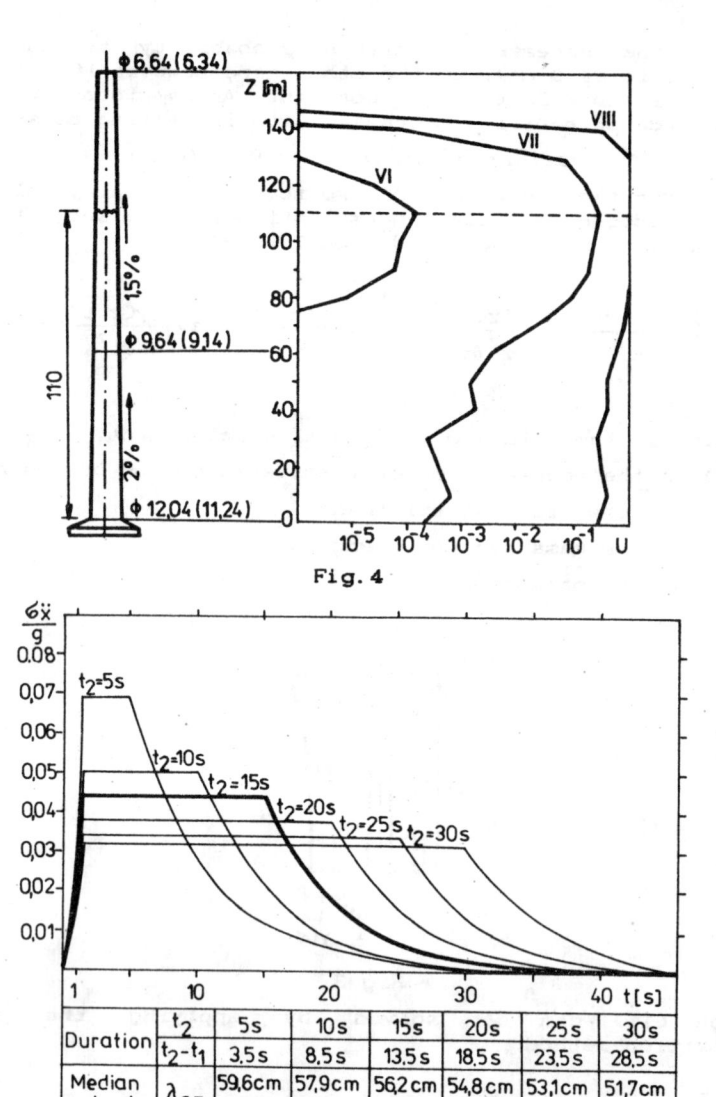

Fig. 5

Duration	t_2	5s	10s	15s	20s	25s	30s
	t_2-t_1	3,5s	8,5s	13,5s	18,5s	23,5s	28,5s
Median peak value	$\lambda_{0,5}$	59,6cm 6,0%	57,9cm 3,0%	56,2cm 0%	54,8cm -2,5%	53,1cm -5,5%	51,7cm -8,0%
First excursion probability	$\lambda=58$ cm	0,54 23%	0,49 11%	0,44 0%	0,39 11%	0,33 25%	0,28 36%
	$\lambda=100$ cm	0,0145 198%	0,0090 85%	0,0049 0%	0,0025 49%	0,0011 77%	0,0005 90%

Applying the method of evolutionary spectra and carrying out a stochastic analysis one can now obtain first passage probabilities for tip displacements or bending moments of a tower-shaped structure. Here the analysis has been performed for a 160m r/c chimney with the assumption of constant Arias intensity and for the envelope 6b. In Fig. 4 the estimation of first passage probability of load capacity of the shaft of the chimney is plotted vs. height. The computations have been carried out for three earthquake intensities and show the greatest probability of failure at about 2/3 of chimney height. This result is confirmed by observations of failures of chimneys during earthquakes.

In figure and table 5 the selected results of crossing analysis for the tip displacements of a 160m r/c industrial chimney are gathered. It can be seen from the table that, as in the previous example, the assumption of constant Arias intensity results in decreasing the first passage probability. Thus the decrease of maximum root mean square acceleration of the excitation process has the greater effect on the probability of failure than the increase of duration.

CONCLUSIONS

The main conclusion of the foregoing analysis is that for constant Arias intensity the increase of strong motion duration decreases the peak response and first passage probability (Figs. 2b, 2d, 4). Thus for constant energy imparted to the structure the level of root mean square excitation influences more significantly first excursion probability and peak response than the duration time. This conclusion holds also for inelastic oscillators, [35].

REFERENCES

1. R. L. Husid, *Analisis de terremotos*, Analisis general, Revista del IDIEM (Santiago, Chile) 8, 1969, 21-42.

2. G. W. Housner, *Measures of severity of earthquake ground shaking*, Proc. U. S. Natl. Conf. Earthquake Eng., Ann Arbor, Michigan (1975).

3. N. C. Donovan, *Earthquake Hazards for Buildings*, National Bureau of Standards, Boulder, Colorado, 1972.

4. M. D. Trifunac and A. G. Brady, *A study on the duration of earthquake ground motion*, Bull. Seism. Soc. Am., 65, 1975, 581-626.

5. R. Dobry, I. M. Idriss and Ng, *Duration characteristics of horizontal components of strong-motion earthquake records*, Bull. Seism. Soc. Am. 68, 1978, 1487-1520.

6. B. A. Bolt, *Duration of strong ground motion*, Proc. 5th World Conf. Earthquake Eng., Rome, 1973.

7. M. W. McCann, H. C. Shah, *Determining strong motion duration of earthquakes*, Bull. Seism. Soc. Am., 69, 1979, 1253-1265.

8. E. H. Vanmarcke and S. P. Lai, *Strong motion duration and rms amplitude of earthquake records*, Bull. Seism. Soc. Am., 70, 1980, 1293-1307.

9. C. Carino and F. Carli, *Some aspects in the qualification of local seismicity*, Proc. 9th, Int. Conf. Struct. Mech. Reactor Technol. Lausanne 1987.

10. R. K. McGuire and T. P. Barnhard, *The usefulness of grund motion duration in predicting the severity of seismic shaking*,

Proc. 2nd U.S. Natl. Conf. Earthquake Eng., Stanford, California, 1979, 713-722.

11. M. D. Trifunac and B. D. Westermo, *Duration of strong earthquake shaking*, Soil Dyn. & Earthquake Eng., **1**, 1982, 117-121.

12. A. Arias, *A measure of earthquake intensity*, in *Seismic Design of Nuclear Power Plants*, (ed. R.I. Hansen), MIT Press, 1970.

13. H. Kameda and K. Kohno, *Effect of ground motion duration on seismic design load for civil engineering structures*, Mem. Fac. Eng. Kyoto Univ. **45**, 1983, 140-184.

14. M. Shahabi and N. Mostaghel, *Strong ground motion duration and effective cycling acceleration*, Proc. 8th World Conf. Earthquake Eng. San Francisco, 2, 1984, 843-850.

15. G. D. Jeong and W. D. Iwan, *The effect of earthquake duration on the damage of structures*, Earthquake Eng. & Structural Dyn. **16**, 1988, 1201-1211.

16. R. W. Clough and J. Penzien, *Dynamics of structures*, Mc-Graw Hill, 1975, 619-626.

17. Z. Zembaty, *A note on non-stationary stochastic response and strong motion duration*, Earthquake Eng. & Structural Dyn. **16**, 1988, 1189-1200.

18. K. Kanai, *An empirical formula for the spectrum of strong earthquake motions*, Bull. Earthquake Res. Inst. Tokyo Univ. **39**, 1961, 85-95.

19. H. Tajimi, *A statistical method of determining the maximum response of a building structure during an earthquake*, Proc. 2nd World Conf. Earthquake Eng., Tokyo, 2, 1961, 781-797.

20. M. Amin and A. H. S. Ang, *Nonstationary stochastic model of earthquake motion*, J. Eng. Mech. Div. ASCE, **94**, 1968, 559-583.

21. A. G. Davenport, *Note on the distribution of the largest value of random function with application to gust loading*, Proc. of the Institution of Civ. Eng., 28, 1964, 187-196.

22. H. Cramer, *On the intersections between the trajectories of a normal stationary stochastic process and a high level*, Ark. Mat., **6**, 1966.

23. M. Shinozuka, Y. Sato, Simulation of nonstationary random processes, Journal of the Engineering Mechanics Division, ASCE, **93**, 1967, 11-40.

24. M. Shinozuka, J.-N. Yang, *On the bound of first-excursion probability*, J. Eng. Mech. Div. ASCE, **95**, 1969, 363-377.

25. M. Shinozuka, J.-N. Yang, *Peak structural response to nonstationary random excitation*, J. of Sound and Vibration, **16**, 1971, 501-517.

26. J.-N. Yang, *Nonstationary envelope process and first excursion probability*, J. of Struct. Mech., **1**, 1972, 231-248.

27. J.-N. Yang, *First excursion probability in nonstationary random vibration*, J. of Sound and Vibration, **27**, 1973,

165-182.

28. E. H. Vanmarcke, *On the distribution of the first passage time for normal stationary random process*, J. of Applied Mech., ASME, 42, 1975, 215-220.

29. J. B. Roberts, *Probability of first passage failure for nonstationary random vibration*, J. of Applied Mech., ASME, 42, 1975, 716.

30. R. Grossmayer, *On the application of various crossing statistics in the aseismic reliability problem*, IUTAM Symposium on Stochastic Problems in Dynamics, (Ed. B. L. Clarkson), 1977, 283-307.

31. L. A. Bergman, J. C. Heinrich, *Petrov-Galerkin finite element solution for the first passage probability and moments of first passage time of the randomly accelerated free particle*, Computer Methods in Applied Mechanics and Engineering, 27, 1981, 345-362.

32. Z. ZEMBATY, *Random vibrations and reliability of tower-shaped structures under seismic excitations*, Ph. D. thesis WSI w Opolu, 1986, (in Polish).

33. Z. Zembaty, *Nonstationary random vibration analysis of the seismic response of chimneys*, Proceedings of 5th International Chimney Congress, Essen, 3-5 Oct. 1984, 103-106.

34. Z. Zembaty, *On the reliability of tower-shaped structures under seismic excitations*, Int. J. of Earthquake Engineering and Structural Dynamics, 15, no. 6, August, 1987, 761-775.

35. Z. Zembaty, *Strong motion duration and inelastic response spectra*, 9th European Conf. on Earthquake Engineering, Moscow, 11-16 September 1990.

LIST OF AUTHORS (A) AND PARTICIPANTS (P)

Prof. Drazen ANICIC (A,P)
Institute of Civil Engineering
Rakusina 1
P.O.Box. 165
41000 Z A G R E B
Y u g o s l a v i a

Prof. Atilla M. ANSAL (A,P)
Istanbul Technical University
Ayazaga
I S T A N B U L
T u r k e y

Dr. M. ARUMUGAM (P)
Structural Engineering Research Centre
CSIR Campus
Taramani P.O.
M A D R A S 600 113
I n d i a

Prof. Miloslav BATA, DrSc. (P)
Fakulta stavebni CVUT
Thakurova 7
166 29 P R A H A
C z e c h o - S l o v a k i a

Doc. Jan BENCAT, CSc. (A)
Stavebna fakulta VSDS
Katedra mechaniky
30. aprila c. 1
010 26 Z I L I N A
C z e c h o - S l o v a k i a

Ing. Vladimir BILY (P)
Fakulta stavebni CVUT
Thakurova 7
166 29 P R A H A
C z e c h o - S l o v a k i a

RNDr. Ivan BROUCEK, CSc. (A)
Geofyzikalny ustav SAV
Dubravska cesta
842 28 B R A T I S L A V A
C z e c h o - S l o v a k i a

Prof. Alberto CASTELLANI (A,P)
Politecnico di Milano
p. Leonardo da Vinci 32
20133 M I L A N O
I t a l y

Prof. Andrzej CHOLEWICKI (A,P)
Building Research Institute (ITB)
Filtrowa 1
00950 W A R S Z A W A
P o l a n d

Prof. Roman CIESIELSKI (A)
Politechnica Krakowska
ul. Warszawska 24
31 155 K R A K O W
P o l a n d

Ing. Alois DANEK (P)
Stavebna fakulta VSDS
Katedra mechaniky
30. aprila 1
010 26 Z I L I N A
C z e c h o - S l o v a k i a

Prof. Ja.M. EISENBERG (A)
TsNIISK
2nd Instituskaya 6
M O S C O W 109428
U S S R

Ing. Ondrej FISCHER, DrSc. (A,P)
UTAM CSAV
Vysehradska 49
128 49 P R A H A 2
C z e c h o - S l o v a k i a

Doc. Dr. Rainer FLESCH (A,P)
BVFA - Arsenal
A-1030 W I E N
A u s t r i a

Prof. Radomir FOLIC (A,P)
Institute of Industrial Building
Faculty of Technical Sciences
21000 N O V I S A D
Y u g o s l a v i a

Mr. L. GALANO (A)
Dept. of Civil Engineering
University of Florence
Via di S. Marta, 3
50139 F L O R E N C E
I t a l y

Dr. B.K. GAYAROV (A)
Scientific Research Institute
for Earthquake Engineering
A S H K H A B A D
Turkmen SSR
U S S R

Dr. E.J. GERASIMOVA   (A)
TsNIISK
2nd Institutskaya 6
M O S C O W   109428
U S S R

Prof. Jacob GLUCK   (A,P)
Faculty of Civil Engineering
TECHNION
H A I F A   32000
I s r a e l

Ing. Tomas HORYNA   (P)
UTAM CSAV
Vysehradska 49
128 49   P R A H A
C z e c h o - S l o v a k i a

Ing. Emilia JUHASOVA, DrSc.   (A,P)
USTARCH SAV
Dubravska c. 9
842 20   B R A T I S L A V A
C z e c h o - S l o v a k i a

Prof. Edward E. KHACHIAN   (A)
ArmNIISA
3 D. Sasuntsky st.
Y E R E V A N   375014
A r m e n i a
U S S R

Dr. Boris KIRIKOV   (A)
TsNIISK
2nd Institutskaya str. 6
M O S C O W   109428
U S S R

Ing. Yvona KOLEKOVA, CSc.   (A,P)
USTARCH SAV
Dubravska c. 9
842 20   B R A T I S L A V A
C z e c h o - S l o v a k i a

Dr. Igor K. KONSTANTINOV   (P)
TsNIISK
2nd Institutskaya str. 6
M O S C O W   109428
U S S R

Mr. Wieslaw KOWALSKI   (A)
Politechnica Krakowska
ul. Warszawska 24
31 155   K R A K O W
P o l a n d

RNDr. Peter LABAK (P)
GfU SAV
Dubravska c. 9
842 28 B R A T I S L A V A
C z e c h o - S l o v a k i a

Mr. M. Aysen LAV (A)
Istanbul Technical University
Ayazaga
I S T A N B U L
T u r k e y

Prof. Edward MACIAG (A,P)
Politechnica Krakowska
ul. Warszawska 24
31 155 K R A K O W
P o l a n d

Dr. M.R. MAHERI (A)
Shams Lane
Negarestani Street.-Gohari Street
Shahab
K E R M A N
I r a n

Ing. Gustav MARTINCEK, DrSc. (A,P)
USTARCH SAV
Dubravska c. 9
842 20 B R A T I S L A V A
C z e c h o - S l o v a k i a

Ing. Rudolf MASOPUST, CSc. (A)
SKODA Praha, k.p.
Vystavba elektraren, k.p.
316 00 P L Z E N - Bolevec
C z e c h o - S l o v a k i a

Dr. Andrej M. MELENTIEV (A,P)
TsNIISK
2nd Institutskaya 6
M O S C O W   109428
U S S R

Dr. Michael.G. MELKUMIAN (A)
ArmNIISA
D. Sasuntsky str. 3
Y E R E V A N 375014
A r m e n i a
U S S R

Dr. Ashot Ts. MINASIAN (A)
ArmNIISA
D. Sasuntsky str. 3
Y E R E V A N 375014
A r m e n i a
U S S R

Ing. Ivan MOTLIK (P)
USTARCH SAV
Dubravska c. 9
842 20 BRATISLAVA
Czecho-Slovakia

Dr. J.M. MUBARAKOV (A)
IMISS UzAN
Akademgorodok
TASHKENT
Uzbekistan
USSR

Mr. John. P. NEWELL (A,P)
WS Atkins Engineering Sciences
160 Aztec West
Park avenue, Almondsbury
BRISTOL
U.K.

Ing. Jan PAVLICKO (P)
Hutny projekt
ul. Bozeny Nemcovej 30
042 18 KOSICE
Czecho-Slovakia

Ing. Ladislav PECINKA, CSc. (A,P)
Ustav jaderniho vyskumu
250 68 REZ u PRAHY
Czecho-Slovakia

Mr. J. PERZYNSKI (A)
Centre for Building Systems R&D
WARSZAWA
Poland

Dr. Kypros PILAKOUTAS (A,P)
Dept. of Civil and Struct. Eng.
The University of Sheffield
PO Box 600
Mappin Street
SHEFFIELD S1 4DU
U.K.

Prof. Vaclav PLACHY, DrSc. (P)
Fakulta stavebni CVUT
Thakurova 7
166 29 PRAHA 6
Czecho-Slovakia

Ing. Michal POLAK (P)
Fakulta stavebni CVUT
Thakurova 7
166 29 PRAHA 6
Czecho-Slovakia

RNDr. Dana PROCHAZKOVA, DrSc. (A,P)
GEOSCI
Faltanova 565
149 00 P R A H A 4
C z e c h o - S l o v a k i a

Prof. Dragan RADIC (A)
Institute of Civil Engineering
Rakusina 1
41000 Z A G R E B
Y u g o s l a v i a

Mr. G. P. ROBERTS (A)
WS Atkins Engineering Sciences
160 Aztec West
Park avenue, Almondsbury
B R I S T O L
U.K.

Dr. Chamidulla SAGDIEV (A)
IMISS UzAN
Akademgorodok
T A S H K E N T
U z b e k i s t a n
U S S R

RNDr. Vladimir SCHENK, DrSc. (P)
GfU CSAV
Bocni ul.
141 31 P R A H A 4 - Sporilov
C z e c h o - S l o v a k i a

Dr. S.G. SHAGINIAN (A)
ArmNIISA
D. Sasoutsky str. 3
Y E R E V A N 375014
A r m e n i a
U S S R

Prof. B. SIMEONOV (A)
Institute of Earthquake Engineering
and Engineering Seismology
University Kiril and Metodij
S K O P J E
Y u g o s l a v i a

Prof. Nikolai N. SKLADNEV (P)
TsNIISK
2nd Institutskaya 6
M O S C O W 109428
U S S R

Mr. M. STERNIK (A)
Faculty of Civil Engineering
TECHNION
H A I F A 32000
I s r a e l

RNDr. Bohumil SVOBODA, CSc.   (A,P)
Stavebni geologie
GEOTECHNIKA
Godovicka 4
152 00   P R A H A   5
C z e c h o - S l o v a k i a

Dr. Tadeusz TATARA   (A)
Politechnica Krakowska
ul. Warszawska   24
31 155   K R A K O W
P o l a n d

Ing. Milan TICHY   (P)
USTARCH SAV
Dubravska cesta 9
842 20   B R A T I S L A V A
C z e c h o - S l o v a k i a

Ing. Dagmar TOTHOVA   (P)
Stavebna fakulta TU
Svermova 9
042 00   K O S I C E
C z e c h o - S l o v a k i a

Dr. Alexander G. TYAPIN   (A)
MEI
Slaviansky boul. 1-119
M O S C O W   121352
U S S R

Doc.Ing. David UKLEBA   (A)
ISMIS GAN
Z. Ruchadze Str. 1
T B I L I S I 380093
G e o r g i a
U S S R

Prof. Andrea VIGNOLI   (A)
Dept. of Civil Engineering
University of Florence
Via di S. Marta, 3
50139   F L O R E N C E
I t a l y

Ing. Marian VRABEC   (P)
USTARCH SAV
Dubravska c. 9
842 20   B R A T I S L A V A
C z e c h o - S l o v a k i a

Mr. E. WADECKI   (A)
Centre for Building Systems R&D
W A R S Z A W A
P o l a n d

Mr. Huseyin YILDIRIM (A)
Istanbul Technical University
Ayazaga
I S T A N B U L
T u r k e y

Dr. Zbigniew ZEMBATY   (A,P)
Wyzsa Szkola Inzynierska
ul. Katowicka 48
45 954 O P O L E
P o l a n d

Ing. J. ZDAREK (A)
Ustav jaderniho vyskumu
250 68   R E Z u P R A H Y
C z e c h o - S l o v a k i a

Dr. A.M. ZHAROV (A)
TsNIISK
2nd Institutskaya 6
M O S C O W   109428
U S S R

FROM CEREMONIES

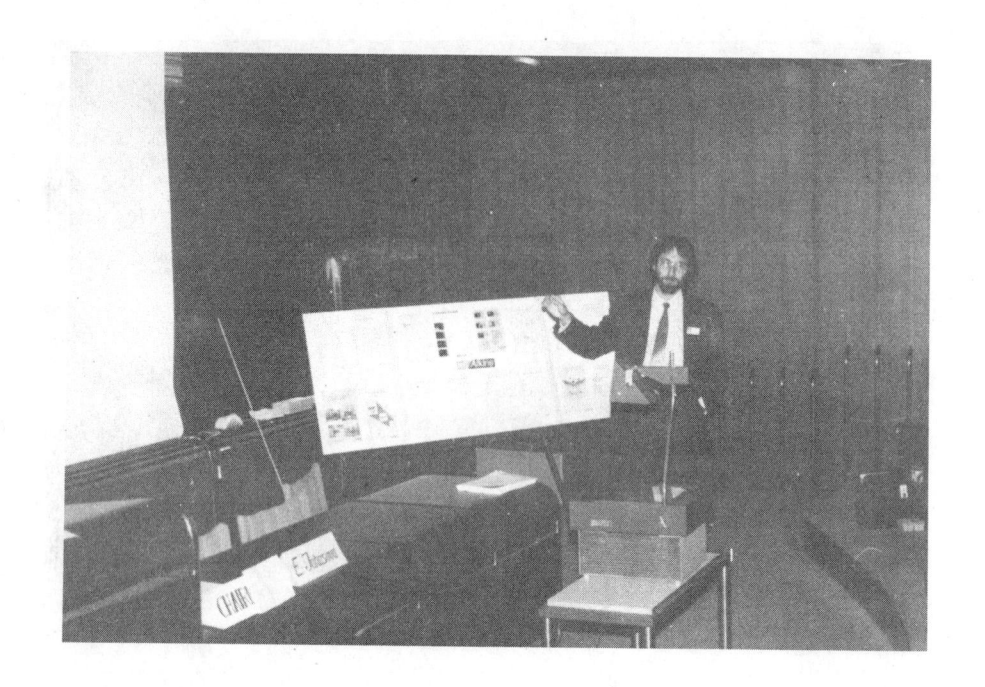

IN FULL WORK

DISCUSSIONS IN THE LECTURE HALL

DISCUSSIONS IN THE EVENING TIME

AND NEAR THE END